North Sea Passage Pilot

BRIAN NAVIN

6th March 1987

Brian Navin.

Imray Laurie Norie & Wilson Ltd
St Ives Cambridgeshire England

Published by
Imray, Laurie, Norie & Wilson Ltd
Wych House, St Ives, Huntingdon,
Cambridgeshire, PE17 4BT, England.

British Library Cataloguing in Publication Data
Navin, Brian
    North Sea Passage Pilot.
    1. Pilot guides – North Sea
    I. Title
    623.89'2916336   VK815

    ISBN 0 85288 102 9

CAUTION
Whilst every care has been taken to ensure accuracy, neither the Publishers nor the Author will hold themselves responsible for errors, omissions or alterations in this publication. They will at all times be grateful to receive information which tends to the improvement of the work.

PLANS
The plans in this guide are not to be used for navigation. They are designed to support the text and should at all times be used with navigational charts.

The scales have been chosen to allow the best coverage in relation to page size and consequently on certain small scale plans numerous buoys and other marks have been omitted for clarity.

The technical data in this work is correct to January 1987.

Set in Plantin by Cromwell Graphics Ltd, St Ives, Huntingdon direct from the Author's and Publishers' disks.

Printed at Tabro Litho Ltd, St Ives, Huntingdon

North Sea
Passage Pilot

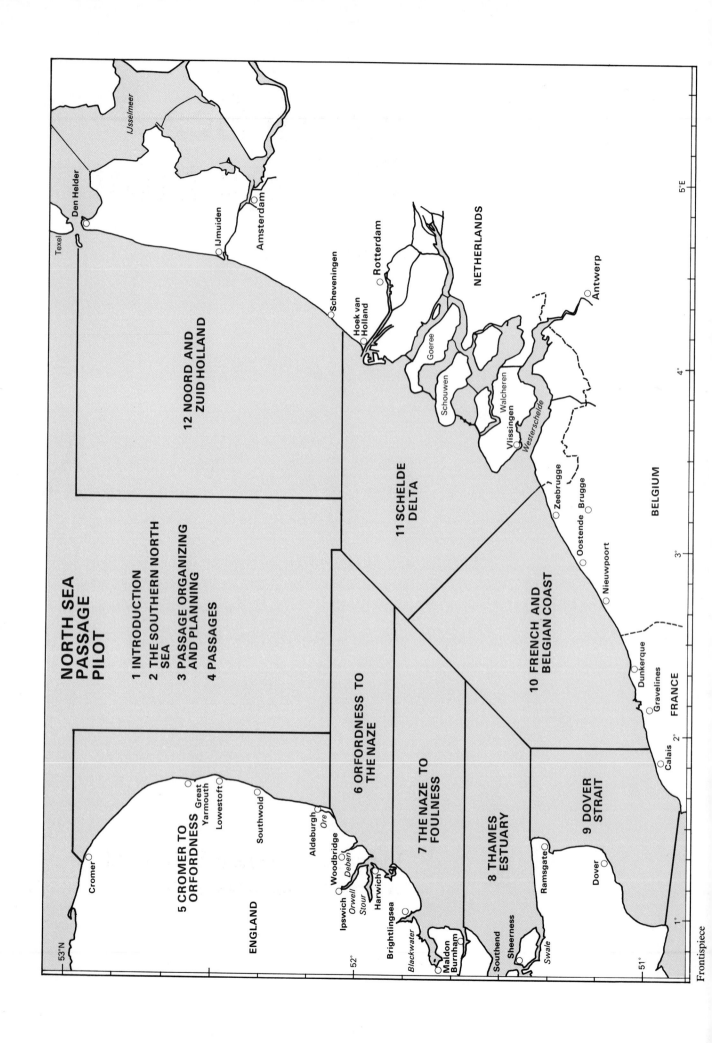

NORTH SEA PASSAGE PILOT

1 INTRODUCTION

2 THE SOUTHERN NORTH SEA

3 PASSAGE ORGANIZING AND PLANNING

4 PASSAGES

5 CROMER TO ORFORDNESS

6 ORFORDNESS TO THE NAZE

7 THE NAZE TO FOULNESS

8 THAMES ESTUARY

9 DOVER STRAIT

10 FRENCH AND BELGIAN COAST

11 SCHELDE DELTA

12 NOORD AND ZUID HOLLAND

ENGLAND

NETHERLANDS

BELGIUM

FRANCE

Cromer
Great Yarmouth
Lowestoft
Southwold
Aldeburgh
Ore
Woodbridge
Deben
Ipswich
Orwell
Stour
Harwich
Brightlingsea
Blackwater
Maldon
Burnham
Southend
Sheerness
Swale
Ramsgate
Dover
Calais
Gravelines
Dunkerque
Nieuwpoort
Oostende
Brugge
Zeebrugge
Vlissingen
Walcheren
Westerschelde
Schouwen
Goeree
Hoek van Holland
Rotterdam
Scheveningen
Amsterdam
IJmuiden
Texel
Den Helder
IJsselmeer
Antwerp

53°N
52°
51°

1°
2°
3°
4°
5°E

Contents

For my wife Barbara who made it all possible.

Preface

Writing this book was easy, but the research has taken half of my lifetime, and draws heavily from the lifetime work of many others before me.

A wealth of background information and charts is available from the various national hydrographic offices in the countries around the southern North Sea, to all of whom, and particularly the UK Hydrographic Department, I am deeply grateful. My thanks are also extended to the many harbour, tourist and public authorities who gave me advice, literature, photographs, and harbour plans, in particular, Great Yarmouth Port and Haven Commissioners, Harwich Harbour Board, Associated British Ports' Lowestoft office, Hoo Marina, Gillingham Marina, St Katharine Yacht Haven, the Port of London Authority, Ramsgate Harbour Office (Thanet District Council), Dover Harbour Board, Yacht Club de la Mer du Nord (Dunkerque), the Netherlands VVV tourist offices at Amsterdam, Rotterdam, Delft, Middelburg, Den Helder, Maassluis, IJmuiden, and the Rijkswaterstaat Deltadienst (Zierikzee) who provided so much information on the Delta Plan. In addition my thanks are extended to the staffs of the multitude of yacht marinas and mooring places I have repeatedly used throughout the area over many years.

On a personal level the book dates back to the London School of Economics Sailing Club in 1954 – 7 where we frequently capsized Firefly dinghies on the Welsh Harp at Hendon, in bitterly cold winters, dressed in T-shirts, shorts and plimsolls. Whilst this was definitely a good grounding for North Sea cruising, also at the time the opportunity arose for three of us from the Club to crew in the 35-foot *Maid of Pligh* (no prizes given for guessing its construction, and I believe she still sails from the Crouch) in the North Sea Race from West Mersea to Oostende via the Smith's Knoll and Goeree light vessels. It blew a gale, we broached, overrode and tore a spinnaker to pieces, and I remember being sequentially seasick, constipated, frightened, tired, exhausted and finally nodding off over the official dinner in the North Sea Yacht Club. We then had a gentle cruise in good weather down-Channel to Fécamp and across to Gosport. Since then I have been a convinced cruising man, looking for the settled window in the weather in which to dash across the North Sea for some unhurried coastal and waterways cruising on the other side. I have lost contact with those LSE friends of the mid-fifties, but if any still sail I am sure they would be glad to exchange a yarn or two.

In addition to LSE Sailing Club I would also like to thank my friends at the Cruising Association for entertainment, inspiration and information from their extensive library and from the CA Handbook. Also thanks to Bradwell Quay Yacht Club for mooring and yet further inspiration, and to whose members I think Barbara and I are conspicuous by our absence on cruising projects, although we do assist in propping up the bar occasionally, particularly in winter.

On a practical level I would like to thank my friend Arthur Somers for reading and commenting on the manuscript, but above all for his advice, encouragement and interest in our projects. Many thanks are also due to the editors of *Practical Boat Owner*; originally Denny Desoutter, and more recently George Taylor, who have frequently allowed me to 'practice' pilotage journalism on their magazine audience.

Barbara and I are also deeply grateful to all those who have crewed with us throughout the area – often coming at short notice, returning by ferry, and above all living in cramped conditions aboard various boats of 8 metres in length and less; more specifically in the earlier years our children Robert and Sarah, and in the past three years John Houbert, Peter Giles, Peter Burton, Peter Skinner, Ron Woolhead, and Dominic Jones. Also many thanks for the advice and various skilled jobs done on *Teazle* by the staff of Rice and Cole's boatyard at Burnham where I used to moor, and have more recently laid up each winter; also to Bradwell Marina, where we frequently load and unload, and to John Morrison without whose skills the engine might have seized in the 1986 season.

Finally many thanks to the editorial and design team at Imray, Laurie, Norie and Wilson for their highly professional work on a complicated reference book requiring high quality, detailed cartography and up-to-date source checking.

Brian Navin
Southminster
October 1986

Acknowledgements

The Publishers are grateful to Elizabeth Cook who compiled the index.

The plans of harbours on the English coast are based on Admiralty charts with the permission of the Hydrographer of the Navy.

The plans on the continental coast are based on official charts published by the Hydrographic Service of the Royal Netherlands Navy, the Belgian Dienst der Kust-Hydrografie and the French Service Hydrographique de la Marine.

I. GENERAL

1. Introduction

There has been a growing need for a cruising guide to passage making between the wide range of southern North Sea cruising grounds. Local coast and inland cruising guides which are available for most of these areas do not cover adequately the daunting prospect to the relative newcomer of a full day's dead-reckoning navigation across the world's busiest shipping lanes including some intricate inshore navigation at each end.

Weekend and long summer holiday cruising of the northern European coasts, is becoming increasingly popular and interest in more enterprising cruising involving 20 to 150-mile nonstop passages to and from interesting new inland and estuary cruising grounds is reflected in the growing traffic across the southern North Sea by yachtsmen from all of the surrounding countries. This guide, together with a suitable set of charts and a nautical almanac, provides the essential information for these medium-distance trips, together with all the necessary departure and approach pilotage information for entering the 8 major coastal areas and 32 harbours and havens in the region.

How to use this book

The three chapters in Section I cover the general background to North Sea passage making, for example: official requirements and regulations, traffic separation schemes, navigation aids, tide and weather information, passage planning and the log of an actual passage.

The nine chapters of Section II are a cruising guide to the area, with Chapter 4 summarising 23 North Sea passages, and Chapters 5 to 12 giving information about 8 coastal areas in an anticlockwise direction round the southern North Sea from Cromer to Den Helder; each including several approach route plans as well as data on each harbour within the area.

If you are interested only in a quick planning reference to a specific passage together with the areas at each end then:

1. Turn to page 38 in Section II (Chapter 4), *Passages page reference*, and extract the passage number in which you are interested.

2. Turn to the appropriate page in Chapter 4 to read about that passage. From *Approach routes*, extract the chapter and route numbers for the coastal areas to be crossed.

3. Turn to the appropriate chapter to read about the required route under *Approach routes and tidal timing*.

4. At the beginning of the same chapter you will find detailed information under the following headings:
   Charts
   Tidal atlases
   Tidal streams
   Tidal differences and ranges
   Lights
   Radiobeacons
   Coast radio stations
   Major fixed daylight marks

5. Harbour information is listed after the route summaries under the following subheadings:
   Port radio
   Entry signals
   Customs
   Entrance
   Mooring and facilities

In addition to the above information you will also require: an up-to-date *Reed's Nautical Almanac*, and in Holland, an up-to-date ANWB *Almanak voor watertoerisme, deel 2* (plus a pocket Dutch dictionary). For legal reasons it is essential that you also carry a copy of the ANWB *Almanak voor watertoerisme, deel 1*, as it includes copies of all the shipping regulations in force in the Netherlands including the Inland Waters Police Regulations and Shipping Regulations on the Westerschelde. It is, to say the least, unfortunate that these Dutch publications are not currently available in English.

Tidal streams and differences

At the beginning of each of the pilotage Chapters 5 to 12 is a list of times of changes in direction in tidal streams at various points near the coast obtained from the Admiralty Sailing Directions, and also a list of HW time differences from HW Dover of the various harbours, obtained from the UK Hydrographic Department, which was up to date at time of going to press.

It must be stressed that these times are averages and approximations, and in practice meteorological conditions can appreciably affect them. For convenience of use the times of changes in direction of tidal streams quoted are in most cases based on HW at each of two places and the implied difference in time of HW between these places may not correspond with the latest HW differences between the same places also listed at the beginning of the chapter. For example at the beginning of Chapter 6 the times of tidal stream changes imply that HW Harwich is HW Dover +30 minutes, whereas the tidal difference for Harwich listed on the same page is HW Dover +50 minutes. This discrepancy underlines the levels of approximation involved, and the more accurate tidal stream time is likely to be that based on the nearest of the two places, in this case Harwich.

Bearings and directions

The bearings given are all in 360° notation and are true. Directions are indicated by the usual abbreviations; W for west, NE for northeast, WSW for west-southwest etc.

Key to abbreviations and symbols

Throughout the text and plans in this guide familiar abbreviations have been used for convenience. Generally these follow standard conventions, those of the Admiralty are listed in the relevant Admiralty publication (i.e. regarding light or radio information). Admiralty *5011, Symbols and Abbreviations* and a dictionary should be consulted for explanations not listed below. See above for direction abbreviations.

Ch	channel (radio)	card	cardinal
CTMG	course to make good	F	fixed light
DTp	Department of Transport	Fl	flashing light
DWR	deep water route	Fl( )	group-flashing light
EP	estimated position	Iso	isophase light
ETA	estimated time of arrival	Lt	light
HW	high water	LtF	light float
LW	low water	LtHo	lighthouse
RC	marine radiobeacon	LtTr	light tower
RDF	radio direction finding	LtV	light vessel
RG	radio direction-finding station	Oc	occulting light
RYA	Royal Yachting Association	Oc( )	group-occulting light
TSS	traffic separation scheme	B	black
hr(s)	hour(s)	G	green
kn	knot(s)	R	red
m	metre(s)	RW	red and white
M	nautical mile(s)	W	white sector (with other colours)
min(s)	minute(s) of time		no letter if white light only
No(s)	number(s)	Wh	white (as of a structure)
⊖	Customs	Y	yellow/amber/orange
⚓HrMr	harbourmaster		
►SC	Sailing Club		
►YC	Yacht Club		

General

Electronic position-fixing equipment

Longer distance cruising is at a 'watershed' since microcircuit technology is making available electronic navigation aids priced within the reach of most small-boat skippers. The equipment is small enough to fit alongside the chart table of the smallest passage-making yacht, and is easy – perhaps too seductively easy – to operate. The legal enforcement of the traffic separation schemes, the development of offshore oil and gas fields, and increasing governmental regulation of offshore areas is placing a premium on the need for a yachtsman to have a continuous accurate position-fix in the busy area covered by this book, and the new equipment provides the experienced yachtsman with an invaluable additional check on his normal navigation estimates. But it is an aid and not a substitute for seamanship, thorough navigational knowledge, and above all for the keeping of a proper log of the passage and making frequent conventional estimates of position. At the same time the new instruments are creating potentially dangerous situations and the day may be near, if it has not already occurred, when some mathematically minded armchair-yachtsman of limited experience, made overconfident by the existence on his boat of an electronic position-fixing instrument and a chart, but without a set of tide tables or a weather forecast, strikes out across the Dover Strait or the North Sea, maybe even sailing against the traffic in the wrong lane of a traffic separation scheme. No apology therefore is needed for including in this book some general guidelines for the North Sea passage-maker about crew experience and equipment requirements in addition to the normal pilotage information.

Requirements of skipper and crew

Before attempting a passage involving long periods out of sight of land and at night, the skipper must ensure that he himself, crew and boat are adequate for the job. Above all this 'team' should be completely competent to carry out the passage alone without the assistance of other vessels, and this equally applies when cruising in company. Cruising in company is not necessarily a bad thing, providing it does not encourage slipshod skippers joining a group and putting other boats at risk as well as their own by needing assistance due to their own shortcomings.

The minimum complement of crew is three: a competent skipper, a second in command, and one other active and useful crew member. On larger yachts at least one additional member is needed to enable 2 distinct watches to be established. All crew members should be free from seasickness or at least a potential sufferer should be confident of medication against a complaint which in its worst form can remove him from the active crew list for most of the brief period likely to be involved in these passages. A *Stugeron* tablet taken 3 to 4 hours in advance of the passage and at 3 to 4-hourly intervals during the passage is an effective remedy for most, but not all, people and there are many other remedies on the market which can be tested in advance of the trip.

It is becoming increasingly important in Europe for a skipper to have some sort of 'certificate of competence' in the use of his vessel. However, except for fast motorised craft in Holland the governments in the area covered by this book do not yet make this compulsory, but clearly some kind of certificate can help in any legally difficult situation which might arise. The Royal Yachting Association issues a *Helmsman's (Overseas) Certificate of Competence* which in turn requires one of the RYA's basic national certificates or a special declaration from a club or school. For the passages described in this book the skipper and his crew members should have reached the higher standards, if not necessarily possessing the certificates, required by three of the certificate courses in the RYA's National Cruising Scheme (Sail). For many cruising yacht owners paper qualifications are the antithesis of what cruising is about, but these certificate standards do provide a minimum objective to aim for, discouraging too hasty a leap into the dark at the expense of other crew members and of the rescue services.

The skipper should have reached a standard approaching that of the RYA/DTp *Yachtmaster Offshore Certificate*; i.e. an experienced yachtsman, competent to skipper a cruising yacht on any passage which can be completed without the use of astronavigation, and having at least 50 days/2500 miles sea time, with 5 passages over 60M, of which 2 as skipper and 2 overnight passages. This may be regarded as slightly difficult in that to achieve the two 60M passages as skipper it might be argued that a UK southeast-coast yachtsman would need to cross the North Sea, and on this argument an experience somewhere between this and the lower grade Coastal Skipper would be more appropriate, but on the other hand it is possible to carry out 60M or longer passages within the inshore areas for example across the Wash or across the Thames Estuary and along the south coast, and it is advisable to do so before striking out on a long cruise across the traffic separation schemes themselves.

The passages covered by this book can range from 5 to 33 hours or more, and for the single-hander or the person who regards his crew as passengers they can be extremely wearing, particularly towards the end, when entrance to the destination harbour requires complete concentration; so there should be at least one other crew member with experience at minimum reaching that of RYA Day Skipper/Watch Leader, i.e. who can skipper a cruising vessel by day in his home coastal waters and with a minimum seatime experience of 10 days, 200 miles and 8 night-hours cruising. Finally the third member of crew should not be a pure passenger but of at least RYA Competent Crew standard, i.e. with enough knowledge of personal safety, seamanship, and helmsmanship to be a useful member of the crew. Fuller details of the certificate courses are listed at Appendix I.

Both skipper and crew should be physically fit; the prospect of a heart attack or seizure in mid North Sea is a daunting one. Eyesight standards should be reasonable, if necessary with spectacles, remembering too that at night colour blindness can be a disadvantage in distinguishing and interpreting lights. If contact lenses are worn, a pair of waterproof goggles are also useful for bad weather. Finally if any potential crew member feels competent only to sail in inshore waters and inland waterways there is always the wide range of ferry (and also air) services across the North Sea from which crew can join or leave; the routes from Harwich, Sheerness, Ramsgate and Dover provide reasonably priced foot-passenger services to Hoek van Holland, Vlissingen, Zeebrugge, Oostende, Dunkerque, and Calais.

Requirements of the boat

For many a skipper an equally difficult question is whether his boat is sound enough for making any of these passages. It is outside the scope of this book to discuss the design features of the ideal cruising yacht, if such a thing exists, or the minimum requirements for a well-found offshore cruising vessel, but a warning on the state of repair of the boat and its engine is in place. The strength of deck and rigging fitments, the hull, keel and keel bolts, condition of sails, and the reliability of the engine should be thoroughly checked in each winter lay-up. Sad though it may seem to the sailing purist, a reliable engine is essential for avoiding danger, conforming to the traffic separation schemes and manoeuvring into harbours.

Whilst there are recognised statutory and recommended minimum standards of equipment for cruising vessels there is still the question of what additional useful equipment to take. A summary of the minimum UK Department of Transport 'recommendations' for yachts of below 13.7m (45ft) in length is listed in Appendix II. Larger vessels are statutorily required to carry a wide range of safety equipment, fire appliances, and other equipment. The smaller vessels should carry at least the 24 items of equipment listed. But how far should one go beyond these recommendations? A VHF radio receiver/transmitter, with the appropriate operator's and installation licences is very much to be recommended in an area with such a comprehensive coverage of port and coast radio stations, Coastguard Rescue Centres and heavy shipping traffic; nowhere in this area of the North Sea is a vessel likely to be out of range of another ship or shore VHF station, a confidence-inspiring situation for the lone yacht on a night passage. Whether to invest in radar is probably the most difficult question of all: in a small, short-handed yacht with a lively movement in heavy seas, the need to concentrate on a small screen is often of less importance than lookout, steering and yacht-handling, but in calm fog conditions it can be worth its weight in gold, although some of the cheaper automatic radar warning devices are more practicable for the smaller vessel; in the hierarchy of equipment a full-blown radar set is probably left to the end. A question increasingly asked is the extent of navigational aids needed. A normal range of charts, publications and chart instruments is of course essential, but what about an electronic position-fixing system and which one to fit – RDF, Decca, Satnav, Consol, or Loran? Is a sextant needed, and if so how much astronavigation is needed? These navigational considerations are dealt with in Chapter 3.

Official requirements

The boat should be insured before making the types of passages covered in this book, and most UK marine policies automatically cover the English Channel and the waters of the North Sea, but it is worth checking on the extent of your coverage for inland waterways, if you intend to cruise at all inside the extensive waterways of Holland, Belgium, France or the UK. It is advisable to check the personal insurance of crew members, and take out some form of travel/accident insurance. It is also worth checking at a tourist office about any vaccination regulations or EEC health services requirements.

To minimise any legal problems of customs and other types of clearance in foreign ports, the boat should be registered with its national registration authority, be appropriately marked and carry aboard the registration documents. In the UK there are two types of register, and the RYA can advise, but essentially the Part 1 Register is for larger vessels and special cases, is expensive, and administered by official Port Registrars. The Small Ships Register, administered by the RYA on behalf of the government, is the easiest method of registering.

Increasing drugs smuggling as well as the rabies epidemic on the continent are causing a considerable tightening up of customs procedures in all countries. Yachtsmen of all nationalities should ensure that they conform to local customs procedures before they embark on a foreign trip:

UK   Always fly the Q flag when entering the UK (failure to do so can result in a fine of up to £400), and if the vessel is not boarded it is the responsibility of the skipper to find the customs authorities himself. In the UK they can be contacted by telephone for instructions (see under relevant ports for addresses and telephone numbers). Departure on a foreign passage should always be reported in advance to the customs office nearest the departure port on the first copy of a triplicate form (*C1328*), retaining two copies for later submission to customs at the port of return. For UK yachtsmen these copies are also useful for clearance with the customs authorities at foreign ports. Foreign yachtsmen also have to fill in appropriate forms.

France   Fly the Q flag at the first port of entry if you have anything to declare. If the authorities do not board, then find them (see under relevant ports for address and telephone numbers). If you are carrying paying passengers (there is a wide interpretation of this) you will be regarded as a commercial craft.

Belgium   Fly the Q flag at the first port of entry until customs officers board. If they do not board, then inquire at the marina or YC for the necessary address to contact.

Netherlands   Fly the Q flag at the first port of entry and report to the nearest customs post (see under relevant ports for addresses and telephone numbers). This is very strictly enforced and you will be provided with a *Verklaring* certificate allowing you to stay up to a certain date. Also make sure you obtain immigration and passport clearance, for which you may need to visit another office.

The vessel should carry aboard all the necessary certificates and operating licences for radio equipment.

Typical North Sea summer cruising

There are a number of interesting cruising areas in the southern North Sea most of which have dense concentrations of boats and facilities. On the English side there are: Dover and Ramsgate; the Medway and Swale; the Thames to St Katharine's Dock, London (also a route for vessels with removable masts into the English inland waterways); the rivers Crouch and

Roach, Blackwater and Colne, Orwell and Stour, Deben and Ore; and Southwold, Lowestoft, Great Yarmouth and the Norfolk Broads. In France and Belgium there is a string of North Sea harbours from Calais to Zeebrugge with entrances to another extensive inland waterway system. In Holland is the densest concentration of high quality yachting facilities of all: the Delta itself leading to yet another inland waterway system, the North Sea ports of Scheveningen and above all of IJmuiden and Den Helder with access to the IJsselmeer and the Frisian canals, lakes, Waddenzee and the islands.

The typical North Sea cruising holiday consists of 2 to 3 weeks and is of two types: a one-day each, outward and return passage between two of the above cruising grounds and day coast-hopping or cruising inland during the bulk of the remaining period; or coast-hopping throughout possibly with a Dover Strait and/or Thames Estuary crossing. More experienced English, Dutch and German yachtsmen – the latter almost as numerous in the Netherlands as the Dutch – are more likely to try the 'quick' way between the Delta or Noord Holland and the Norfolk, Suffolk and Essex coast. The French and Belgians have less need and tend to do more of the coast-hopping but the Dover Strait in season is crowded with all nationalities of cruising yacht, whilst the Oostende–UK east coast passage is also a popular way of avoiding the tricky Thames Estuary crossing.

Chapters 4 to 12 of this guide cater for all possible combinations of holiday passages as well as the coasting and saltwater river/estuary element of these cruises, but do not cover the freshwater inland waterways beyond the first locks and bridges. For example the gateway to the Broads – Great Yarmouth and Breydon Water – is examined and the routes up the great rivers to London and Antwerp, but not the IJsselmeer or the canals of Belgium and Holland. However, it proved impossible to resist forays down the Rotterdamsche Waterweg and down the North Sea Canal to Amsterdam.

2. The southern North Sea

The position of the traffic separation schemes (TSSs) in relation to a yacht's departure and destination points, is a critical factor dictating the direction of any passage across the southern North Sea. The boundaries of the TSSs are determined by the depths of water in the region, and these boundaries in turn determine the positioning of many of the major lights and radiobeacons, which are another significant factor in passage-making. Some knowledge of the problems of deep-draught ships and of the thinking behind TSSs, as well as a thorough knowledge of the International Collision Regulations, therefore, can be of considerable practical use to the yacht skipper. Admiralty chart *5500 English Channel Passage Planning Guide* gives a very useful snapshot of the considerations deep-sea masters should give in passage-planning for this area, as well as a few direct guidelines for yachts and crossing vessels.

Traffic separation schemes (TSSs) and deep water routes (DWRs)
See also Fig. 1 and Appendix III

TSSs concentrate through traffic in the region into approximately 10M-wide bands in two opposing 5M-wide lanes, with a number of branching and shallow-area variations on this theme. In 1977 a Dover Strait traffic analysis showed that on average 1 ship every 6 minutes passed in either direction along the Strait's traffic lanes, i.e. 1 every 12 minutes in a single lane. A yacht can take an hour at 4 to 5kn to cross a 5M-wide traffic lane, so on average could pass or be passed by 5 ships within the lane during this period followed by another 5 ships in the next lane. But statistical averages can deceive and ships can be widely spread and pass at distances up to the full 5M width of the lane, or sometimes traffic tends to bunch so there can be a series of vessels virtually in line across a yacht's track followed by a long gap. The same survey showed that an average of 1 vessel, including ferries, hovercraft and hoverfoils, crossed the lanes every 8 minutes, but again the average conceals variations such as higher frequencies in the summer season. North of the turn-off for Europoort and heading for the Texel Separation Scheme through traffic as well as ferry traffic reduces considerably, probably to about half or less than in the Dover Strait.

The southern North Sea is wedge-shaped: at the southwest corner the Dover Strait is 20M wide from South Foreland to Cap Gris Nez and at the northern exit from Cromer to Texel is 125M. On the western side from South Foreland, across the Thames Estuary and round the convex bulge of Norfolk to Cromer the distance is 110M; and on the eastern side following the concave bay formed by the coastline of Belgium and Holland from Cap Gris Nez to Den Helder the distance is 170M. Usually TSSs are confined to narrow straits and turning points round headlands, but since only a small proportion of the southern North Sea is deep enough for safe navigation by long-distance commercial vessels TSSs and also DWRs are specified extensively in the region by the International Maritime Organisation (IMO). TSSs are subject to occasional revision, as official experience of their operation develops and depths change, so yachtsmen should always use the latest, corrected large-scale charts when passage-making and consult the weekly Admiralty *Notices to Mariners*.

In general the controlling depth for the fairways in the TSSs is 21m so they are generally outside the 20m depth contour (see Fig. 1), although there are a few shallower patches within some lanes mostly marked by light buoys. The schemes consist of two opposite-going lanes of traffic leaving each other port to port, with a traffic separation zone (TSZ) in between. The main destinations of incoming North Sea traffic are Antwerp, Europoort and Hamburg and although the eastern side of the North Sea is the shortest route to these ports, since it is also the shallowest side, the schemes are well out in the middle to allow clearance of the 20m contour. The main TSS leads northeast through the Dover Strait with a subsidiary offshoot eastwards past the West Hinder LtV towards the approaches to the Westerschelde, and a second eastern offshoot near the Noord Hinder LtV to Europoort. The junction point of the latter,

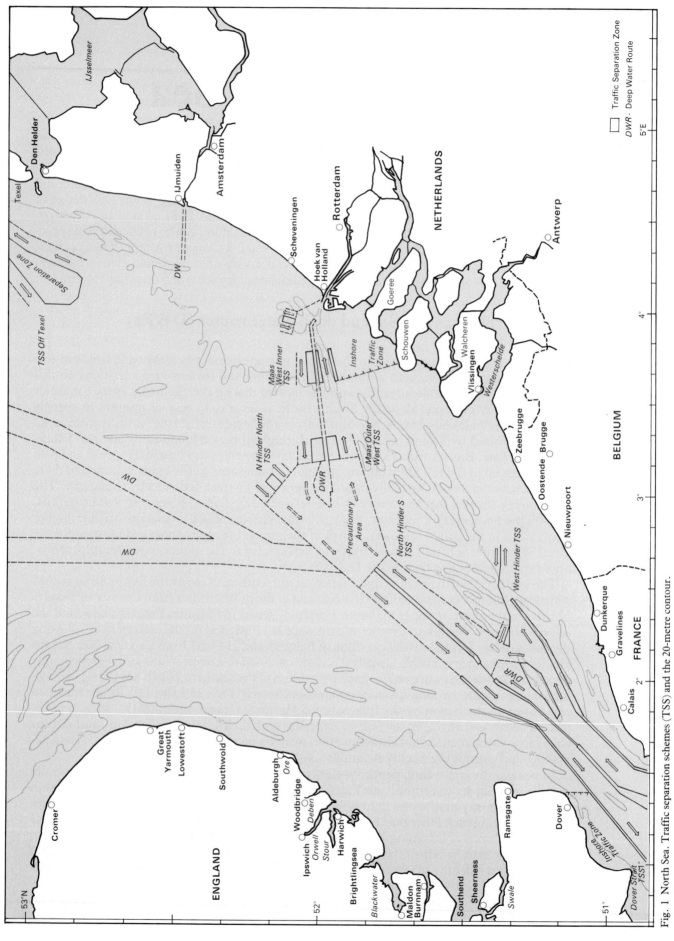

Fig. 1 North Sea. Traffic separation schemes (TSS) and the 20-metre contour.

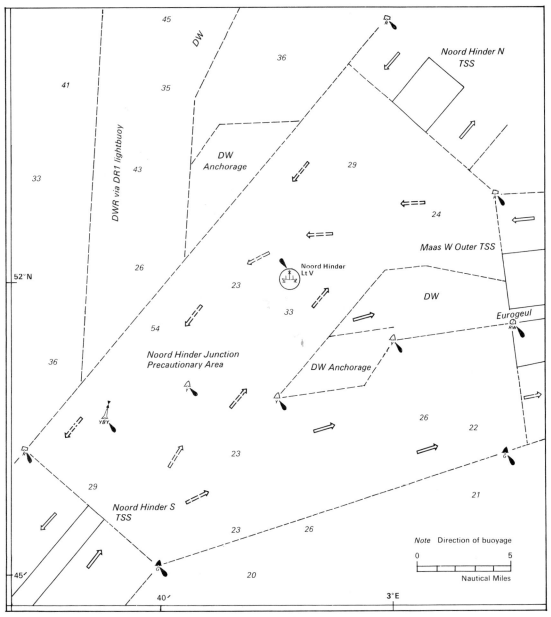

Fig. 2 Noord Hinder Junction Precautionary Area

Fig. 3 Approaches to Europoort

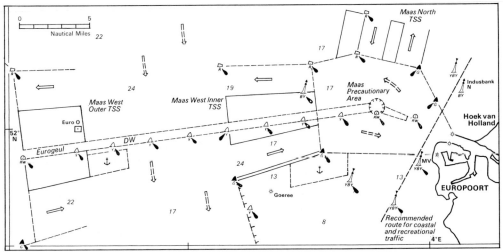

the Noord Hinder Junction Precautionary Area, is the maritime equivalent of a traffic roundabout, with a 1M diameter circle round the Noord Hinder LtV acting effectively as a traffic 'island' albeit somewhat complicated by the nearby DW vessel anchorage as well as the entrance to the Eurogeul DWR; it is an area in which vessels should be particularly careful and is to be avoided by yachts if at all possible. There is a second such area, the Maas Precautionary Area, just off Europoort, where another 1M diameter circle, Maas Center marked on its S side by a RW Iso.4s light buoy, acts as the roundabout. Yachts heading into the Nieuwe Waterweg for Rotterdam or coast-hopping past Europoort cannot avoid entering this latter precautionary area and should keep to the landward side of it crossing the traffic at the roadstead into Europoort at as close to a right angle as practicable. The officially recommended procedure is formidable: call Maasmond Radar (VHF Ch 2) before crossing, give name of vessel, position, course, maintain listening watch, follow a track close west of a line joining buoys MV, MVN (south of the entrance) and Indusbank N (north of the entrance), and cross under power and in company when possible. (Fig. 3)

The only other TSS in the region is off the Texel TSS north of the most northern yacht passage covered in this book. Yachts heading for Den Helder should aim to approach well to the south of the Texel LtV to give a wide berth to this scheme.

Near coastal areas special inshore traffic zones (ITZs) are designated outside the traffic lanes. Rule 10(d) of the International Collision Regulations states: 'Inshore traffic zones shall not normally be used by through traffic which can safely use the appropriate traffic lane within the adjacent traffic separation scheme. However, vessels of less than 20m in length and sailing vessels may under all circumstances use inshore traffic zones'. Effectively carte blanche is given for the smaller motor yacht and all sailing vessels not to use the traffic lanes for passage-making except in a crossing situation, the reasoning for this being the speed differential between these vessels and those in the lanes, and in the past, although to an increasingly lesser extent nowadays with the more widespread use of Decca, the greater difficulty of small vessels in fixing their position far from land. There are three ITZs in the area, one on each side of the Dover Strait on the English and French coasts but west of the South Foreland and Cap Gris Nez so effectively outside the scope of this book, and one south of Europoort and off the two dammed-up entrances of the Delta tributaries, the Haringvliet and Grevelingenmeer.

The Admiralty chart symbol for wrecks dangerous to surface navigation is based on a depth of 28m or less, and allowing for the accepted 'static underkeel allowance' of at least 4m, the Hydrographic Department are assuming that heavily laden vessels may in certain extreme circumstances reach as much as 24m (78ft in draught). The Netherlands Hydrographic Office publish a special *Deep Draught Planning Guide* and a chart for vessels with a draught between 20.7m (68ft) and 21.4m (70ft). Therefore 20m as a minimum for the traffic lanes is clearly not sufficient for the deeper draught vessels and the IMO has designated additional deeper lanes (DWRs) which are generally, but by no means always, of 30m depth and over. Fig. 2 shows the beginning of one of these DWRs with soundings. Some DWRs are within the normal lanes but separated from the other shallower draught traffic, e.g. there is one northwest of the Sandettié Bank TSZ, whilst another forms the approach to and also the Eurogeul route itself into Europoort and is in the middle of the TSZ between the two shallow-water lanes. The most extensive DWR is outside the shallower water schemes and leads northwards from the Noord Hinder Precautionary Area, with an additional northeastern branch leading into the German Bight. For obvious reasons these separate DWRs are not divided into two lanes although of course the Steering and Sailing Rules of the Collision Regulations apply here as in the TSSs.

International Collision Regulations (ICRs) – avoiding action

Clearly, two major priorities for a passage-making yacht are a radar reflector, preferably oversize and correctly mounted so as to give warning well in advance to crossing ships, as well as bright, reliable navigation lights. Secondly the skipper should be thoroughly conversant with the ICRs, and is advised to brush up on these immediately before the passage, particularly Rule 10 dealing with traffic separation schemes (see Appendix III) and the rules dealing with lights which are easy to forget yet so necessary at night in the TSSs. Observance of these rules is mandatory and prosecution can follow noncompliance.

There is no longer the old ambiguity of the tidal-set problem in the requirement to cross a traffic lane 'as nearly as practicable at right angles to the general direction of traffic flow' since the *Dover Strait Pilot* now states unequivocally that in the opinion of the English and French Governments: 'To follow this advice, low powered vessels and sailing vessels should therefore not make allowance for the tidal stream while crossing, if by so doing they will not have a heading nearly at right angles to the traffic flow'. Admiralty chart *5500* states that 'by keeping a vessel's heading at right angles, whatever the tidal stream, her time in the traffic lanes will be reduced to a minimum and a true right angle aspect will be presented to shipping following the lanes'. So the routine for the crossing vessel, whenever 'practicable' (to quote the ICR's oft-repeated expression), is to steer on a heading at an exact right angle to a lane but estimating in advance the likely course to be made good with the tide so as to check on likely hazards, buoys, shoals etc., and working out an estimated position at safe intervals. Yachtsmen have been prosecuted and fined for not complying with Rule 10, so other than in taking action to avoid collision and in the most difficult circumstances such as storm or breakdown, a yacht should head at right angles to the lane she is crossing. This rule applies officially only to TSSs but although not mandatory for DWRs, yachts are well advised not to navigate in these areas except to cross at right angles. Deep-draught tankers and bulk carriers are some of the least manoeuvrable vessels and slowest to stop so the objective again should be to cross these DWRs as quickly as possible. A right-angled heading in most situations ensures that the lane is crossed in the shortest possible time. I also believe in using or at least having the engine idling whilst crossing and if under sail as well as engine then under mainsail only so that traffic can see the black motoring cone on the forestay. However, a yacht is not prohibited from crossing the lanes under sail alone, providing she crosses at right angles; sailing yachts have been fined for tacking across the lanes at oblique angles to the traffic.

A major consideration for a vessel on a right-angled heading across the lane is the avoiding action to take when crossing the tracks of ships using the lane. The problem is that although Rule 10 does not override the other Steering and Sailing Rules for crossing vessels avoiding collision, unfortunately an element of ambiguity is introduced in 'stand on' situations by paragraph 10(j) which warns that small vessels and sailing vessels shall not impede the safe passage of through vessels. A yacht under engine not only has an advantage of manoeuvrability, but in crossing the second lane avoids ambiguity in that she has to give way to vessels coming from starboard, although in the first lane the ambiguity of which vessel should stand on, holding its course, does arise; however it is certainly not unseamanlike for a low-powered crossing vessel to give way to through vessels from either direction, provided that she takes positive avoiding action well in advance. Remember, it is difficult if not impossible to gauge the draught and manoeuvrability of passing ships. It is advisable to take a hand bearing of an approaching vessel as soon as her lights are seen at night or in daylight as soon as she appears hull down over the horizon, taking repeat bearings at not more than 3 minute intervals from then on, and if the vessel is getting larger or the lights brighter to regard a steady bearing – a change of 5° or less – as a potential collision situation requiring appropriate avoiding action. In giving way, reducing speed rather than changing heading is one possibility as well as stopping or heaving to, but if changing heading, for example to go astern of the approaching vessel, then the change of course should be large and obvious to the other vessel. Crossing both lanes under sail alone theoretically places the yacht in the position of stand-on vessel in many situations, but the rule of 'no impedence by small vessels' produces an element of ambiguity throughout and it cannot be assumed that all merchant navy skippers are familiar with the sailing manoeuvres of yachts. Discussions of this type always end with the unsatisfactory statement that in the end all depends on the seamanship of skipper or helmsman in a given situation.

Lights
See also Fig. 4

The small-yacht navigator with or without electronic position-fixing equipment tends to confirm his reckoning with as many visual fixes as possible, and in particular to use visual departure and arrival points which are well offshore to reduce the range of possible inaccuracy on passage and ensure an accurate course in the approaches to his final destination. Light vessels and floats, Lanbys and the occasional light tower are particularly useful, since at night

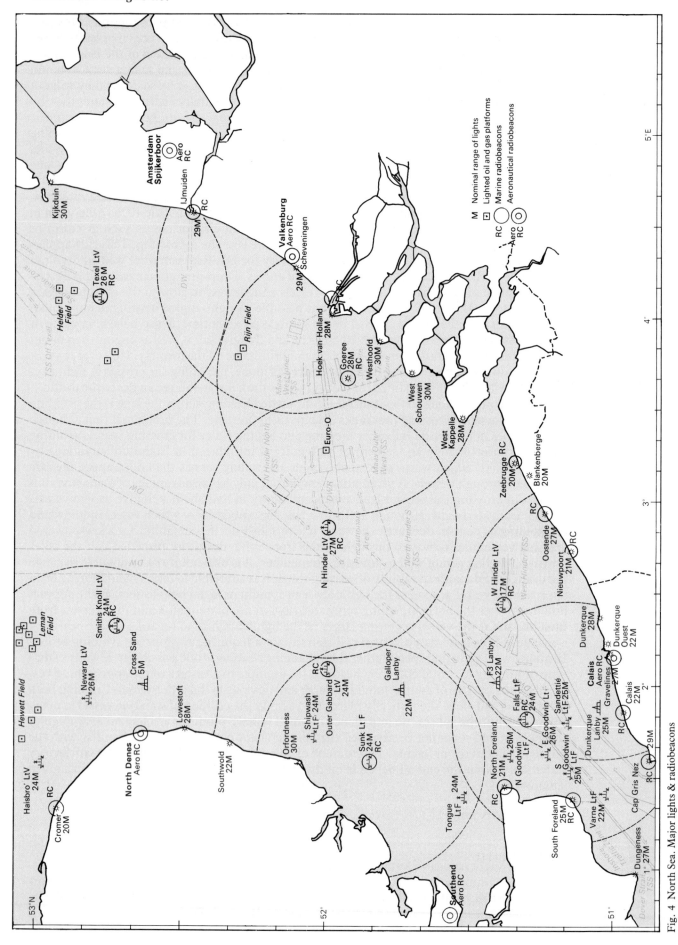

Fig. 4 North Sea. Major lights & radiobeacons

although the dipping distance of the average LtV from a yacht cockpit or foredeck is only about 11M the nominal range quoted on the charts is usually over 20M and in good atmospheric visibility their loom over the horizon may be seen well beyond 11M. Nominal range is the maximum luminous range at which a light can be seen with an atmospheric visibility of 10 nautical miles. In BBC shipping forecasts 'good visibility' is over 5M, whilst the station weather reports following the forecast quote actual visibility distances; useful if you are in the vicinity of one of the stations.

There are 21 of these lights in the region; 10 have radiobeacons. As befits their importance in the TSS the Noord Hinder LtV (27M) and the Goeree LtTr (28M) have the longest range. Somewhat surprisingly the West Hinder (17M) marking the TSS branch to the Westerschelde has the shortest, whilst most of the rest are from 22M to 26M. No less than 8 of the 21 are bunched in or near the Dover Strait, and another 9 off the shoal-ridden East Anglian coast and Thames Estuary providing excellent passage departure or arrival points. The north Dutch coast is poorly served with only the Texel LtV (26M).

There are 24 major onshore light locations: 9 on the English and 15 on the continental coast. These vary between 20M and 30M nominal range with a large number at the upper end of the range; Lowestoft 28M, Orfordness 30M, Cap Gris Nez 29M, Dunkerque 28M, Westkappelle 28M, West Schouwen 30M, West Hoofd 30M, Hoek van Holland 28M, IJmuiden 29M, Kijkduin 30M. The elevations of these lights are in some cases appreciable so the dipping distances are usually above 13M and up to 19M, being particularly long on the Dutch coast and enabling ideal night-time or dawn approaches.

Radiobeacons and radio aids
See also Fig. 4

There are four groups of radiobeacons in the region:

1. Dover Strait area (5 from 20M to 50M range)

2. A most useful central group for northern crossings stretching from Cromer down to Goeree and with good range (6 of 50M range)

3. Zeebrugge & Nieuwpoort (5M range only)

4. Hoek van Holland, IJmuiden, Eierland (Texel Island) (all 20M range)

In addition the Texel LtV in the north and Dungeness and Cap Gris Nez lights in the south are useful isolated members of other groups intruding into the region.

There are also three isolated stations with continuous transmissions: North Foreland (50M), Sunk LtF (10M only) and Dunkerque Lanby (10M but temporarily inoperative at time of going to press).

There are 5 aerobeacons in the area: 2 on the English side, Great Yarmouth (15M) and Southend (20M); and 3 on the continental coast, Calais/Dunkerque (15M), Valkenburg/Scheveningen (25M), and the most useful of all Amsterdam/Spijkerboor with 75M range.

There are 2 of the experimental VHF radio lighthouses with continuous digital transmission on VHF Channel 88, on either side of the Dover Strait at N Foreland and Calais, but with a horizon-limited range of only 20M. As these are experimental, one should expect service disruptions.

Finally for the few yachts with radar, about 30 of the lights and smaller buoys have Racons (radar transponder beacons) which trigger a flash on a radar screen indicating bearing and distance off, and in a number of cases, not all, a morse identification letter. The range of the majority of these is around 10M and a complete list with their characteristics is available in *Reed's* or the Admiralty *List of Radio Signals, Volume 2*.

Radio and coastguard stations
See also Fig. 5 and Appendix IV

Most yachts are nowadays fitted with VHF radio and so can communicate with other stations up to a maximum effective range which depends on the height above the horizon of each station; about 10M in the case of other vessels, and 40–60M in the case of land-based stations.

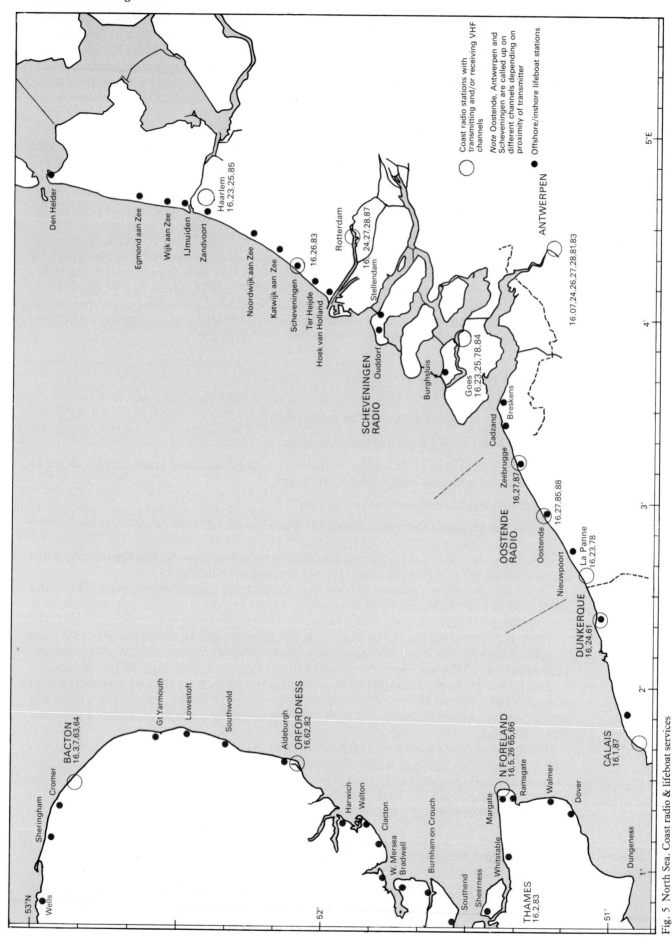

Den Helder

Egmond aan Zee
Wijk aan Zee
IJmuiden
Zandvoort
Haarlem
16,23,25,85

Noordwijk aan Zee
Katwijk aan Zee
Scheveningen
Ter Heijde
16,26,83
Hoek van Holland

Rotterdam
16,24,27,28,87
Stellendam

SCHEVENINGEN
RADIO

Ouddorp

Burghsluis

Goes
16,23,25,78,84

Breskens
Cadzand

Zeebrugge
16,27,87

OOSTENDE
RADIO

Oostende
16,27,85,88

Nieuwpoort
La Panne
16,23,78

DUNKERQUE
16,24,61

CALAIS
16,1,87

ANTWERPEN

16,07,24,26,27,28,81,83

○ Coast radio stations with
transmitting and/or receiving VHF
channels

Note Oostende, Antwerpen and
Scheveningen are called up on
different channels depending on
proximity of transmitter

● Offshore/inshore lifeboat stations

5°E

4°

3°

2°

Wells
Sheringham
Cromer
BACTON
16,3,7,63,64

Gt Yarmouth
Lowestoft
Southwold
Aldeburgh
ORFORDNESS
16,62,82

Harwich
Walton
Clacton

W. Mersea
Bradwell

Burnham on Crouch

Southend
Sheerness
Whitstable
Margate
Ramsgate
N FORELAND
16,5,26 65,66

Walmer
Dover

Dungeness

THAMES
16,2,83

53°N

52°

51°

1°

Fig. 5 North Sea. Coast radio & lifeboat services

14

This means that shore stations can be reached at sea throughout the area with the exception of a limited north-central area of doubt. However, even here in the DWRs and the approaches to the Texel TSS there is usually a ferry or a merchant or fishing vessel within range listening on Channel 16, or communicating on other channels.

The coast of the region is covered by a network of 40 inshore and offshore lifeboat stations, plus summer auxiliaries, operating within rescue networks, and a MAYDAY call on Channel 16 reaches stations either direct or is relayed by other shore stations or shipping to bring help rapidly. Again the area between the Norfolk and Noord Holland coasts is the more difficult since a vessel can be up to 55M from shore-based help, nearly 4 hours even at an average 15kn, so rescue by helicopter or by nearby ships coordinated by the coastguard services are more likely options here. In the area between Harwich and Vlissingen the potential distance is up to 40M and nearer to the Thames Estuary and Dover Strait narrows to about 10M maximum, although the very heavy traffic in this area also means a good chance of assistance from passing ships.

Yarmouth, Thames (based at Walton-on-Naze) and Dover are the three H.M. Coastguard Maritime Rescue Centres on the English east coast, and on any west–east crossing it pays to participate in their 'Yacht and Boat Safety Scheme', leaving complete details and a photograph of your vessel with your nearest centre, appointing a friend as agent and logging in a position report with the nearest station on departure from the English coast and on approach. The Coastguard Rescue Centres are of particular use, as on request from a vessel in distress a centre will transmit the bearing of the vessel *from* the direction-finding antenna; and many of them have subsidiary remote-controlled DF stations for this emergency use. Also useful is the Channel Navigation Information Service provided by Dover Coastguard and CROSSMA, its French equivalent at Cap Gris Nez; the VHF channel number of a broadcast is announced in advance in English on Channel 16, and information is given on request to vessels.

There are nine UK and continental coast radio stations (see beginning of Chapters 5–12; details also in *Reed's Nautical Almanac* and the Admiralty *List of Radio Signals, Volume 1*) in the region. These are primarily used for public correspondence, weather reports and forecasts, and distress coordination. There are 4 on the English coast: Bacton, Orfordness, Thames and North Foreland. Some of the 5 on the continental coast have substations: Calais, Dunkerque, Oostende (also La Panne, Zeebrugge), Antwerp, and Scheveningen (also Amsterdam, Goes, Hoek van Holland, IJmuiden, Rotterdam, Vlissingen as well as others).

Even more widespread are the port radio stations (also detailed in *Reed's Nautical Almanac* and the Admiralty *List of Radio Signals, Volume 6*). There are 52 of these, 24 on the English and 28 on the continental coasts, and some with subdivisions (see under each harbour section of Chapters 5–12) with separate call signs. Most are for port operations only, but three stations, Thames Navigation Service (subdivided Gravesend and Woolwich), Channel Navigation Information Service (subdivided Dover and Gris Nez), Schelde Information Service (subdivided Vlissingen and Zandvliet), put out regularly timed information broadcasts about hazards, weather, traffic movements etc. Also IJmuiden and Amsterdam on the Noordzeekanaal both broadcast at specified times when visibility is less than 1000m. All these services are worth listening to if your vessel is in the vicinity, but listening on any one of the ordinary port operations channels provides a very useful guide to the state of traffic movement providing the language can be understood, and sometimes it is in English; certainly most of the stations have operators who will respond to a request in English which is definitely advisable when entering or leaving harbour.

Tides, sandwaves and tidal surges

Accurately predicting tidal depths and tidal sets and drifts is particularly important in navigating this shallow heavily trafficked region, but 100% precision is impossible due to a number of unpredictable factors.

Throughout the region the undersea banks and the channels between them tend to follow the general direction of the tidal scour into and out of the southern North Sea 'wedge', and a high proportion are concentrated into the bottleneck presented by the Thames Estuary and Dover Strait. North of the Dover Strait some of the shoal patches are composed of 'sandwaves' which are series of waves or ripples of sand and shingle at right angles to the tidal flow which form and reform, move appreciably, and can be up to 20m above the seabed. There is limited hard evidence on their seasonal, long term and lateral movement to enable their predictability. Sandwaves of significance to deep-draught vessels and of course to any navigator using his depth sounder, are noted in the *Dover Strait Pilot* as follows:

1. Off Longsand Head Lt buoy, which should be given a wide berth and should never be cut behind.

2. At the NE end of Sandettié Bank towards Fairy Bank as well as encumbering the DWR between MDW buoy and the separation zone 2M NW.

3. In the SW-going lane near a 17.7m patch (51°33′N 2°14′E in 1983).

4. Sandwaves rising up to 17m above the seabed extend up to 4M SSW from the Tail of the Falls and in 1982 the shoalest depths in this vicinity were 18.5m to 21.5m.

To compound the problem of estimating depth, strong or prolonged winds, and unusually low or high barometric pressures can cause positive or negative tidal surges raising or lowering the height of tide from its predicted level. Winds blowing with the ebb can lower LW and prolong the ebb, and those blowing with the flood can raise HW and prolong the flood, whilst winds blowing against the tides can have the opposite effects. The wedge shape of the region can cause severe tidal surges from NW and N winds, particularly if they are of storm force from intense slow-moving depressions crossing to the north of the area. A storm-surge wave can raise tidal heights by as much as 3m; the January 1953 floods on the UK east coast and in the Netherlands were caused by this phenomenon. Persistent southerly winds or high pressure can produce much less predictable negative tidal surges in shallow-water areas, reducing predicted tidal heights by up to 2m, and this can be particularly hazardous to very deep-draught shipping; the Thames Estuary is particularly prone to these.

Although it is useful to know of these dangers in case the problem arises, the small yacht tends to avoid bad weather and dash across the North Sea with good shipping forecasts so tidal streams and depths can be predicted with acceptable accuracy but allowance for margins of error must always be made particularly on trans-North Sea passages involving the longer distances and crossing shoals.

Tidal streams
See also Fig. 6 and Appendix VI

The average yacht navigator is well versed in making tidal passages following the stream inshore, but the occasional North Sea passage crosses a series of reversing tidal streams requiring considerable care in estimating position.

In general the tide starts moving down the North Sea into the Dover Strait at around HW Dover –5 (hours), and turns northwards at HW Dover +1 (see Fig. 6). On the north Norfolk coast it tends to turn about an hour earlier than these times and on the opposite north Dutch coast about an hour later. In the central area spring tides seldom exceed 2kn, but on the Norfolk coast 'bulge' the streams tend to be much stronger than in the Dutch 'bight': at HW Dover –4 the East Anglian S-going stream averages 1.9kn (neaps) to 3.4kn (springs); on the Dutch coast north of Europoort it reaches its maximum up to 3hrs later at HW Dover –1 when the average is 0.8 to 1.5kn, about half the rate of the Norfolk coast. The N-going stream reaches its fastest across the whole of the northern part of the region at about HW Dover +3 at similar rates to the S-going though a little slower on the East Anglian coast: 1.7 to 3kn.

In the Dover Strait itself the tide turns about 2hrs earlier than in the northern part of the area, the SW-going stream starting approximately HW Dover +5 and the NE-going HW Dover –1. The funnelling effect of the Strait creates much stronger rates: the SW-going reaches 1.8 to 3.3kn in the middle of the Strait at HW Dover –4; the NE-going reaches 1.7 to 3.1kn HW Dover +2.

Tidal diamonds, tidal atlases and extreme tidal streams

Although the tidal diamonds on the appropriate Admiralty or Imray charts provide tidal set and rate information to use during a passage, this information cannot be fully appreciated

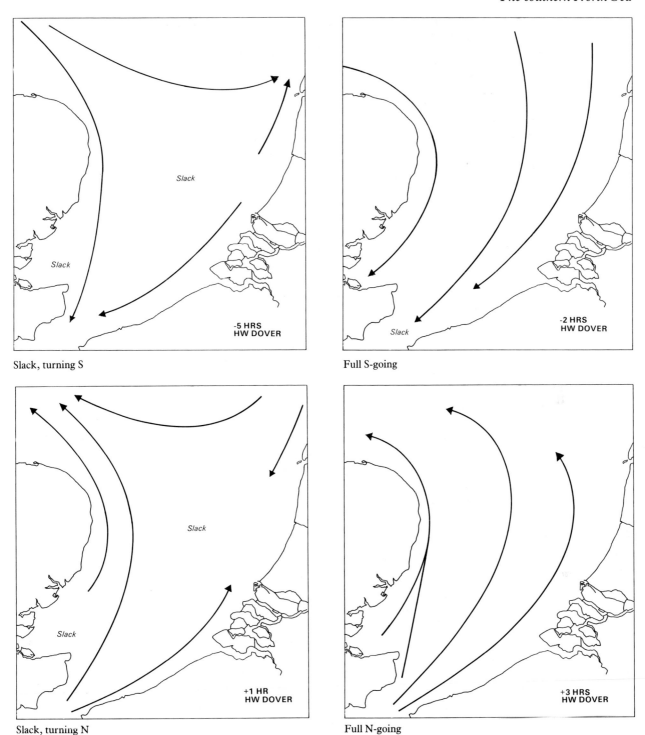

Slack, turning S

Full S-going

Slack, turning N

Full N-going

Fig. 6 Tidal streams (see also Appendix VI for detailed diagrams)

without the use of tide tables and tidal atlases. *Reed's Nautical Almanac* provides adequate tide tables and also tidal charts but it is well worth investing in the three more robust Admiralty *Tidal Stream Atlases* for use in planning, and also on passage when actual times of day can be pencilled on the corners of each hourly page for quick reference. These Admiralty atlases cover:

1. *Thames Estuary* (with co-tidal chart) (based on HW Sheerness)

2. *Dover Strait* (based on HW Dover)

3. *North Sea – southern portion* (based on HW Dover)

All three include, inside the front cover, a table for 'computation of rates' of tidal stream, not only between the mean or average spring and neap ranges of rate included in the atlases and in the tidal diamond information on Admiralty charts, but also for the more extreme spring and neap ranges which at certain times of year, and throughout some years, vary considerably above and below the mean. These tables work on the assumption that the rates vary with the range of tide. Dover mean springs range is 5.9m but can reach as much as 6.6m, and at Dover entrance in these conditions the rate can reach as much as 4.6kn compared with a mean of 4.1kn; off the Norfolk coast a 3.4kn mean can be associated with 3.8kn in practice. At the other extreme Dover mean neaps range of rate is 3.3kn but can be as small as 2.1kn in practice, producing on the northern Dutch coast tidal sets as low as 0.5kn *maximum*.

For the inlets of the Schelde Delta and the Schulpengat on the Dutch coast it is also advisable to buy the appropriate *Stroomatlassen* (listed in Appendix VII), since the Admiralty tidal atlases do not adequately cover these areas.

Very often it is a matter of simple mental interpolation to estimate tidal rates from an atlas or tidal diamond using the *average* springs and neaps rates found there. However, it is always advisable in advance of the passage to check the actual Dover range of tide on the day in question to see if it is significantly different from the average, and if so to make a few estimates of rates at various times and places on the route using the interpolation table in the atlases.

Tidal times and depths
See also Appendix VI

Estimating depths is critical for approach navigation but also can be extremely useful in estimating position from soundings whilst crossing the longitudinal North Sea banks well offshore. Estimating times and heights of local HW and LW and interpolating to correct your soundings to chart datum in coastal locations within range of the local secondary port (Reed's or Admiralty) is a major part of a yachtsman's training, but few skippers bother with reducing to soundings offshore using co-tidal charts. In the northern part of the area offshore, where the tidal range is only 1m or less (see Fig. 7) it is less critical, but accurate estimation is important in the area bounded by the Dover Strait and a line from Harwich to the Schelde and including the Thames Estuary; this area is littered with shoals and has a tidal range which at extremes can vary from less than 2m to well over 6m. To depth-navigate with any confidence here it is necessary to convert to chart datum using Admiralty chart *5057*, *Dungeness to Hoek van Holland: Co-tidal and Co-range Chart*, or in the Thames Estuary using the co-tidal and co-range chart included at the back of the appropriate Admiralty Tidal Atlas. Another co-tidal and co-range Admiralty chart *5059* covers the whole of the southern North Sea from Dover Strait to Denmark and northeastern England on a very small scale, and will probably remain in your chart drawer unless you are sailing well offshore to Scandinavia or Scotland. Fig. 7 is a very small scale diagramatic representation extracted from the two Admiralty charts.

Fig. 7A of the HW time interval illustrates how the tidal wave created behind the moon's meridian passage spins in an anticlockwise direction around a nodal point which lies somewhere about halfway between Great Yarmouth and IJmuiden. The lines represent equal Mean High Water time intervals referred to this meridian passage; i.e. HW occurs at the same time along each line. The tidal wave is pulled up behind the moon and at the time when the moon is crossing the Greenwich meridian and also 12hrs later when the moon is near the international dateline meridian on the other side of the earth, the HW line is roughly 90M behind the meridian (100M at 0hrs and 80M at 12hrs) lying from the above nodal point to an area near Oostende, 1hr later it has swung round to the Schelde Delta, 2hrs later to IJmuiden, 6hrs later to Den Helder, by which time a wave coming down the northern North Sea pushes down towards Great Yarmouth (7hrs later), and progressively moves to the Thames Estuary which it crosses 12hrs after the first HW, i.e. when the moon is near the international dateline meridian on the other side of the earth; during this process another wave from the English Channel reaches the Dover Strait 11hrs after the Greenwich passage and continues on to meet the North Sea wave in the Thames Estuary at the time of the tidal day's second HW. Fig. 7B is almost the mirror image of Fig. 7A showing the LW trough's similar anticlockwise swing a little over 6hrs behind the HW wave producing the intervening two LWs.

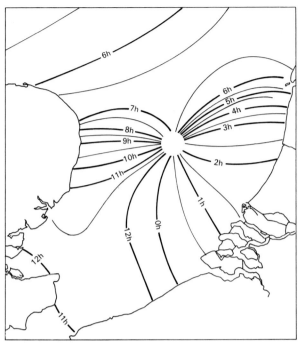

7A Mean high water interval.
Lines of equal mean high water time interval referred to the time of the moon's meridian passage at Greenwich.

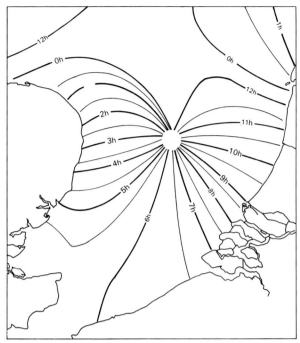

7B Mean low water interval.
Lines of equal mean low water time interval referred to the time of the moon's meridian passage at Greenwich.

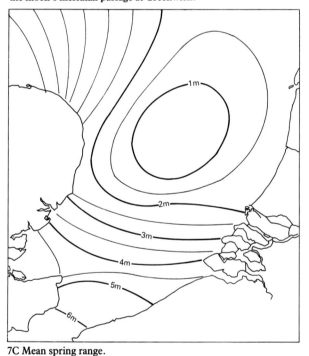

7C Mean spring range.
Lines of equal mean spring range in metres.

7D Mean neap range.
Lines of equal mean neap range in metres.

Fig. 7 Co-tidal and co-range diagrams

These movements can be appreciated more easily by using HW at Dover as the base time: local HW on the continental coast is progressively later; Dunkerque/Oostende area is approximately +1hr, Vlissingen +2, Hoek van Holland +3, and IJmuiden/Den Helder area is over 4hrs later. On the UK east coast Orfordness has HW at exactly the same time as Dover, while Great Yarmouth has it over 2hrs earlier, so that from Orfordness southwestwards along the north side of the Thames Estuary as well as from North Foreland westwards along the south side of the Estuary local HWs are progressively later than Dover. On the north side: Woodbridge Haven +0hr40mins, Walton-on-the-Naze +0hr42, Whitaker Beacon +1hr05, Southend +1hr30. On the south side: Ramsgate +0hr20, Margate +1hr, Shivering Sand +1hr10, Sheerness +1hr36, approximately. In the case of the harbours inside the rivers and creeks of the two coasts times of HWs are later, and London Bridge is +2hr52, Antwerp +3hr42, and Dordrecht +5hr05, all approximate.

Figures 7C and 7D show lines of equal average neaps and springs ranges, the difference in tidal height between HW and LW. The bottleneck effect increases the range of tide progressively towards the Thames Estuary and Dover Strait: Great Yarmouth 1m neaps/1.9m springs range, Oostende 3m/4.8m, Ramsgate 2.6m/4.5m, Sheerness 3.3m/5.5m, Calais 3.8m/6.2m. Similarly the range up the rivers increases, London Bridge no less than 4.2m/6.6m, and Antwerp 4m/5.4m. On the open shores of the north and in the artificial coastal harbours the range is very small: Lowestoft a tiny 1.1m/1.9m, IJmuiden 1.2m/1.8m, Hoek van Holland 1.5m/1.9m. Finally of course there is a wide area of open sea with 1m and less range around the tidal 'node' between Norfolk and Noord Holland.

Reduction to soundings using co-tidal charts

There is a detailed explanation of this on Admiralty chart *5057*; Fig. 8 gives a brief example. To correct a depth sounding out at sea in the area of chart *5057*, the position must be estimated within a reasonable accuracy, say 1–2M. The method used is then essentially the same as estimating heights and times of tide at a secondary port, except that the chart is used to obtain the differences at the estimated position from the standard port instead of the tide tables; the four stages are (1) find HW and LW times and heights at the nearest standard port for the day, (2) use the chart to extract the differences at the position, (3) apply these to the standard port data to obtain HW and LW times and heights at the position, (4) interpolate to obtain height at the position for the given time.

This one and only method of reduction to soundings offshore is complicated, very approximate, and the number of steps involved lends itself to error. It is advisable on any passage to work out in advance on a comfortable stable desk at home the HW and LW times and heights for a small number of waypoints on the intended route constructing a small tidal curve diagram for each to clip into the logbook. Alternatively electronic calculator users may wish to pre-program part of the method, but this will still involve inputting tidal position data extracted from Admiralty chart *5057* en route or storing data for a limited number of waypoints.

Position
51°40′N 2°E. Approximately 32M from Harwich and 43M from Oostende on a line between.

Time   1200 BST

Finding range of tide at the position

Ranges	Harwich (metres)	Position (metres)	
Mean springs range	3.6 (chart)	3.4 (chart)	
Mean neaps range	2.1 (chart)	1.9 (chart)	* Estimate by inspection of MSR and MNR and Har-
Actual range (see below)	2.8 (Reed's)	2.6*	wich actual figures in this table.

Finding HW and LW heights and times at the position

	HW		LW		
	Time	Height (metres)	Time	Height (metres)	
Harwich	1101	3.7	1647	0.9	(Reed's) Range 2.8*
difference	+20 min.		+30 min.		(chart)
Position	1121	3.4*	1717	0.8*	Range 2.6*

* Estimate of position's heights: 2.6 range at position divided by 2.8 range at Harwich represents 0.93. Apply this factor to Harwich actual heights gives 3.4 and 0.8.

Finding height at 1200 BST (times above have all been corrected to BST on extraction from tide tables).

1200 BST is +39 minutes on HW at the position. Factor for HW +40 minutes from tidal diagram on Chart *5057* is 0.95.

0.95 times 2.6 range is 2.5
2.5 on LW of 0.8 gives:
3.3 metres height of tide at the position

This can now be applied to the depth sounding obtained from instruments to convert to chart datum.

Fig. 8 Reduction to soundings using Admiralty co-tidal chart *5057*

Weather, wind and fog
See also Fig. 9 and Appendix v

Weather and wind conditions, as every North Sea yachtsman will wholeheartedly agree, are variable in the extreme and seldom remain the same for more than a few hours on any passage. Across the seasons pressure averages 1015 millibars throughout the region but with the generally eastward passage of depressions and intervening high pressure ridges through or to north and south of the region it can fluctuate hourly falling as low as 950 millibars in a deep depression and bringing a series of rapid changes in wind strength and direction. Less frequently more stable conditions may last for up to a week or more due to the spreading of high pressure air from (usually) the Azores region of the Atlantic in summer (June to August) and Siberia in winter (December to February) with pressure sometimes reaching 1050 millibars, and usually with lighter more variable conditions prevailing except sometimes at the edges of the system where the isobars may be compressed by adjacent depressions creating strong persistent winds. Spring (March to May) and autumn (September to November) are the most variable seasons.

In terms of wind strength and direction an early cruising season from April to August, providing thick winter underwear is worn at the beginning of the season, is preferable to a late one say from July to November. The early season approach tends to suffer from slightly more danger of fog, which is least prevalent in July, August, September and October, but a reduction in air pollution in the last 25 years due to legislation has reduced the fog problem considerably.

The average frequency and strength of winds from each direction gives a useful guide to holiday timing, providing the skipper is prepared to wait for the right conditions over a few days. On short holidays it often pays to sail with the wind direction, choosing the objective when the day arrives, and there are many interesting destinations to strike out for in wide-ranging directions from most harbours in the region. The highest frequency of winds is from SW to NW with a N to NE component mainly from February to May. As the coasts on each side trend in these directions winds near the shore tend to be stronger and rough conditions are often encountered at the end of a passage: on the continental side both W to SW and N to NE winds create rough conditions and on the English side including the Thames Estuary NE to E winds produce the worst lee shore conditions; near to and through the Dover Strait both SW and NE winds are funnelled often creating rough conditions which sometimes persist building up high seas. In autumn and winter there is a high frequency of winds from the SW quadrant, i.e. from south round to west, (approximately 60% frequency) so sailing from the English coast to Holland is usually a 'downhill' run but very difficult for the Dutchman (or returning Englishman) in the opposite direction; crossing the wind between England and France is more difficult but feasible; getting down to and through the Dover Strait can be very difficult indeed. Gale force and above winds are also much more frequent in autumn and winter and rare in spring and summer. Spring and summer generally are less of a problem with a better spread of wind-rose directions and lighter winds; March to May brings winds from the NE quadrant with a similar frequency to those from the SW quadrant, giving the Dutchman at least an equal opportunity of a downwind run to the UK east coast or Thames Estuary as his opposite number has in sailing from England to Holland. In summer wind frequency from the NE quadrant declines a little and more is concentrated into the whole of the western semicircle but with more chance of calms and the need for supplementing with engine power. Finally throughout the yachting season in spring, summer and autumn winds from the SE quadrant are very infrequent making it more difficult to sail direct from Belgium to England except reaching or close-hauled.

Fog, a hazard to be avoided in the traffic separation schemes if at all possible, is rare in the sailing season – 5% frequency (1½ days a month) – and is often a brief coastal feature in quiet anticyclonic conditions due to night-time temperature inversion on the land and fog drifting onto the sea but clearing during the morning. There tends to be a slightly higher fog frequency nearer to the Dover Strait particularly on the French/Belgian coast stretching up as far as the Delta (see Appendix v). This stems from a tendency for warm moist air to spread through the Dover Strait from the west, but even this means only 2–4 days fog per month early and late in season and 1–2 days in summer at places such as Calais and Oostende.

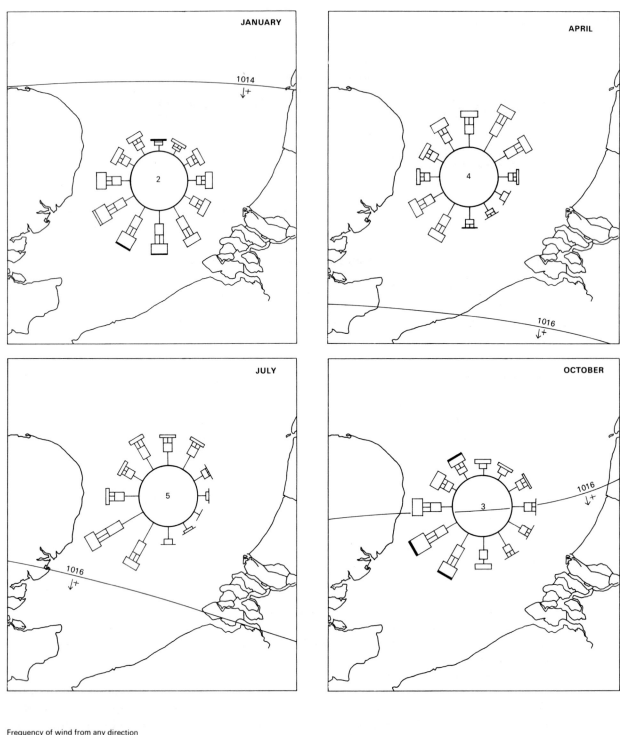

Frequency of wind from any direction
Subdivided by Beaufort Force

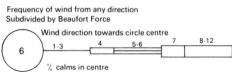

Wind direction towards circle centre

1-3   4   5-6   7   8-12

% calms in centre

Fig. 9 Wind distribution by season

3. Passage organising and planning

The average yacht skipper is unlikely to make frequent North Sea passages and before every such departure there is no shame in being keyed up and slightly nervous; overconfidence may just be blissful ignorance. In addition there is little time to relax during the short period of the crossing despite the fact that he needs to conserve his energies to cope with any problems which may arise and to be ready for the final approach pilotage. It is wise therefore to be prepared for all reasonable contingencies before motoring out of the marina or dropping the mooring. Equipment and engine will no doubt have been thoroughly checked, but there are a number of potential problem areas worth extra attention, especially weather forecasting, navigation lights, liferaft/dinghy, roller-furling sails, the watchkeeping system and finally the route plan.

Weather forecasts, NAVTEX

The weather for a short crossing can be carefully chosen and a day or two's delay to allow bad weather to pass is never a loss. Shipping forecasts up to 24hrs ahead tend to be reliable although particularly unstable weather patterns with fast-moving deep depressions sometimes leave the forecasters guessing beyond 12hrs ahead. A careful check of the weather maps found in *The Guardian, The Times* or *The Daily Telegraph* newspapers on a few days running up to a passage gives an impression of the tendency of the weather pattern. For yachtsmen within commuting distance of London the Meteorological Office's Weather Centre on High Holborn sells an up-to-date yachtsman's weather forecast package, consisting of the latest actual met. chart and shipping forecast together with forecast charts for 5 subsequent consecutive 24hr intervals, and blank forms and charts for filling in the subsequent shipping forecasts. This 4-day (i.e. beyond the current day) view can be combined with subsequent BBC shipping forecasts to enable the navigator to maintain an updated met. sketch chart near the chart table, either on the forms in the package, or preferably on a pad of The Royal Meteorological Society/Royal Yachting Association's well proven *Metmaps* available from the RYA or good chandlers. If the passage is delayed, later versions of the actual met. chart plus a 24hr forecast can continue to be obtained from the above newspapers to assist in reinterpreting the 5-day forecast.

The yacht which has room for a NAVTEX set on the chart table can obtain instant printouts of the latest forecasts. This extremely small and compact radio telex receiver is permanently tuned to a special fixed frequency transmission and prints out messages immediately on receipt, or alternatively if required, stores them and prints them out on request. The service provides weather messages and forecasts, gale warnings, ice reports, coast navigation and NAVAREA warnings, initial distress information messages, pilot service messages, and electronic navigation aid warnings. The beauty of NAVTEX for the undermanned yacht is that it does not need constant radio monitoring, although inevitably it is yet another drain on a sailing yacht's finite battery resources. The smallest NAVTEX I set however consumes only 0.2 amps on standby and 1 amp during the brief printing periods which can be reduced by selecting only certain information for printout. The system is sponsored by the International Maritime and International Hydrographic Organisations and organised through 16 NAVAREA coordinators appointed by the coastal states and operating as collection and distribution services. It is limited currently to northwest European waters (NAVAREA I) but will eventually be available to vessels sailing within 200M of all coasts worldwide. The printouts still need interpreting on a weather map pad.

Navigation lights

Crossing a TSS at night without lights as a result of lack of power supply or bulb/contacts/ fuses failure is every skipper's nightmare. Prior to the cruise a trip down to the local garage for a slow charge of the batteries over at least 10hrs does wonders for morale. Contacts for all the navigation lights and their fuses should be cleaned, and spare bulbs and fuses, and a spare anchor or tricolour light on a long length of electrical wire with clip-on ends for direct attachment to the battery should also be available as a last resort as well as white flares. Finally there should be more than one battery. Two good capacity batteries can either have a junction switch for use alternately or run simultaneously, or they can operate through a blocking diode system which charges both batteries automatically by an 'overflow' system; one can be used to start the engine, and the other for lights and electronic aids.

If worried about lighting capacity it is well worth remembering the formula: watts divided by volts = amperes. A modest 60 ampere-hour capacity 12 volt battery can probably be relied on to supply about 45a.h. in practice, i.e. two thirds, so a 10-watt masthead tricolour light needing a supply of 0.83 amps (10 watts divided by 12 volts) would be alight for 54hrs on this battery, i.e. 45 divided by 0.83. Even four 10-watt lights – sidelights, sternlight and steaming light – take only 3.3 amps (40÷12) giving over 13hrs (45÷3.3) lighting time; ample for a summer North Sea crossing, when darkness lasts for 6–8hrs, even in the unlikely event of not using the engine. It is always advisable to carry a hydrometer (syphon type), which is very useful at the other end of a passage preparing to return when the engine has been used sporadically and the state of charge of the battery is in doubt. If the specific gravity of the battery acid has fallen as low as 1.15 then it needs recharging to the full level of 1.28.

Finally it is also useful to remember that the Collision Regulations have been revised to allow a craft less than 12m in length under power to have an all-round steaming light to act as a combined forward and stern light (in addition of course to the usual green and red sidelights) which saves the drain of one light bulb on the battery compared with the old system. Under sail of course the single masthead tricolour light creates a minimal battery drain, so the ideal economical masthead unit is a combined tricolour and all-round white light which serves both functions and can also act as an all-round white anchor light when needed.

Liferaft or dinghy

Whether to have liferaft or dinghy is an agonizing question for the penniless cruising yachtsman. The DTp recommendations for yachts up to 13.7m (45ft) leave four options open (see Appendix ii):

1. A DTp-approved liferaft, capacity all members of crew, carried on deck or in a locker with direct access to deck, and must be serviced annually.

2. Solid/collapsible dinghy, permanent (not inflatable) buoyancy, secured oars and rowlocks, on deck or in tow.

3. Inflatable dinghy, 2-compartment with one fully inflated, secured oars and rowlocks, on deck or in tow.

4. Inflatable dinghy, 2-compartment, need not be on deck if yacht has enough permanent buoyancy to float when swamped, with 115 kilos (250lbs) added weight, or in tow.

It is unlikely (and inadvisable) that owners following the expensive first option would not also have an inflatable or solid dinghy; the thought of a liferaft not opening on pulling the lanyard is another cruising nightmare. The fourth option is interesting in that increasingly boats are becoming available with polyurethane foam buoyancy and claimed unsinkability. Even in the case of conventional craft if well away from hazards it is advisable to remain as long as possible with the vessel even in an apparently sinking condition, since it may float for a considerable period, rather than putting oneself and one's crew too hastily at the mercies of a flimsy inflatable or a solid dinghy. It is also worth remembering that serviced liferafts can be hired for short periods or for varying term longer lease. Finally whilst it is essential to have an annual service for a liferaft it is equally important to service annually, usually at considerably lower cost, the inflatable dinghy.

Remote control sail-handling equipment

Removal of the need to go on the foredeck can be a particular boon on longer distance passage making. Headsail roller furling and quick-reefing mainsail systems operated mainly from the cockpit or for brief periods on deck are very useful for short-handed cruising. But a roller headsail can become a major liability if it jams in bad weather. Readership correspondence in yachting magazines shows that sailing in circles with a jammed headsail is sometimes the only physically possible way of furling in rough weather, but the thought of doing this in heavy seas in the Noord Hinder TSS is somewhat daunting! So again the message is to make sure the equipment is in reliable working order before the start of the passage; the main problem with bearings, top and bottom, is saltwater ingress. A storm jib with a wire luff should also be available to rig on a spare halyard as a jury headsail.

Watchkeeping

Compared with a well crewed offshore racer discipline in a smaller often family crewed cruising yacht is usually more relaxed and the watchkeeping system looser. This is all the more reason for the skipper to lay down in advance some minimum watchkeeping rules to avoid the exhausting situation of being on watch himself for long periods with everyone else asleep. The problem on these short passages with limited crew numbers is the shortage of time to adapt to a new sleeping timetable, and few people get much sleep during off watches which tend to be short and frequent. Some cruising vessels start the watchkeeping system at around 2000 or 2100 to take the crew through the night only, but this does not avoid the need to ensure a constant lookout and to allocate helmsmen for the rest of the passage, so watchkeeping is advisable directly on leaving the mooring. A good system for a crew of three on overnight passages is the three/three or three/two method; three 3hr or 2hr watches (depending on the crew's preference as some prefer short lookout/helming periods) rotated for each member, including the skipper, as follows:

- 2 or 3hrs ON watch (at the helm if not on autohelm)
- 2 or 3hrs OFF watch (lying down essential even if not asleep)
- 2 or 3hrs STANDBY (cooking, helping with sail changes, or relaxing)

This order should be strictly kept as the on-watch member must be allowed to 'switch off' and sleep immediately on finishing his watch. The standby member can prepare a hot drink and food half an hour before he goes on watch. Inevitably the skipper of this three-man crew will find himself navigating or on lookout 'out of hours' depending on likely dangers, but the opportunity of a complete scheduled 'off' watch for him is still essential. With a four-man crew it is probably better if the navigator/skipper still operates this system, exempting himself to be continually on watch, snatching sleep during longer non-navigational periods. Above all the skipper has a duty to himself and his crew to juggle the three somewhat contradictory balls of conserving his own strength, putting the crew at their ease, and remaining vigilant. With a yet larger crew of 5 or more a 4hr watchkeeping system can be operated with two watches of two persons each within which helm/lookout/cooking can be rotated but again the skipper/navigator should exempt himself. But the best laid watch systems usually degenerate a little on a short passage, being strictly maintained initially and during the hours of darkness but slipping a little as the other side is reached and spirits raise, even though the weather often blows up when closing these 'wind-parallel' coasts. This is yet another cue for the skipper to be on his toes since he needs all his pilotage skills to get into that unfamiliar harbour.

Avoiding navigational error

Knowledge of astronavigation can be useful but is not an indispensible tool even on the longer 1 to 1½-day passages in the North Sea. However, the prospect of an engine breakdown during light weather and drifting for considerably longer periods persuades many a skipper to put yet another reassuring set of safety equipment into an accessible locker; namely a sextant and a set of sight reduction tables (*AP3270* or *NP401*), which, together with the little-used nautical ephemeris pages of *Reed's Nautical Almanac* and a digital watch provide all the necessary information for estimating astronavigational position lines. On such short partly overnight passages in a region where the sky can cloud over for long periods, reliance on sun-run-meridian altitude navigation or rounds of star/planet position lines in two brief twilight periods is impractical. The most the amateur navigator can hope for is an occasional near-vertical morning or afternoon sun position line confirming distance run on E–W passages or a meridian altitude from sun or Pole Star to confirm maintenance of track on E–W passages or distance run on southerly passages.

In the final resort the navigator must depend on thorough experience, culled over several years under varying weather conditions, of conventional coastal navigation techniques requiring accurate extraction and use of published tidal information and accurate chart drawing, combined with a properly corrected and swung steering compass, a hand-bearing compass, an echo sounder, and possibly, but not indispensibly, the help of one of the various electronic postion-fixing systems. These systems should always be regarded as adjuncts to the basic chartwork and logbook entries. There are frequent claims by characters propping up yacht club bars that eventually these position-fixing systems will become so reliable, that conventional navigating techniques will become as unnecessary as several hundred feet of lead line now that echo sounders are so reliable. But an echo sounder provides a single position line and if it goes on the blink there is always a variety of other necessary position-line data (times, distances, log readings, fixes etc.), worked up to various stages on the chart and in the log, to help one get back to harbour. In contrast the new electronic position-fixing systems provide a single pinpoint latitude/longitude position as well as a built-in clock, which theoretically removes the need for any chartwork and log entries, but if they go wrong and there happen to be no visual objects from which to fix, and no logged records have been kept then the navigator is completely lost in a featureless void. It cannot be stressed enough that these systems should be regarded as an adjunct to and not a replacement for conventional chartwork, logbook entry/calculations and records.

Passages of 60 to 120M out of sight of land require a thorough appreciation by the navigator of the concept of an area of error, usually a four-sided quadrilateral or diamond-shaped figure around each of his position-fixes and EPs, which enables him to set the safest subsequent course to steer for his destination that will give a reasonable clearing distance from potential dangers. The size of this diamond of error is dependent on two potential errors and two physical factors.

Potential errors:

1. Distance run by the log; a potential 5% error is not unusual in extreme sea conditions even with the best-calibrated equipment.

2. Angle of the bearing taken or of the estimated track direction; 5°is not an unusual error in reading a hand-bearing or steering compass or in taking an RDF bearing. Estimating a helmsman's course steered, leeway, and tidal set and drift especially in exceptional weather conditions can produce errors even larger than 5° in ground track estimation particularly if errors happen to work in a similar direction rather than compensate.

Physical factors:

1. The longer the distance from the object of bearing or the starting point of a course the wider the ultimate distance off track.

2. The further away from 90°, either obliquely or acutely, is the angle of cut of bearings and track lines then the longer and narrower (i.e. further away from a square) is the quadrilateral or diamond of error resulting from one of the above errors in distance or angle measured.

A series of reciprocal tidal sets tend to cancel out over a long distance leaving a small net factor which usually can be reasonably estimated. Steering error and leeway cause the largest element of error. Over a 100M passage a beam or slightly ahead-of-beam wind possibly combined with helmsman's correction can sometimes mean a 5–10M error downwind of the track. Simple geometry indicates that a constant 5° error over 100M can mount to 8–9M inac-

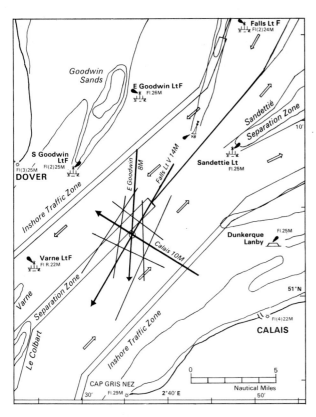

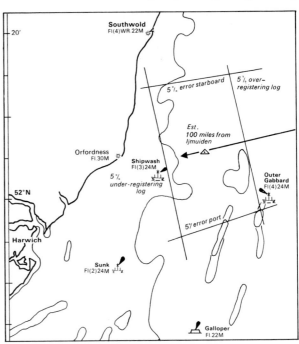

Fig. 10A Diamond of error. 5 degrees on each side of the assumed bearing is an easy error to make with RDF. In the Dover Strait TSS two such bearings 8M and 10M from sources produce a large diamond of error.

Fig. 10B Diamond of error. Distance and course. This yacht is 100M from IJmuiden aiming to pick up either the Orfordness, Shipwash or Outer Gabbard lights. In the absence of previous fixes from (say) the Rijn Oil Field or the loom of the Noord Hinder light or from RDF, the vessel could be anywhere within this 15M by 10M quadrilateral. In practice many navigators would have a strong indication of whether they were in the N or S section by the direction of the weather and their interpretation of helmsman's error. Soundings would also help.

Harwich harbour and looking up the River Orwell

Photo Lester McCarthy, Motor Boat & Yachting

27

curacy. RDF bearings and very high objects, such as chimneys or hills, from which bearings can be taken at considerable distances suffer particularly from distance inaccuracy; an acceptable error of 3° for a good RDF receiver means a 1M error of position if the beacon is 20M away.

Electronic position-fixing equipment

RDF, the first cheap electronic position-fixing system for yachts and now reaching reasonably high levels of accuracy, is rapidly losing popularity to the new Decca and Satnav systems. The worldwide Omega system is expensive, less widely available and less reliable than the others, Loran C does not cover this region, whilst Consol coverage is no longer adequate, so the real choice is between two systems; Decca and Satnav.

The Decca Navigator System operated by Racal–Decca is currently the best of the two with accuracy of the order of 0.1 to 0.2 nautical miles. This currently covers the whole of northwestern Europe and Scandinavia, but excludes Iceland and most of the Mediterranean east of Gibraltar. It also covers a number of isolated areas: Newfoundland, Japan, South Africa, Persian Gulf, northwest and northeast Indian coasts and northwest Australia. Chains of master and slave stations transmit on different frequencies generating a set of three curved (hyperbolic) position-line lattices to provide a fix for vessels on special hyperbolic lattice-overlaid Admiralty charts and also on similar special fishing charts published by Imray, Laurie, Norie and Wilson Ltd. Provided a broad estimate of position relative to the appropriate chains is available, a three-line position fix can be placed on the lattice chart. These charts will be phased out in the near future since more recently Decca receivers have become available, particularly those of a compact size for yachtsmen, which convert these hyperbolic position lines into a single position-fix in latitude and longitude on a digital readout. It is necessary to start off a passage by keying in an estimated position to enable the receiver to select the most appropriate station chains. Laying off a fix on a large chart by latitude and longitude on a small yacht's chart table can often be difficult and prone to error, but, depending on the type of equipment used, there are a wide range of other functions available to assist chartwork. There are a number of receivers on the market; for example the Decca *Yacht Navigator III* marketed by Racal–Decca itself operates at less than 0.2 amps. These are some of its functions:

1. Storage of up to 25 waypoint positions on a route plan.

2. Instant display of distance off track and to which side, and bearing and distance to the next waypoint.

3. An audible signal is given when passing a waypoint, i.e. bisecting the courses between successive waypoints, and the next waypoint is then automatically locked in.

4. Instant display of ETA and distance to go to any selected waypoint.

5. Estimate of true course and speed over the ground over a selected interval of between 2 and 99 minutes.

6. Instant store of a position in a man overboard situation together with instant display of course and distance to return to that position.

7. A self-checking function in the event of difficulty in position updating which tests whether the receiver or the Decca chain being used is at fault.

8. Maintenance of all stored information (including an automatic clock and date function) when switched off for up to six months.

Laying off Decca latitude/longitudes and waypoint bearings/distances on the chart, interspersed with the laying off of conventional courses, tidal sets, fixes and EPs, provides a very powerful, accurate and continuous (20-second updating, unlike Satnav) tool for long-distance passage making.

Satnav is in a development phase. Currently receivers use the US Navy Navigation Satellite System which depends upon a series of somewhat unsatisfactorily spaced satellites, and since positions may be obtained only when a satellite is sufficiently above the receiver's horizon fixes are obtained at intervals varying from 90 minutes in low latitudes to 30 in high though much longer intervals may occur. Accuracy varies with receiver from 0.1 to 0.2 of a nautical mile. Receivers vary but provide additional functions somewhat similar to those of the Decca receivers described above. A new satellite system, Navstar GPS, is currently being established with 21 satellites in 6 evenly spaced orbits with continuous fixing ability and a higher level of accuracy than the USNNSS, but this will take until the early 1990s to develop. When this does happen Decca systems will probably be phased out, but the navigational technique of using Decca receivers is equally applicable to Satnav.

Automatic steering, radar

It is hardly necessary to list the advantages of electronic, or mechanical, automatic steering equipment for single-course steering over long periods, but while the usual advantage cited is the removal of fatigue and eyestrain for a short-handed crew leaving them free for proper lookout there is no doubt that for the navigator the removal of varying levels of helmsman's error is its major attribute, providing the equipment works efficiently and does not need constant manual adjustment.

Fog is an equally, if not more dangerous situation to cope with than gales during a long passage and every effort should be made to pick weather which removes the chance of this arising whilst crossing the TSSs. Increasingly compact and cheaper radar equipment is becoming available for yachts, although not so far practicable for the smallest vessels. For this equipment to be of any use the operator has to have had considerable practice in interpreting the information, which means good visibility daylight practice to check results before risking bad visibility. On a short-handed yacht on a short passage the problem arises of handling yet another piece of electronic equipment which uses up significant battery capacity. More limited audible-warning radar systems which give bearings of the signals emitted from oncoming ships' radars at varying levels of sound, and available at reasonable prices for quite small vessels, can be a useful and more practical alternative. However, the compactness, quality of output, and simplicity of usage of the more sophisticated yacht radar systems is constantly improving, and there is no doubt that their use will spread to eventually reach the usage levels of electronic position-fixing equipment.

Tidal planning and the traffic separation schemes (TSSs)

For yachtsmen used to short-distance coastal cruising where familiarity minimises the number of tidal calculations, logbook and chart entries, a North Sea passage requires a little more planning. There are three planning objectives:

1. To minimise the overall distance and time taken in order to be confident of forecast weather conditions.

2. To minimise hazards en route (shoal patches, TSSs and DWRs, the need to enter a destination estuary on the flood etc.) in the given weather conditions.

3. To simplify the navigation and thus the chances of error, e.g. minimise the number of course changes, minimise distance between major departure and arrival fixes on each coast reducing the number of EPs required as opposed to accurate position fixes.

The passages examined in the next chapter vary from 22 to 147M, and at the assumed 4½kn average for an auxiliary sailing vessel require anything from 1 to 5 reciprocal tidal sets often directly across the vessel's track. A motor cruiser might double this average speed but still require ½ to 2½ tidal sets and careful position estimation. A second complication is that passages are usually across the wind, resulting often in appreciable leeway. A third complication is the need to change course to cross the TSSs and DWRs at right angles.

The following 7-stage passage-planning routine is therefore recommended before setting off, preferably done on the day before the passage but thoroughly re-checked with the weather forecast on the day of the passage.

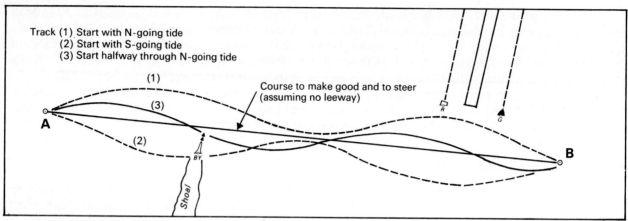

Track (1) Start with N-going tide
(2) Start with S-going tide
(3) Start halfway through N-going tide

Course to make good and to steer
(assuming no leeway)

Fig. 11A Tests of three start times

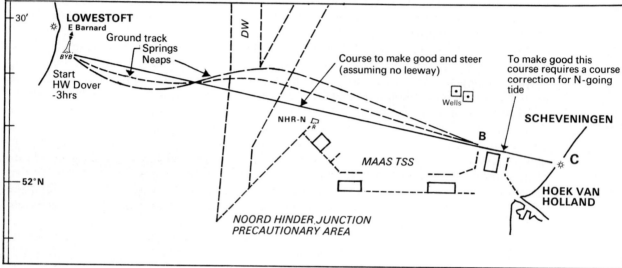

Fig. 11B Lowestoft to Scheveningen (Chapter 4, Passage 21)

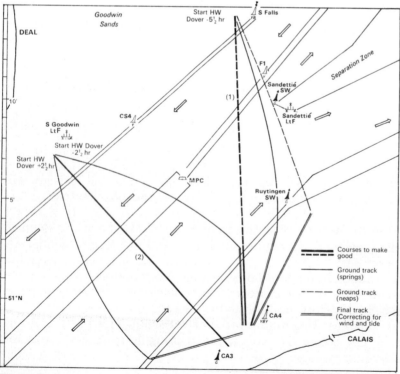

Fig. 12 Course correcting in the Dover Strait

Photographs opposite

1. At Zierikzee

2. The motoring cone about to be hoisted in Dutch waters where the regulations are tightly enforced

3. Typical of North Sea traffic – the roll-on roll-off container ship

4. The yacht harbour at Den Helder

5. The old harbour at Maassluis on the Nieuwe Waterweg.
   Photo: VVV Maassluis

1

3

2

4

5

1. Rule off on the chart the most important long-distance course to be made good (CTMG), correcting if necessary for a right angle across the shipping lanes. Some people prefer to do this latter correction when on passage.

2. Measure distance of main CTMG and approximate distances to and from the departure and arrival harbours.

3. List times of HW Dover on day(s) of passage and also for the local HW/LWs at the departure and arrival harbours.

4. Estimate an average speed for the passage assuming reasonable but not excessive use of engine to keep up to the average.

5. Estimate the approximate ground tracks of the vessel for several different 'test' start times along the main CTMG. Four alternative times from the main departure point, at most, are usually sufficient: at each of the turns of the tide northwards and southwards, and halfway through each N and S tide. With practice this estimation can be rapid and easy since each pair of the above tracks are usually mirror images on opposite sides of the CTMG. Each track is estimated by assuming the CTMG is the course steered/wake course (i.e. also assuming no significant leeway). Each ground track can easily be approximated by pencilling in EPs at approximate 6-hourly intervals at the points where the tidal streams turn and the vessel is usually farthest away or nearest to the required CTMG. Each 6-hourly EP involves totalling up 6 tidal drifts and estimating a single tidal set from the tidal diamonds or the tidal atlas.

6. Estimate times of departure and arrival for each 'test' track in order to check the state of tides in the departure and approach areas.

7. Given the expected weather conditions and potential leeway drift select the start time which minimises hazards en route and provides the best tides in the approach and departure areas.

With electronic position-fixing equipment a series of waypoints along the route and/or marking nearby hazards can be loaded into the machine and marked on the chart. As well as displays of bearings, distances off, and lat/long fixes most machines also have a function which displays 'distance off track to the next waypoint' and also the side of the track to which the vessel has deviated, and if waypoints have been marked on the main CTMG this function can give an instant picture of the vessel's deviation in relation to the above passage plan.

Fig. 11A illustrates the 'test' start approach for a simplified situation in which 4 tides occur and there are two hazards en route, a shoal to the south of the CTMG and the beginning of a TSS to the north. Line 1 shows the estimated ground track starting at 'A' at the commencement of the N-going tide which keeps the vessel to the north of the CTMG throughout, which of course would be augmented if there were a southerly wind. Line 2 is a mirror image course to the south of the CTMG starting at 'A' at the beginning of the S-going tide. Line 3 starts halfway through the N-going tide and wanders backwards and forwards across the CTMG, and of course a 4th line could be drawn (not on the figure) which is a mirror image of line 3 starting halfway through the S-going tide. Of the three lines charted 3 seems the safest since 2 clips the shoal and 1 comes very close to the TSS. The 4th potential line might be better if you wish to keep well away from the buoy marking the shoal, but a sighting could provide a useful position fix.

Tidal streams never produce a precise symmetrical pattern in reality. Fig. 11B shows a practical example from Passage 21 (Chapter 4) with actual tides laid off. Assuming 4½kn average speed the distance implies a little more than 3 tides. A ground track has been selected starting at the East Barnard buoy at HW Dover −3hrs leaving 3hrs of S-going tide ensuring the yacht is well to the north of the major TSSs throughout except nearing Scheveningen where a N-going tide helps to keep the vessel away from the Maas north TSS but requires a course correction to reach the final destination. The course steered and CTMG cross the DWR where it fans out into two lanes 30M from the East Barnard buoy but at an angle which is acceptably close to right angles to the traffic, although with an exceptional spring tide it might cross both DWRs further north over an unacceptably long period of time. A start 1 or 2hrs earlier might be better at springs, although this would bring the vessel closer to the NHR-N buoy and close to the ends of both TSSs.

The Dover Strait is a difficult passage which, despite its brevity requires careful timing since at exceptional spring tides the streams can run at over 4 knots in places. Fig. 12 examines two routes (included in Passages 1, 2, 3, and 4 in Chapter 4) from England to France; the reverse routes would be different. CTMG 1 on this figure, between the South Falls and CA4 buoys, is usually taken en route to or from Ramsgate and the Thames Estuary and outside Goodwin Sands. The 15M from South Falls to CA4 is a 3hr plus crossing, half a tidal set. A course heading at right angles to the TSS (no leeway correction) and starting 1M west of South Falls buoy, 5½hrs before HW Dover on a SW-going tide, and averaging 4½kn makes a track which at average springs exits the lane close west of the Ruytingen SW buoy and at average neaps to the east of the buoy; and from these points a course correction for tide is needed to round CA4 buoy and push a slackening tide in the buoyed channel to Calais. At exceptionally weak neaps on this route the SW-going tide may not be enough to push the vessel outside the Sandettié LtV so the bank may have to be crossed at its western end, or alternatively by starting the TSS crossing further west of the South Falls buoy this hazard can be avoided.

CTMG 2 starting 1M SW of the South Goodwin LtF is 13M/3hrs to CA3 buoy, and two timings and tracks are charted. The northerly one starts at 2½hrs after and the southerly at 2½hrs before HW Dover. In both cases the position of exit from the TSS must be estimated and a course correction applied to reach CA3 and the channel buoys. The northerly track uses the last 3hrs of NE-going to slack tidal stream, reaching the channel as the tide is turning against, while the southerly uses the last 3hrs of SW-going stream reaching the channel with the beginnings of a fair tide. The southerly route is not only much better because of the fair tide in the channel, but also because yachts from Ramsgate and the Thames Estuary can use the preceding 4hrs of slack and S-going tide in the Downs.

This type of advance planning not only enables an optimum start time to be selected but thoroughly familiarises the navigator with the route and its potential hazards so he can react more rapidly if the weather does deteriorate. It is considerably helped by the use of one of the Decca position-fixing receivers, where advance inputting of the coordinates of a series of waypoints requires careful scrutiny of the chart along the route. The section below examines this.

Lessons from an actual long-distance passage
See also Chapter 4 Passage 20

A long-distance passage at an acute or obtuse angle to the TSSs requires a substantial course correction to cross the lanes on the right course heading; the actual passage discussed below illustrates this as well as summarising points made in this Chapter. The rhumb line on this passage crosses a DWR, and although it is not mandatory to cross a DWR on a heading at right angles to the traffic, in this case it is definitely advisable, since the crossing is at a dangerously acute angle which could impede slow-moving and slow-stopping deep-draught ships, as well as covering a considerable distance across the junction of two DWRs.

The passage from Den Helder to Levington (River Orwell) was made on 30–31 May 1986 in the author's vessel *Teazle* a 26ft LOA, 4½ft draught, Offshore Eight GRP sloop with a long, bulbed fin keel and a 9hp Farymann diesel. The vessel was equipped with all the usual safety apparatus including a hired liferaft. In addition to the standard instruments of compass, VHF radio and echo sounder, it was also equipped with a Decca *Yacht Navigator III* (*DYNIII*) a *NAVTEX I* and an *Autohelm 1000* (push-button control). There were three crew operating a 'three/two' watchkeeping system (2hrs on, 2 off, 2 on standby).

ADVANCE PLANNING INFORMATION FOR THE PASSAGE

Total distance 150M

Likely duration 33hrs, 5–6 tidal sets at 4½kn

Tides (all times BST)
Dover: 30 May HW 0452, 1710; 31 May HW 0602, 1819
Den Helder: 30 May LW 0608, HW 1226 (*Stroomatlassen* based on LW Den Helder obtainable from the Dutch *Almanak deel 2* or Admiralty *Tide Tables*)
Harwich: 31 May LW 1226, HW 1905

Tidal streams (all times BST)
2–1 days before neaps. Dover actual range 4m (average ranges: neaps 3.3m/springs 5.9m)
Schulpengat: ingoing starts 30 May 0800, outgoing 1400
Shipwash LtF: S-going starts 31 May 1230
Harwich Ent: outgoing starts 31 May 1900

Waypoints list for entry into Decca

A straight-line course between waypoints 2 and 10 was calculated from the *DYNIII* by inputting these two waypoints only, enabling a line to be drawn from each end of this course without the need for a long straight edge; always a problem on a small-scale chart over a long distance. This initial 'sailplan' was then cleared from the *DYNIII* and lat/longs of all the waypoints inputted, and double-checked. Since waypoints 2–10 were on the straight line the course between each, obtained from the *DYNIII*, had to be the same, providing an automatic crosscheck of those inputs. The waypoints were as follows:

WP1   Kap Hoofd Point
WP2   SG, RW buoy
WP3   Abeam of TX1, G buoy
WP4   Abeam of Texel LtV
WP5   MO8, Y buoy
WP6   S end of Brown Ridge
WP7   Entry of DWR boundary
WP8   Departure from DWR boundary
WP9   Halfway between WP8 and WP10
WP10   Shipwash LtF
WP11   Landguard Point

Accurate input of waypoints is critical otherwise the ETA and waypoint bearing information obtained from the Decca on passage is nonsense and then only the actual lat/long can be relied upon. Choice of waypoints is a personal one, but physical objects are best if possible, although between two long-distance landfalls it is much easier to handle waypoints on the same line whether they mark an object or not, since 'distance off track' then gives a good running estimate of tidal set en route. It is essential to use Decca for charting position rather than as an automatic homing device, otherwise it is possible to collide with waypoint objects or ships heading for the same waypoints. Recently charted positions provide the necessary information to deviate from the inputted course to avoid unforeseen danger, and leave one able to reckon where the yacht is when the Decca fails. The above list for example included a number of visual objects, but in all cases en route they were used for bearing and distance position, often checked by the lat/long position.

On this passage it was necessary to push the tide out of the Schulpengat in order to have a chance of obtaining a full S-going tide at the exhausting end of the passage from the Shipwash LtF onwards. This was not difficult as it was near to neaps, and was certainly well advised as the high pressure ridge with which we started the passage gave way to strong SW winds at the end requiring the fair tide, although of course this built up uncomfortable seas off the Suffolk coast.

The log entries (Fig. 13B) summarise the 34hr passage, although a much larger number of positions were recorded on the chart, and the diagram (Fig. 13A) shows the actual track with its 5½ tidal sets distorted by the DWR course correction. The problem on this passage was light, though favourable, winds requiring motor-sailing for the first 15hrs, veering to SW winds for the last 12hrs, when the pace was beginning to tell, again requiring motor-sailing. Non-use of the engine would have required probably an extra day's passage. For 90% of the passage the *Autohelm 1000* was left completely alone, after putting the vessel on the required true course by crosschecking between hand-bearing and steering compass. Apart from hand bearings of visual objects and checking the steering compass occasionally these were the only occasions when the compass was consulted. En route at frequent intervals the true course and speed made good over the ground was taken from the *DYNIII* to check how the tide was setting the vessel, whilst positions were noted frequently from the Decca using both lat/long as well as bearing and distance from the next waypoint.

It is essential to enter a log reading and time with every charted position, and these should be made at least hourly. In the event of an electrical equipment or battery failure this information is essential in order to take over dead-reckoning from the Decca. Visual fixes are also a constant useful check, but on this trip these were only available in the early and later stages: the Texel LtV, production platforms/rigs, yellow buoys marking wellheads, the Shipwash

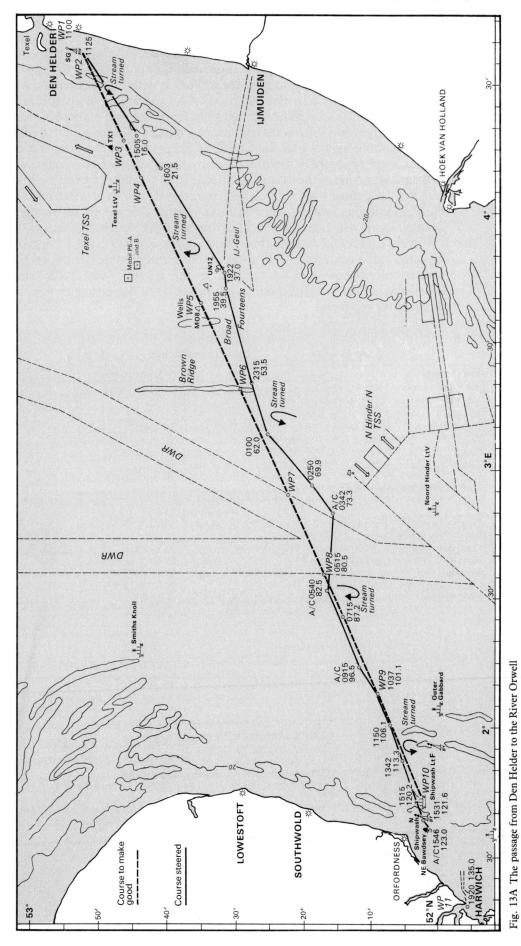

Fig. 13A The passage from Den Helder to the River Orwell

LtF and the many buoys and landmarks on the Suffolk coast. Soundings, such as those near the Brown Ridge and homing in on and leaving the coasts, were occasionally useful.

Oil and gas platforms are being established, removed and towed constantly from location to location in the North Sea, and it is difficult if not impossible to keep charts up to date for this. The essential thing is to note the wellheads without installations which are marked by yellow buoys, and we were not surprised by the appearance of a platform in one of these locations but not yet on the chart. At night platforms are well marked by varying combinations of a 15M range Mo(U) ('you are running into danger') flashing light, a subsidiary lesser range Mo(U) flashing light at each corner of the platform, 2 F.R lights, and a Mo(U) Horn. The *NAVTEX I* was particularly useful in locating the latest platforms, telexing out on a day prior to the passage a list of the extant installations. Its two other uses came in listing floating dangers such as lighted buoys which had come adrift, and in printing out a substantial piece of the BBC shipping forecast, although some hours after the actual broadcast and without the station reports which are so helpful in drawing up a sketch met. map.

Finally it is useful to note that whilst on the English coast a departure position can be reported to the nearest H.M. Coastguard Maritime Rescue Centre; on the continental coast similar information can be radioed to the nearest coast radio station, in this case Scheveningen Radio (Huisduinen area Ch 16), not forgetting to sign-off from Orfordness Radio at the other end.

Voyage from _Den Helder_ to _Levington Marina_ Date _30/31 May_					
Time	Remarks	Log	Course	Wind	
1015	Den Helder Ent. – Engine on / Full Sail			NNW3	
1025	TR reported Scheveningen Radio			"	
1100	WP1 (Kap Hoofd) 0.3 to port			"	
1125	SG / WP2 Log streamed A/C 245° True	0	245°		
1505	WP3 / Tx1 Abeam 2.0 to Starboard	16.0	"	"	
1603	WP4 / Texel Abeam 3.3 to Starboard	21.5	"	NW2	
1922	UN12 (yellow buoy) close to port	37.0	"	"	
1955	WP5 (M08 buoy) 4.3 to Starboard	39.5	"	0	
2100	Topped up diesel tank			"	
2315	WP6 (Brown Ridge) 2.1 to Starboard (Sounding confirmed)	53.5	"	" / S2	
0100	Engine off / Full Sail 1.0 South of track plan	62.0	"	S 3/4	
0250	WP7 (DWR entry) 3.9 to Starboard	69.9	"	"	
0342	A/C 286° True to cross Deep Water Route	73.3	286°		
0515	WP8 (DWR departure) close to	80.5	"	"	
0540	A/C 245° True for Shipwash Light Vessel	82.5	245°	SW2	
0715	Engine on / Full Sail. On track	87.2	"	SW3	
0915	A/C 230° True. Sails off. Track plan 0.8 to port	96.5	230°	"	
1037	WP9 close (en route for Shipwash Lt V)	101.1	"	"	

Continued Voyage from _____ to _____ Date _____					
1150	Track planned 0.3 to port	106.1	"	SW4	
1342	1 reef in main sail / Roll in jib. Track planned 0.8 to port	113.3	"	SW5	
1515	WP10 (Shipwash LtV) 0.4 to port	120.2	"	"	
1531	N Shipwash buoy close port	121.6	"	"	
1546	NE Bawdsey buoy close Starboard A/C 245°	123.0	245°	"	
1615	TR reported Orfordness Radio / Link call home				
1715	Cutler buoy close Starboard	128.5	"	"	
1844	Platters buoy / Crossing of Harwich Channel started	133.7	–	SW3	
1920	WP11 (Landguard Pt) 0.4 to starboard / Log taken in / Engine off	135.0	–	"	
2015	Arrive Levington Marina	–	–	"	

Fig. 13B Log extracts from the author's passage from Den Helder to the River Orwell

II. PILOTAGE

4. Passages

Using the passage plans and descriptions

23 suggested passages across the southern North Sea are examined below. They provide a framework for the passage-maker and are not immutable channels cut through the water, even though thousands of yachts may make the same tracks. Weather, particularly wind strength and direction and sea state, will always determine the actual routes taken at time of sailing and during passage; some of the passages cross banks which in bad weather should be avoided, and to take any passage dead to windward with the possibility of beating across a traffic lane albeit under engine, is not a decision likely to be made. These plans aim to provide simple passages, prominent marks, minimal hazards, minimal time within the traffic lanes, and minimal potential confusion from turning traffic by avoiding the major 'precautionary areas'.

The following guidelines should be used for the plans and passage descriptions:

Plans   These sketches are not to be used for navigation. Courses and bearings are not given and should be measured by the navigator on his own charts.

Courses on the plans   These are courses to make good (CTMGs) uncorrected for tides between departure and destination points and not the actual ground tracks which vary with time of start, tidal streams and weather conditions. Bold lines are the CTMGs between major marks charted. Pecked lines are:
   1. CTMGs across TSSs where the course to steer should be as close to a right angle to the traffic as is practicable.
   2. CTMGs in approach and departure areas which require following and crossing buoyed and unbuoyed channels not drawn in detail on the plan, but examined in detail regionally in Chapters 5 to 12 under *Approach routes and tidal timing*.

Distances   Nautical miles along the CTMGs charted with no allowance for tides.

Average speed   All plans and related timings are based on an assumed average of 4½ knots uncorrected for tides. This represents an average-sized auxiliary sailing yacht which sails whenever possible and motors if necessary to keep up this average. The navigator of a yacht which averages a significantly different speed from this, say half a knot or more, should recalculate the listed duration and tidal sets before planning his passage.

Tidal streams   Tidal times are expressed in hours, rounded to the nearest half hour. Passages are planned to achieve a favourable tidal set near hazards and in the coastal approaches. Once again the navigator of a yacht which averages a significantly different speed from 4½ knots should re-estimate start times to achieve this objective. This can be done usually by thinking about the implications of the speed variation on the various parts of the route and subtracting or adding a few hours to the listed start times after comparison of the recalculated 'duration' with the listed figure.

Marks   Only major departure, turning and arrival marks are listed in detail. Groups of channel buoys are described in general terms with some further explanation in Chapters 5 to 12 under *Approach routes and tidal timing*.

Approach clearance   Except where otherwise stated, the conventional clearance of 1M on the approach course is assumed for large sea marks and buoys. In practice, however, it is usually easier to work on the chart with lines between the actual objects, and in any case in narrow channels or areas with many buoys and beacons a 1M clearance is impossible. The major thing to keep in mind is not the charted course but the actual position of the vessel and the position of dangers in relation to its track. Above all keep a constant lookout

PASSAGES – PAGE, DISTANCES, TIMES & TIDAL SETS

Page No	Passage No	Passage	Distance (n. miles)	Hours at 4½ kn	Tidal sets at 4½ kn
		DOVER & RAMSGATE TO CALAIS			
39	1	Dover Went. – Calais via CA6	22	5	1
41	2	Dover Went. – Calais via S Goodwin Lt F & CA6	23	5	1
41	3	Ramsgate – Calais via S Goodwin Lt F & CA6	32	7	1+
		THAMES ESTUARY TO CALAIS			
43	4	*Southend Pier – Calais via Princes Channel	67	15	2½
44	5	*Southend Pier – Calais via Four Fathoms Channel	65	14½	2½
		ESSEX RIVERS TO CALAIS			
45	6	†R. Crouch (Inner Crouch buoy) – Calais	60	13½	2+
		HARWICH TO CALAIS			
47	7	Harwich (Landguard Pt) – Calais	71	16	2½
		RAMSGATE TO DUNKERQUE			
49	8	Ramsgate – Dunkerque	44	10	1½+
		ESSEX RIVERS & HARWICH TO BELGIUM & WESTERSCHELDE			
50	9	†R. Crouch (Inner Crouch buoy) – Oostende	92	20½	3½
51	10	†R. Crouch (Inner Crouch buoy) – Nieuwpoort	97	21½	3½
52	11	†R. Crouch (Inner Crouch buoy) – Zeebrugge	99	22	3½
53	12	†R. Crouch (Inner Crouch buoy) – Vlissingen	115	25½	4+
54	13	Harwich (Landguard Pt) – Oostende	77	17	3
55	14	Harwich (Landguard Pt) – Nieuwpoort	82	18	3
56	15	Harwich (Landguard Pt) – Zeebrugge	84	18½	3
57	16	Harwich (Landguard Pt) – Vlissingen	100	22	3½
		ESSEX RIVERS & HARWICH TO OOSTERSCHELDE			
58	17	Harwich (Landguard Pt) – Roompot	92	20½	3½
60	18	†R. Crouch (Inner Crouch Buoy) – Roompot	106	23½	4
		NORFOLK & SUFFOLK COAST TO NOORD HOLLAND			
62	19	Harwich (Landguard Pt) – IJmuiden	126	28	4½
64	20	Harwich (Landguard Pt) – Den Helder	147	33	5½
65	21	Lowestoft – Scheveningen	98	22	3½
66	22	Great Yarmouth – IJmuiden	105	23½	4
67	23	Great Yarmouth – Den Helder	121	27	4½

Note *Distances, times and tidal sets are the same from Queenborough Spit buoy (River Medway) to each of these destinations.

†Distances, times and tidal sets are the same from the Nass Beacon (River Blackwater) to each of these destinations.

Passages page reference

KEY TO PASSAGE PLANS

Boundary of traffic separation scheme

Boundary of inshore traffic zone

Course to make good between major marks ━━━━━

Course to make good ━ ━ ━ ━ ━
a) Along narrow channels not shown in detail. Reference should be made to larger scale plans in Chapters 5 to 12
b) Where yacht must cross traffic at right angles

completely around the horizon; this is particularly important with the increasing use of automatic steering, radar and electronic position-fixing equipment which allow other vessels to precisely home in on the same marks.

Clearance in buoyed channels   Closer approach than 1M to marks in channels is often essential to avoid danger. The assumption here is that the cruising yacht always crosses a channel at right angles, keeps to the extreme starboard edge of a channel in either direction, and if depths permit keeps just outside this edge.

Passage plans   Start times are only recommended; there may be other possibilities depending on the weather and other strategic objectives. The commentaries, however, should help to give the feel for a particular route even if a different start time is selected.

Dover & Ramsgate to Calais
See also Chapter 9

The three passages below cross the world's busiest TSS where successful prosecutions of yachts disobeying the ICRs have originated. These passages also cross the strongest tidal streams in the region and are often a beam reach across a wind funnelled by the Dover Strait exaggerating the sea conditions, particularly with wind against tide. They are passages which require careful homework.

The Strait has heavy commercial through traffic and a considerable ferry and hovercraft cross traffic from outside as well as within the area, so vigilance is essential. The Goodwins, Varne bank, and the Ramsgate and Calais channels are well buoyed, and all three harbours have outer basins in which to lie until near HW for locking through into the inner basins.

Passage 3 from Ramsgate is a variant of Passage 2 joining it at the S Goodwin LtF. The first half of the outward passage can be done in most conditions, being well protected by the Goodwins as well as by the land, and in bad weather Dover is always an alternative refuge if the Strait looks rough. Passage 3 is more likely to be used outwards, and inwards, in SW weather; in NE weather the outward route outside the Goodwins is often taken instead (part of Passage 4), clawing to windward initially via the N Goodwin LtF to make CA6 (R) buoy off Calais more easily. Conversely in strong NE weather the return journey is sometimes better via the inside passage (Passage 3) protected by the Goodwins.

PASSAGE 1   DOVER W ENTRANCE TO CALAIS ENTRANCE VIA CA6

Distance   22M

Duration   5hrs, 1 tidal set

Approach routes   See Chapter 9 Routes 3, 4 and 5

Charts   Admiralty *323, 1352, 1698, 1892*
Imray *C8*

Recommended start time outward
HW Dover –1½

Arrival time outward
HW Dover +3½, HW Calais +3

Recommended start time return
HW Dover +5½, HW Calais +5

Arrival time return
HW Dover –1½

Tidal streams

At both Dover and Calais NE-going starts at HW Dover –1½ and SW-going at HW Dover +4½.

Commentary

Outward   If you get the earliest possible lock-opening before HW out of Wellington Dock it should be possible to approximate this schedule. Head direct for CA6 (R) buoy, alter course in reasonable time before the outer boundary to cross the traffic lanes at right angles, crabbing across with the tide, make a tidal correction for CA6 when well clear of the lanes then follow the channel buoys. There is a long time to wait before the lock into the Port de Plaisance is available.

Return   There is similarly a long wait to start after locking through from Port de Plaisance. Follow the channel buoys, immediately on rounding CA6 alter course to cross the TSS at right angles, and finally make a down-tide correction for Dover W entrance when well clear of lanes. Arrival is conveniently near to HW for locking through into Wellington Dock.

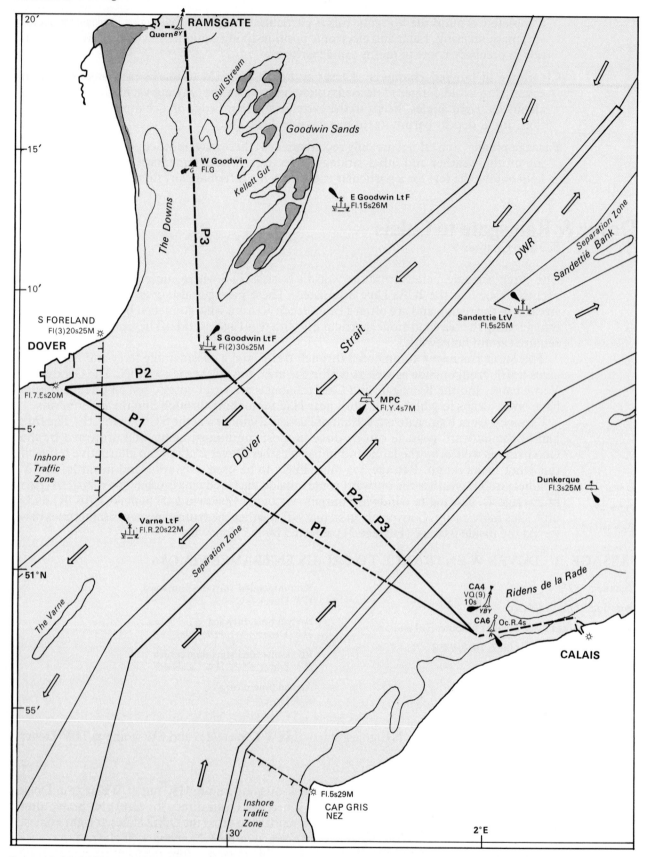

Passages 1, 2 and 3 Dover and Ramsgate to Calais

PASSAGE 2   DOVER W ENTRANCE TO CALAIS ENTRANCE VIA S GOODWIN LTF & CA6

Distance   23M

Duration   5hrs, 1 tidal set

Approach routes   See Chapter 9 Routes 3, 4 and 5

Charts   Admiralty *323, 1352, 1698, 1892*
Imray *C8*

Recommended start time outward
HW Dover –5½

Arrival time outward
HW Dover –0½, HW Calais

Recommended start time return
HW Dover –0½, HW Calais

Arrival time return
HW Dover +4½

Tidal streams

At both Dover and Calais NE-going starts at HW Dover –1½ and SW-going at HW Dover +4½.

Commentary

Outward   You will need to anchor for a few hours at Dover after leaving Wellington Dock. Motor hard up tide to the vicinity of S Goodwin LtF, alter course to right angles to the lanes, and when well clear of the lanes make a tidal correction towards CA6 and the channel buoys. The tide will be slack but it may be a race to enter Calais' tidal lock.

Return   Motor straight out of the lock hard against the tide along the channel buoys, immediately on rounding CA6 alter course to right angles to the lanes, and when clear of the lanes make a tidal correction for Dover and past S Goodwin LtF. Tide will be slack but there are a few hours to anchor to wait for the lock. A good route at neaps.

PASSAGE 3   RAMSGATE ENTRANCE TO CALAIS ENTRANCE INSIDE GOODWINS VIA S GOODWIN LTF & CA6

Distance   32M

Duration   7hrs, 1+ tidal set

Approach routes   See Chapter 9 Routes 1, 4 and 5

Charts   Admiralty *323, 1352, 1827, 1828, 1892*
Imray *C8*

Recommended start time outward
HW Dover +4, HW Ramsgate +3½

Arrival time outward
HW Dover –0½, HW Calais

Recommended start time return
HW Dover –2, HW Calais –1½

Arrival time return
HW Dover +5, HW Ramsgate +4½

Tidal streams

Off Ramsgate N-going starts at HW Dover –1½ and S-going at HW Dover +4½, i.e. at about the same time as off Dover and Calais. At Calais NE-going starts at HW Dover –1½ and SW-going at HW Dover +4½.

Commentary

Outward   Follow the Ramsgate dredged channel along its extreme S edge. From the Quern (N card) buoy (leave to starboard), head S with the tide down to the W Goodwin (G) buoy in the Gull Stream, then head towards the S Goodwin LtF leaving about 1M to port on the approach course, then alter course to right angles to the lanes, and when well clear of them make a tidal correction towards CA6 and the channel buoys. The tide will be slack and it may be a race to enter the tidal lock.

Return   Leave Calais as soon as the lock opens. Motor hard against the tide along the channel buoys, immediately on rounding CA6 alter course to right angles to the lanes, and finally when clear of the lanes make a down-tide correction for the S Goodwin LtF. Then take the tide to the Ramsgate entrance channel via the W Goodwin and Quern buoys. Cross the channel at right angles and follow its extreme N edge into Ramsgate. A long wait to lock through into the inner harbour.

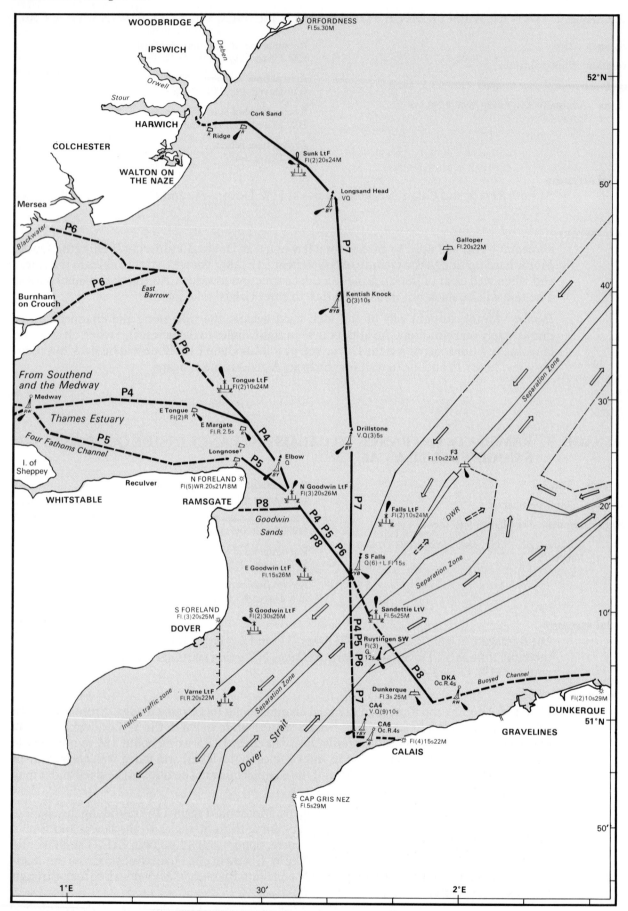

Passages 4 to 8 Essex and Kent rivers to Calais and Dunkerque

Thames Estuary to Calais

Distances, times, tidal sets and most of the commentary are the same from Queenborough Spit buoy in the Medway as from Southend Pier. See Chapter 8, *Rivers Medway and Swale. Entrance and river passage* for the pilotage from the Medway to the Medway (RW) buoy.

These are very popular passages although the beginner tends to break the journey at Ramsgate, particularly on return where the tides are more difficult. A third of each passage crosses the Dover Strait with the problems discussed for Passages 1, 2 and 3, whilst the other two-thirds follows the Thames Estuary channels and sandflats. Again winds tend to be SW or NE, into and out of the Estuary rather than across, and, as in the Dover Strait, the tidal stream often coincides with wind direction, and wind-over-tide conditions frequently occur. The worst conditions are generated by E to N winds due to their much longer fetch. Since the passage is a dogleg, part of it is usually to windward and part freed off.

Part of Passage 4 in the Estuary is out of sight of land but it is well marked and in addition to the immense number of lighted and unlighted buoys is never far from one of the litter of towers, wrecks and beacons (see Chapter 8, *Major fixed daylight marks*).

PASSAGE 4 SOUTHEND PIER TO CALAIS ENTRANCE VIA PRINCES CHANNEL AND OUTSIDE GOODWINS

Distance 67M

Duration 15hrs, 2½ tidal sets

Approach routes See Chapter 8 Route 4 and Chapter 9 Routes 2, 4 and 5

Charts Admiralty *323, 1185, 1352, 1607, 1892* Imray *C1, C8, Y7*

Recommended start time outward
HW Dover –3, HW Sheerness –4½

Arrival time outward
HW Dover –0½, HW Calais

Recommended start time return
HW Dover –2, HW Calais –1½

Arrival time return
HW Dover +0½, HW Sheerness –1

Tidal streams

Outward From Southend tides start flowing E out of the Estuary at HW Sheerness (HW Dover +1½). North Foreland, 7 to 8hrs away from Southend Pier, is the critical point, and here the tidal stream first 'turns the corner' S towards Calais 3½hrs later at HW Sheerness +3½ (HW Dover +5). At Calais NE-going (unfavourable) starts at HW Dover –1½. Outwards therefore it pays to start well in advance of HW and push the tide in order to get a full set from North Foreland to Calais.

Return NE-going (favourable) tide starts at HW Dover –1½, and at North Foreland, 7 hrs distant, the tide turns ingoing/westwards into the Estuary 11hrs later at HW Dover –3, so it is impossible to avoid motoring hard to push some foul tide albeit a weakening to slack stream for 3 to 4hrs across to Princes Channel. Arriving at North Foreland before HW Dover +5 is critical; arrive any later and a strong S-going stream is encountered.

Commentary

Outward First 4hrs will be against the stream, probably motoring. Cross Yantlet Channel from Southend Pier at right angles to the channel, follow extreme S edge of channel and make another right-angled crossing of the outer end of Medway dredged channel in the vicinity of the Medway (RW) buoy, yachts from the Medway enter the route at this point, then follow extreme S edge (R buoys) of channel to Red Sand Tower, E Redsand (R) buoy, the Princes Channel (R) buoys to the E Tongue (R) buoy. Thence take the strong favourable stream to the E Margate (R) buoy, Elbow (N card) buoy and the N Goodwin LtF keeping to the E in all cases, correct your course across the tide towards S Falls (S card) buoy choosing a distance from this buoy based on the strength of tidal stream (see Chapter 3), a right-angled course to the traffic lanes will crab across to CA4 (W card), CA6 (R) and then turn to follow the channel buoys into Calais, in time for locking through to the inner harbour.

Return Lock out of Calais inner harbour as early as possible on the rising tide. It may be better to lie in the outer harbour in order to start promptly and avoid the possibility of missing the tide at North Foreland and having to run into Ramsgate. Follow the channel buoys to CA6 rounding to cross the traffic lanes at a right angle; the NE-going tide will push the vessel out of the lanes at a point, depending on strength of tide, between the S Goodwin LtF and S Falls (S card) buoy. Then take a tide-corrected course E of the N Goodwin LtF, Elbow (N card) and E Margate (R) buoys, by which time some hard motoring will be required to catch a favourable tide back along the Princes and Yantlet channels, this time keeping to the extreme N edge of the channels and crossing the end of the Medway and Yantlet channels at right angles near the Medway (RW) buoy, yachts going to the Medway exit at this point.

PASSAGE 5   SOUTHEND PIER TO CALAIS ENTRANCE VIA FOUR FATHOMS CHANNEL AND OUTSIDE THE GOODWINS

Distance   65M

Duration   14½hrs, 2½ tidal sets

Approach routes   See Chapter 8 Route 5 and Chapter 9 Routes 2, 4 and 5

Charts   Admiralty *323, 1185, 1352, 1607, 1892* Imray *C1, C8, Y7*

Recommended start time outward
HW Dover –3, HW Sheerness –4½

Arrival time outward
HW Dover –0½, HW Calais

Recommended start time return HW Dover –2, HW Calais –1½

Arrival time return
HW Dover +0½, HW Sheerness –1

Tidal streams

Tides are generally weaker over this route across the Kentish Flats than in Princes Channel so with a moderate draught it is a much better route to take if the weather is right. Both outwards and return the shallowest part of the route across the Flats is conveniently on a rising tide.

Outward   From Southend tides start flowing E out of the Thames Estuary at HW Sheerness (HW Dover +1½). North Foreland, 7 to 8hrs away from Southend Pier is the critical point, and here the tidal stream first 'turns the corner' S towards Calais 3½hrs later at HW Sheerness +3½ (HW Dover +5). At Calais NE-going starts at HW Dover –1½ and SW-going at HW Dover +4½. Outwards therefore it pays to start well in advance of HW in order to get a full set from North Foreland to Calais, and pushing the first 4hrs of tide is not difficult across the Kentish Flats.

Return   NE-going (favourable) tide starts at HW Dover –1½, and at North Foreland, 7hrs distant, the tide turns ingoing/westwards into the Estuary 11hrs later at HW Dover –3, so it is impossible to avoid motoring hard to push some foul tide albeit a weakening to slack stream for 3–4hrs across to Princes Channel. Arriving at North Foreland before HW Dover +5 is critical; arrive any later and a strong S-going stream is encountered. Weaker streams will be encountered after rounding North Foreland than those on the route across to Princes Channel.

Commentary

Timings are the same as for Passage 4; the leg in Passage 4 between the Medway (RW) buoy and the N Goodwin LtF is the only difference. The main buoys are all lighted to aid night sailing.

Outward   First 4hrs will be against the stream, almost certainly motoring. Cross Yantlet Channel from Southend Pier at right angles to the channel, follow extreme S edge of channel and make another right-angled crossing of the outer end of Medway dredged channel in the vicinity of the Medway (RW) buoy, yachts from the Medway enter the route at this point. Then round the Spile (G) buoy to the S, cross the Flats to pass between the Hook Spit (G) and E Last (R) buoys (N of the twin Reculver Towers landmark, see Chapter 8), follow the Gore Channel buoys out to the yellow buoy marking the end of a pipeline 6 cables N of Longnose (R) buoy, round this yellow buoy to its N then cross to the N Goodwin LtF keeping to its E. Then correct course across the tide towards S Falls (S card) buoy choosing a distance from this buoy based on the strength of tidal stream (see Chapter 3), a right-angled course to the traffic lanes will crab across to CA4 (W card), CA6 (R) and then follow the channel buoys into Calais, in time for locking through to inner harbour.

Return   Lock out of Calais inner harbour as early as possible on the rising tide. It may be better to lie in the outer harbour in order to start promptly and avoid the possibility of missing the tide at North Foreland and having to run into Ramsgate. Follow the channel buoys to

CA6 (R) rounding to cross the traffic lanes at a right angle; the NE-going tide will push the vessel out of the lanes at a point, depending on strength of tide, between the S Goodwin LtF and S Falls (S card) buoy. Then take a tide-corrected course to round the N side of the N Goodwin LtF, then across to the yellow buoy marking the end of a pipeline 6 cables N of Longnose (R) buoy. Round this yellow buoy to its N, and follow the Gore Channel buoys to pass between the Hook Spit (G) and E Last (R) buoys (N of the twin Reculver Towers landmark, see Chapter 8). Then set a course to close S of the Spile (G) buoy, then towards the Medway (RW) buoy, yachts going to the Medway exit at this point. Cross the Medway dredged channel at right angles, then follow the outside of the S edge of Yantlet Channel and cross at right angles opposite Southend Pier.

Essex rivers to Calais

Distances, times, tidal sets and most of the commentary are the same from the Nass Beacon in the Blackwater as from the Inner Crouch buoy. Chapter 7, in particular Route 1 (first half), describes the pilotage from the Nass to the Swin Spitway. Most of the notes applying to Passages 1 to 5 apply to Passage 6.

There is the problem of crossing a number of Estuary swatchways; despite the frequent possibility of beam reaches crossing the longer stretches, the need to zigzag around the banks and cross the shipping channels at right angles means frequent course changing, beating and use of engine. Estimating tidal depths is critical in crossing the banks and if in doubt do not hesitate to detour. Chapter 7, first half of Route 1, describes the route from the Blackwater which joins this passage at the Whitaker (RW) buoy after passing through the Spitway channel, and is the same in overall distance and timing.

PASSAGE 6   RIVER CROUCH (INNER CROUCH BUOY) TO CALAIS ENTRANCE

Distance   60M

Duration   13½hrs, 2 plus tidal sets

Approach routes   See Chapter 7 Route 1, Chapter 8 Route 6 and Chapter 9 Routes 2, 4 and 5

Charts   Admiralty *323, 1183, 1352, 1607, 1892, 1975, 3741, 3750*
Imray *C1, C8, C30, Y6, Y7, Y17*

Recommended start time outward
HW Dover –1½, HW Burnham-on-Crouch and Bradwell (Blackwater) –2½

Arrival time outward
HW Dover –0½, HW Calais

Recommended start time return
HW Dover –5, HW Calais –4½

Arrival time return
HW Dover –4, HW Burnham-on-Crouch and Bradwell (Blackwater) –5

Tidal streams

Outward   These start running out of the Essex rivers with local HW (HW Dover +1). At North Foreland the tidal stream first 'turns the corner' S towards Calais 4hrs later; HW Burnham +4 (HW Dover +5). At Calais NE-going starts at HW Dover –1½ and SW-going at HW Dover +4½. Outwards therefore it pays to start well in advance of HW and push the tide in order to get a full set from North Foreland to Calais.

Return   This direction is a very difficult tidal route; if in any doubt it pays to break the journey at Ramsgate and cross the Estuary the following day (see Chapter 8, Route 6). NE-going tide at Calais starts at HW Dover –1½, but in this return direction the tide in N Edinburgh Channel is critical, starting to flow outwards, i.e. adverse, at HW Dover +2½, only 4hrs later. Since 8 to 9hrs are needed to reach Edinburgh Channel it is essential to start from Calais early with 3 to 4hrs of SW-going stream, and even then some engine power may be needed through the N Edinburgh. The ingoing stream to the Essex rivers starts at local LW (HW Dover –5).

Commentary

Outward   Follow Whitaker Channel to the Whitaker (RW) bell buoy, pushing about half of a tidal stream (the engine may be needed); at springs or at any time when there is doubt about a vessel's capability of averaging 4½ knots, it is best to leave earlier than the above suggested time. By Whitaker buoy the tide will be starting to ebb out of the Estuary so the engine may be needed again to round NE Middle (R) buoy to the N, and to head for Barrow No 7 (G), sounding round the edge of East Barrow Sand en route, then a dogleg pushing the tide down the extreme W edge of Barrow Deep. Cross the channel at right angles when opposite to Barrow No 8 (R) buoy continuing between this buoy and the SW Sunk beacon directly across Sunk Sand remaining on this course at right angles across Black Deep towards the NW Long Sand beacon. 8 cables NW of the beacon in 20m soundings near the edge of the channel (keep clear of the remains of *Radio Caroline* which are close to the edge of the channel at this point) alter to a S course, sounding carefully across a 3m corner of Long Sand (this is reasonably close to HW and thus a good time to cross) to the entrance buoys to Edinburgh Channel. The tide is now favourable for the remaining 8 or 9hrs to Calais. Edinburgh Channel should be followed along its extreme starboard edge, and then head for the E Margate (R) buoy passing between Tongue LtF and Tongue Sand Tower, then Elbow (N card) buoy and N Goodwin LtF keeping to the E in all cases. Correct course across the tide towards S Falls (S card) buoy, choosing a distance from this buoy based on the strength of tidal stream (see Chapter 3), a right-angled course to the traffic lanes will crab across to CA4 (W card), CA6 (R) and then follow the channel buoys into Calais, in time for locking through to the inner harbour.

Return   At this start time from Calais there are still 3½hrs of SW-going stream left pushing the vessel towards Varne LtF and Dover whilst crossing the TSS. The NE-going stream usually starts on the English side outside the lanes, and it is generally quicker to take the route inside the Goodwins at this stage rounding South Foreland and following the Gull Stream channel buoys to Broadstairs Knoll (R) buoy, then round North Foreland and across E of the E Margate buoy between Tongue Sand Tower and Tongue LtF to N Edinburgh Channel. North Foreland should be rounded by about HW Dover +2 with some remaining fair or cross-tide to Edinburgh Channel, and from North Foreland a burst of engine power is advisable to reach N Edinburgh No 1 (S card) buoy before the worst of the adverse tide through the channel. From the N end of the Edinburgh Channel it pays to be cautious on the falling tide, following a similar route to that outward; NE along the edge of Black Deep, crossing the Deep at right angles using the NW Long Sand beacon as a back bearing from 20m soundings 8 cables NW of the beacon in the vicinity of, but giving a wide berth to, the *Radio Caroline* wreck close to the edge of the channel, crossing Sunk Sand close S of the SW Sunk Beacon and then following the E edge of Barrow Deep, then from midway between Barrow No 6 (W card) and No 4 (R) set a tide-corrected course for midway between the Heaps (R) buoy and NE Middle (R) buoy, giving a wide berth to East Barrow and N Hook Middle Sands. Then to the Whitaker (RW) buoy and follow Whitaker Channel to the Inner Crouch (RW) buoy. If the timing is right the fast ebbing tide will enable the vessel to reach the Whitaker buoy/Spitway channel an hour or two before LW in time to take a slack to adverse stream and the first hour of flood into the Crouch. Timing on this route is very much dependent on the amount of engine power available; with a powerful engine life is much simpler, you can start later from Calais, motor through the Edinburgh and arrive at the Whitaker Spit at around LW for a full flood down Whitaker Channel.

Harwich to Calais

Although there is not the complication of a zigzag course to avoid shoals, most of the other comments applying to Passages 1 to 6 apply here. In addition Harwich and Felixstowe are both extremely busy ferry and container ports with a major dredged channel approach, whilst between Sunk LtF and North Foreland there are very few seamarks or landmarks. There are a number of widely separated lit cardinal marks en route but in daylight these are often difficult to sight. On a first-time cruise in daylight, without the help of lighted buoys and without electronic position-fixing equipment, after passing the Kentish Knock (E card) buoy the skipper may prefer to head inshore in order to sight North Foreland and the N Goodwin LtF before heading for the S Falls (S card) buoy and the Dover Strait TSS.

PASSAGE 7   HARWICH (LANDGUARD POINT) TO CALAIS ENTRANCE

Distance   71M

Duration   16hrs, 2½ tidal sets

Approach routes   See Chapter 6 Route 3 and Chapter 9 Routes 2, 4 and 5

Charts   Admiralty *323, 1183, 1352, 1491, 1607, 1610, 1892, 1975, 2052, 2449, 2693*
Imray *C1, C8, C30, Y6, Y7, Y16*

Recommended start time outward
HW Dover –4, HW Harwich –5

Arrival time outward
HW Dover –0½, HW Calais

Recommended start time return
HW Dover –5, HW Calais –4½

Arrival time return
HW Dover –1½, HW Harwich –2½

Tidal streams

Streams start running into and out of Harwich entrance at LW and HW Harwich, and in the middle of the Estuary approaches between Long Sand Head and Kentish Knock they turn S at HW Dover –6 (HW Harwich +6), and N at HW Dover (HW Harwich –1). At Calais NE-going starts at HW Dover –1½ and SW-going at HW Dover +4½.

Commentary

Outward   4hrs of SW and S-going tide at the start, means crossing the tide following the S outer edge of the Harwich Channel buoys to Pitching Ground (R) buoy across to Cork Sand (R) buoy passing very close to its S, across N of the Roughs Tower, Sunk LtF and Longsand Head (N card) buoy, and then with the tide down to the E of Kentish Knock (E card) buoy, meeting an adverse stream before reaching the latter buoy. Then 6hrs tide against, not as strong in the approaches to the Estuary as in the Estuary channels, to a position between Drillstone (E card) buoy and S Falls (S card) buoy when the SW-going stream takes over. Choosing a distance from S Falls buoy based on the strength of tidal stream (see Chapter 3), a right-angled course to the traffic lanes will crab across to CA4 (W card), and CA6 (R). Then follow the channel buoys into Calais, in time for locking through to the inner harbour.

Return   It will be necessary to lock through into the outer harbour and wait for several hours before leaving. Follow the channel buoys out to CA6 and cross the TSS at right angles. At the above suggested start time there are still 3½hrs of SW-going stream left pushing the vessel towards Varne LtF and Dover whilst crossing the TSS. The NE-going stream should start on the English side when outside the lanes, and it is generally quicker to take the route inside the Goodwins at this stage rounding South Foreland and following the Gull Stream channel buoys to Broadstairs Knoll (R) buoy, then rounding North Foreland and continuing with the N-going fair stream to the vicinity of Kentish Knock (E card) buoy when the tide turns into the Estuary and is against for 1 to 2hrs as far as Longsand Head (N card) buoy, where finally the course is cross-tide; N of the Sunk LtF, Roughs Tower and Cork Sand (R) buoy for the last 4hrs to Harwich keeping well to the S of the Harwich Channel and entering Harwich across The Shelf, with a flooding tide for a further 2½hrs into the Orwell (see Chapter 6).

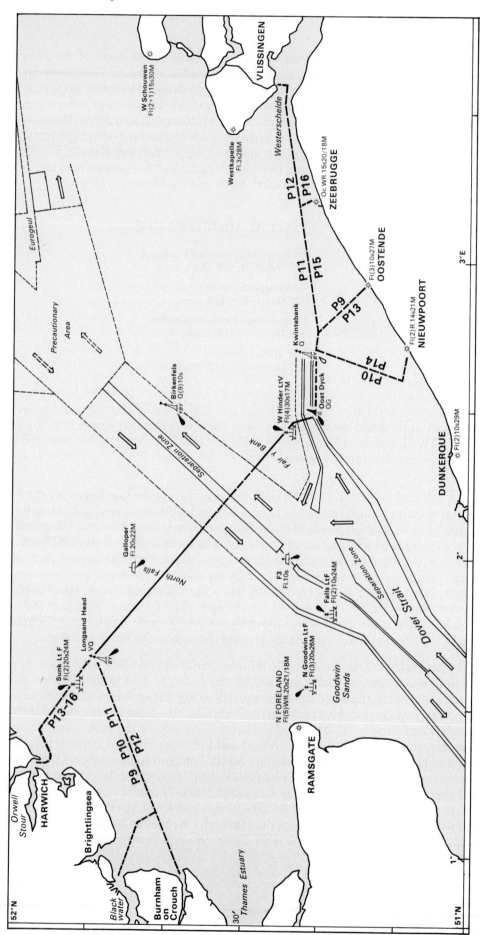

Passages 9 to 16 Essex rivers to the Belgian coast and Vlissingen

Ramsgate to Dunkerque

A direct route out to the N Goodwin LtF and straight to Dunkerque Port Est is not recommended since it crosses the Sandettié, Ruytingen and Dyck banks, which are uncomfortable and dangerous in anything above Force 4. It also crosses a wide section of diverging TSS lanes causing complicated course changes. The route recommended below is a dogleg so in NE or SW weather there are easy and difficult stretches in each direction. All the traffic, weather hazards and other advice described for Passages 1 to 3 apply.

PASSAGE 8   RAMSGATE ENTRANCE TO DUNKERQUE ENTRANCE

Distance   44M

Duration   10hrs, 1½+ tidal sets

Approach routes   See Chapter 9 Routes 2 and 5, Chapter 10 Route 1

Charts   Admiralty *323, 1350, 1827, 1828, 2449*
Imray *C30*

Recommended start time outward
HW Dover –6, HW Ramsgate +6

Arrival time outward
HW Dover +4, HW Dunkerque +3

Recommended start time return
HW Dover –5, HW Dunkerque –6

Arrival time return
HW Dover +5, HW Ramsgate +4½

Tidal streams

As this route is a dogleg via the Dunkerque Lanby it is essential to catch the turn of tide N to NE in the vicinity of this mark on both outward and return routes. The stream turns NE-going at this point at HW Dover –1½. At both Ramsgate and Dunkerque NE-going starts at approximately HW Dover –1 and SW-going at HW Dover +5.

Commentary

Outward   Follow the S edge of the Ramsgate dredged channel out to the Goodwin Knoll (G) buoy, then a course corrected for tide to S Falls (S card) buoy and cross the TSS at right angles, with the tide sweeping SW and pushing the vessel outside the Sandettié LtV. Alter course in the vicinity of Ruytingen SW (G) buoy (1M to its NW there is a W cardinal marking 20m and 17.8m wrecks) towards Dunkerque Lanby and with the NE-going tide keep well S of the buoyed channel to the entrance to Port Ouest, making a careful crossing of the channel at right angles giving a wide berth to the N wall of this harbour, and then following the starboard side of the Passe de l'Ouest buoyed channel to Dunkerque, Port Est entrance, and the yacht basin within.

Return   Follow the starboard side of the Passe de l'Ouest buoyed channel to Port Ouest entrance making a careful crossing of the channel at right angles giving a wide berth to the N wall of the harbour, then a course well S of the approach channel towards the Dunkerque Lanby. Again a slack then NE-going tide from the vicinity of Dunkerque Lanby and Ruytingen SW (G) buoy will give a good northerly tidal drift on the right-angled heading across the TSS, pushing the vessel well to the N of the E Goodwin LtF towards the N Goodwin LtF, then nearing the final cross-tide approach from Goodwin Knoll (G) buoy to the Ramsgate channel on a slackening tide. Follow the extreme N starboard edge of the channel into the harbour.

Essex rivers & Harwich to Belgium & Westerschelde

Distances, times, tidal sets and most of the commentary are the same from the Nass Beacon in the Blackwater as from the Inner Crouch buoy. Chapter 7, in particular Route 1 (first half), describes the pilotage from the Nass to the Swin Spitway.

These are good routes for the prevailing wind direction which is across the main course between Long Sand Head and West Hinder, but with the inevitable complications on the doglegs at both ends. Shipping is light in the Crouch and Blackwater approaches and along King's Channel, but near Harwich (see Passage 7 above) is heavy, if only for a short time out to the Sunk LtF. Traffic is heavy in the two TSS crossings and in the approaches to, and channels into, the Westerschelde.

The West Hinder LtV is a major mark on these routes, and although the bank to its N has 9m and can be crossed near to the LtV in most weather, in Force 5 and over it is advisable to pass S of the LtV.

PASSAGE 9   RIVER CROUCH (INNER CROUCH BUOY) TO OOSTENDE ENTRANCE

Distance   92M

Duration   20½hrs, 3½ tidal sets

Approach routes   See Chapter 7 Route 3 and Chapter 10 Route 3

Charts   Admiralty *125, 1183, 1406, 1610, 1872, 1975, 2449, 3741, 3750*
Imray *C1, C30, Y6, Y17*

Recommended start time outward
HW Dover, HW Burnham-on-Crouch and Bradwell (Blackwater) –1

Arrival time outward
HW Dover –5, HW Oostende –6

Recommended start time return
HW Dover +5, HW Oostende +4

Arrival time return
HW Dover, HW Burnham-on-Crouch and Bradwell –1

Tidal streams

Streams turn into and out of the Essex rivers at local LW and HW (HW Dover +1). At Longsand Head (N card) buoy approximately 30M from the start point, i.e. just over one tidal set distant, they turn SW-going about 6½hrs later at HW Dover –5½, and NE-going at HW Dover +1 (near to local HW). A useful objective, therefore, in either direction on this route is to reach Longsand Head buoy at latest by HW Dover –5½ (near to local LW), and thus have a favourable tide along King's Channel, and Whitaker Channel or Blackwater. On the Belgian coast the stream turns E-going at approximately HW Dover –1 and W-going at HW Dover +5, and the final approaches to Oostende are cross-tide.

Commentary

Outward   Proceed along Whitaker Channel to the vicinity of the Whitaker (RW) bell buoy, continue along King's Channel N of the vicious-looking stumps of Sunk Head Tower and its attendant N cardinal buoy, round Longsand Head (N card) buoy to the N as the tide turns SW-going, and alter course to a heading for West Hinder LtV allowing 1M clearance to the E of the LtV. This heading will cross the Noord Hinder south TSS at right angles, with the SW-going stream pushing the vessel towards but not too close to the F3 Lanby (very useful at night). The stream turns NE-going just before leaving the TSS, and the vessel crosses Fairy Bank reaching the West Hinder LtV on its E side. However, with these tidal sets the track crosses North Falls bank (N end of bank 8.2m minimum sounding) and, beyond the TSS, Fairy Bank (S end of bank 5.5m minimum sounding). In winds of Force 5 and over, therefore, it is safer to sail from Longsand Head buoy close up to the Galloper Lanby, then take a heading at right angles to the TSS, missing N Falls, locate Garden City (W card) wreck buoy and using this position take avoiding action to cross N of or across the N end of Fairy Bank, approaching West Hinder LtV to the N. Whichever side of West Hinder LtV is reached the next course is across and heading at right angles to the West Hinder TSS, and then following the buoyed S edge of the TSS to the Kwintebank (N card) buoy with a continuing NE-going fair stream. The SW-going stream will have started when making the cross-tide approach to Oostende. The near approaches to Oostende are a maze of banks with many buoys, not to be taken lightly in rough weather. Chapter 10, Route 3 examines this approach in more detail.

Return   Leaving Oostende on a weakening NE-going stream the Kwintebank buoy is reached at slack or SW-going, the TSS is crossed at right angles at that point and the N edge of the TSS followed with the favourable tidal stream to West Hinder LtV, then a heading set for Longsand Head buoy (1M clearance E of the buoy) that also crosses the TSS at right angles. The stream turns NE-going as the lanes are entered and the vessel is pushed relatively close to Galloper Lanby from which a course corrected for tide should bring up Longsand Head buoy at the beginning of the flood, and then the course is down King's Channel and Whitaker Channel. Alternatively at Longsand Head a tide-corrected course N of Sunk LtF and the Cork Sand buoy will fetch Harwich with a few hours of rising tide left.

PASSAGE 10 RIVER CROUCH (INNER CROUCH BUOY) TO NIEUWPOORT ENTRANCE

Distance 97M

Duration 21½hrs, 3½ tidal sets

Approach routes See Chapter 7 Route 3 and Chapter 10 Route 2

Charts Admiralty *125, 1183, 1406, 1610, 1872, 1975, 3741, 3750*
Imray *C1, C30, Y6, Y17*

Recommended start time outward
HW Dover, HW Burnham-on-Crouch and Bradwell (Blackwater) –1

Arrival time outward
HW Dover –4, HW Oostende –5 (HW Nieuwpoort is 15mins earlier than Oostende)

Recommended start time return
HW Dover +4, HW Oostende +3

Arrival time return
HW Dover, HW Burnham-on-Crouch and Bradwell –1

Tidal streams

Streams turn into and out of the Essex rivers at local LW and HW (HW Dover +1). At Longsand Head (N card) buoy approximately 30M from the start point, i.e. just over one tidal stream distant, they turn SW-going about 6½hrs later at HW Dover –5½, and NE-going at HW Dover +1 (near to local HW). A useful objective, therefore, in either direction on this route is to reach Longsand Head buoy at latest by HW Dover –5½ (near to local LW), and thus have a favourable tide along King's Channel, and Whitaker Channel or Blackwater. On the Belgian coast the stream turns E-going at approximately HW Dover –1 and W-going at HW Dover +5. The suggested timing gives a favourable tide for Nieuwpoort outwards from Kwintebank buoy and a fairly slack stream to Kwintebank on the return.

Commentary

Outward Proceed along Whitaker Channel to the vicinity of the Whitaker (RW) bell buoy, continue along King's Channel N of Sunk Head Tower and its attendant N cardinal buoy, round Longsand Head (N card) buoy to the N as the tide turns SW-going, and alter course to a heading for West Hinder LtV allowing 1M clearance to the E of the LtV. This heading will cross the Noord Hinder south TSS at right angles, with the SW-going stream pushing the vessel towards but not too close to the F3 Lanby (very useful at night). The stream turns NE-going just before leaving the TSS, and the vessel crosses Fairy Bank reaching the West Hinder on its E side. However, with these tidal sets the track crosses North Falls bank (N end of bank 8.2m minimum sounding) and, beyond the TSS, Fairy Bank (S end of bank 5.5m minimum sounding). In winds of Force 5 and over, therefore, it is safer to sail from Longsand Head buoy close up to the Galloper Lanby, then take a heading at right angles to the TSS, missing N Falls, locate Garden City (W card) wreck buoy and using this position take avoiding action to cross N of or across the N end of Fairy Bank, approaching West Hinder LtV to the N. Whichever side of West Hinder LtV is reached the next course is across and heading at right angles to the West Hinder TSS, and then following the buoyed S edge of the TSS to the Kwintebank (N card) buoy with a continuing NE-going fair stream. The SW-going stream will have started conveniently when making the final approach. The near approaches to Nieuwpoort are a maze of banks with many buoys, not to be taken lightly in rough weather. Chapter 10, Route 2 examines this approach in more detail.

Return Leaving Nieuwpoort on a weakening NE-going stream the Kwintebank buoy is reached at slack or SW-going, the TSS is crossed at right angles at that point and the N edge of the TSS followed with the favourable tidal stream to West Hinder LtV, then a heading set for Longsand Head buoy (1M clearance E of the buoy) that also crosses the TSS at right angles. The stream turns NE-going as the lanes are entered and the vessel is pushed relatively close to Galloper Lanby from which a course corrected for tide should bring up Longsand Head buoy at the beginning of the flood, and then the course is down King's Channel and Whitaker Channel. Alternatively at Longsand Head a tide-corrected course N of Sunk LtF and the Cork Sand buoy will fetch Harwich with a few hours of rising tide left.

PASSAGE 11 RIVER CROUCH (INNER CROUCH BUOY) TO ZEEBRUGGE ENTRANCE

Distance 99M

Duration 22hrs, 3½ tidal sets

Approach routes See Chapter 7 Route 3 and Chapter 10 Route 4

Charts Admiralty *97, 325, 1183, 1406, 1610, 1872, 1975, 3741, 3750*
Imray *C1, C30, Y6, Y17*

Recommended start time outward
HW Dover, HW Burnham-on-Crouch and Bradwell (Blackwater) –1

Arrival time outward
HW Dover –4, HW Zeebrugge –5½

Recommended start time return
HW Dover +4, HW Zeebrugge +2½

Arrival time return
HW Dover, HW Burnham-on-Crouch and Bradwell –1

Tidal streams

Streams turn into and out of the Essex rivers at local LW and HW (HW Dover +1). At Longsand Head (N card) buoy approximately 30M from the start point, i.e. just over one tidal stream distant, they turn SW-going about 6½hrs later at HW Dover –5½, and NE-going at HW Dover +1 (near to local HW). A useful objective, therefore, in either direction on this route is to reach Longsand Head buoy at latest by HW Dover –5½ (near to local LW), and thus have a favourable tide along King's Channel, and Whitaker Channel or Blackwater. On the Belgian coast the stream turns E-going at approximately HW Dover –1 and W-going at HW Dover +5. The suggested timing gives a slightly unfavourable tide to Zeebrugge outwards from Kwintebank buoy and a fairly slack stream to Kwintebank on the return.

Commentary

Outward Proceed along Whitaker Channel to the vicinity of the Whitaker (RW) bell buoy, continue along King's Channel N of Sunk Head Tower and its attendant N cardinal buoy, round Longsand Head (N card) buoy to the N as the tide turns SW-going, and alter course to a heading for West Hinder LtV allowing 1M clearance to the E of the LtV. This heading will cross the Noord Hinder south TSS at right angles, with the SW-going stream pushing the vessel towards but not too close to the F3 Lanby (very useful at night). The stream turns NE-going just before leaving the TSS, and the vessel crosses Fairy Bank reaching the West Hinder LtV on its E side. However, with these tidal sets the track crosses North Falls bank (N end of bank 8.2m minimum sounding) and, beyond the TSS, Fairy Bank (S end of bank 5.5m minimum sounding). In winds of Force 5 and over, therefore, it is safer to sail from Longsand Head buoy close up to the Galloper Lanby, then take a heading at right angles to the TSS, missing N Falls, locate Garden City (W card) wreck buoy and using this position take avoiding action to cross N of or across the N end of Fairy Bank, approaching West Hinder LtV to the N. Whichever side of West Hinder LtV is reached the next course is across and heading at right angles to the West Hinder TSS, and then following the buoyed S edge of the TSS to the Kwintebank (N card) buoy with a continuing NE-going fair stream. The adverse SW-going stream will have started when making the final approach. The approaches to Zeebrugge are not difficult and Chapter 10, Route 4 examines them in more detail.

Return Leaving Zeebrugge and pushing a weakening NE-going stream Kwintebank buoy is left well to the S so that the N edge of the TSS can be followed with the favourable tidal stream to West Hinder LtV, then a heading is set for Longsand Head buoy (1M clearance E of the buoy) that also crosses the TSS at right angles. The stream turns NE-going as the lanes are entered and the vessel is pushed relatively close to the Galloper Lanby from which a course corrected for tide should bring up the Longsand Head buoy at the beginning of the flood, and then the course is down King's Channel and Whitaker Channel. Alternatively at Longsand Head a tide-corrected course N of Sunk LtF and the Cork Sand buoy will fetch Harwich with a few hours of rising tide left.

PASSAGE 12 RIVER CROUCH (INNER CROUCH BUOY) TO VLISSINGEN ENTRANCE

Distance 115M

Duration 25½hrs, 4+ tidal sets

Approach routes See Chapter 7 Route 3, Chapter 10 Route 4 and Chapter 11 Route 1

Charts Admiralty *325, 1183, 1406, 1610, 1872, 1975, 3741, 3750*
Imray *C1, C30, Y6, Y17*

Recommended start time outward
HW Dover, HW Burnham-on-Crouch and Bradwell (Blackwater) –1

Arrival time outward
HW Dover +0½, HW Vlissingen –1½

Recommended start time return
HW Dover +3, HW Vlissingen +1

Arrival time return
HW Dover +3½, HW Burnham-on-Crouch and Bradwell +2½

Tidal streams

Streams turn into and out of the Essex rivers at local LW and HW (HW Dover +1). At Longsand Head (N card) buoy approximately 30M from the start point, i.e. just over one tidal stream distant, they turn SW-going about 6½hrs later at HW Dover –5½, and NE-going at HW Dover +1 (near to local HW). A useful objective, therefore, in either direction on this route is to reach Longsand Head buoy at latest by HW Dover –5½ (near to local LW), and thus have a favourable tide along King's Channel, and Whitaker Channel or Blackwater. Streams in the constricted Scheur/Westerschelde entrance are extremely fast and turn ingoing at HW Dover –3 (HW Vlissingen –5) and outgoing at HW Dover +3 (HW Vlissingen +1), and it is difficult on the return passage to compromise between getting the best stream out of the entrance as well as into the Essex rivers. The best solution on balance is to take the unusual opportunity of a full 9hrs of fair stream from Vlissingen which means pushing the ebb stream for 2 to 3hrs right at the end of the passage, down Whitaker Channel. If you are too exhausted, divert at the Longsand buoy and head cross-tide for Harwich as in Passage 16.

Commentary

Outward Proceed along Whitaker Channel to the vicinity of the Whitaker (RW) bell buoy, continue along King's Channel N of Sunk Head Tower and its attendant N cardinal buoy, round Longsand Head (N card) buoy to the N as the tide turns SW-going, and alter course to a heading for West Hinder LtV allowing 1M clearance to the E of the LtV. This heading will cross the Noord Hinder south TSS at right angles, with the SW-going stream pushing the vessel towards but not too close to the F3 Lanby (very useful at night). The stream turns NE-going just before leaving the TSS, and the vessel crosses Fairy Bank reaching the West Hinder LtV on its E side. However, with these tidal sets the track crosses North Falls bank (N end of bank 8.2m minimum sounding) and, beyond the TSS, Fairy Bank (S end of bank 5.5m minimum sounding). In winds of Force 5 and over, therefore, it is safer to sail from Longsand Head buoy close up to the Galloper Lanby, then take a heading at right angles to the TSS, missing N Falls, locate Garden City (W card) wreck buoy and using this position take avoiding action to cross N of or across the N end of Fairy Bank, approaching West Hinder LtV to the N. Whichever side of West Hinder LtV is reached the next course is across and heading at right angles to the West Hinder TSS, and then following the buoyed S edge of the TSS to the Kwintebank (N card) buoy with a continuing NE-going fair stream. A full 6hrs of adverse stream follows with a final 1 or 2hrs of fair stream into the Westerschelde entrance. The approach route from Kwintebank buoy to Vlissingen is not difficult and Chapter 11, Route 1 examines it in more detail.

Return There are 9hrs of fair stream from Vlissingen taking the vessel (hopefully) to at least the West Hinder LtV, following and keeping just outside the N edge of the TSS. From a position 1M NE of the LtV a course is set for Longsand Head buoy (1M clearance E of the buoy) that also crosses the TSS at right angles. The stream is now NE-going and the vessel is pushed relatively close to the Galloper Lanby at the turn of the tide SW, from which a course corrected for tide should bring up Longsand Head buoy and then be well down King's Channel before an adverse stream sets in for the last 2 to 3hrs to Burnham. Alternatively at Longsand Head a tide-corrected course N of Sunk LtF and the Cork Sand buoy will fetch Harwich at around HW.

PASSAGE 13   HARWICH (LANDGUARD POINT) TO OOSTENDE ENTRANCE

Distance   77M

Duration   17hrs, 3 tidal sets

Approach routes   See Chapter 6 Route 3 and Chapter 10 Route 3

Charts   Admiralty *125, 1183, 1406, 1491, 1610, 1872, 2052, 2693*
Imray *C1, C30, Y6, Y16*

Recommended start time outward
HW Dover +3, HW Harwich +2

Arrival time outward
HW Dover –5, HW Oostende –6

Recommended start time return
HW Dover +5, HW Oostende +4

Arrival time return
HW Dover –3, HW Harwich –4

Tidal streams

Streams start out of and into Harwich at local HW and LW (HW Dover +1 and –5½). On the Belgian coast the stream turns E-going at approximately HW Dover –1 and W-going HW Dover +5. Both the approaches to Harwich and to Oostende are cross-tide, so the above start time is only one of several possibilities.

Commentary

15M shorter than Passage 9 the start is 3hrs later, arrival same time, and the return journey starts at the same time, with arrival 3hrs earlier on a rising tide.

Outward   3hrs of SW and S-going tide at the start means crossing the tide following the S outer edge of the Harwich channel buoys to Pitching Ground (R) buoy, across to Cork Sand (R) buoy passing very close to its S, across N of Roughs Tower, Sunk LtF to 1M E of Longsand Head (N card) buoy. Alter course to a heading for West Hinder LtV allowing 1M clearance to the E of the LtV. This heading will cross the Noord Hinder south TSS at right angles, with the SW-going stream pushing the vessel towards but not too close to the F3 Lanby (very useful at night). The stream turns NE-going just before leaving the TSS, and the vessel crosses Fairy Bank reaching the West Hinder LtV on its E side. However, with these tidal sets the track crosses North Falls bank (N end of bank 8.2m minimum sounding) and, beyond the TSS, Fairy Bank (S end of bank 5.5m minimum sounding). In winds of Force 5 and over, therefore, it is safer to sail from Longsand Head buoy close up to the Galloper Lanby, then take a heading at right angles to the TSS, missing N Falls, locate Garden City (W card) wreck buoy and using this position take avoiding action to cross N of or across the N end of Fairy Bank, approaching West Hinder LtV to the N. Whichever side of West Hinder LtV is reached the next course is across and heading at right angles to the West Hinder TSS, and then following the buoyed S edge of the TSS to the Kwintebank (N card) buoy with a continuing NE-going fair stream. The SW-going stream will have started when making the cross-tide approach to Oostende. The near approaches to Oostende are a maze of banks with many buoys, not to be taken lightly in rough weather. Chapter 10, Route 3 examines this approach in more detail.

Return   Leaving Oostende on a weakening NE-going stream the Kwintebank buoy is reached at slack or SW-going, the TSS is crossed at right angles at that point and the N edge of the TSS followed with the favourable tidal stream to West Hinder LtV, then a heading set for Longsand Head buoy (1M clearance E of the buoy) that also crosses the TSS at right angles. The stream turns NE-going as the lanes are entered and the vessel is pushed relatively close to Galloper Lanby from which a course corrected for tide should bring up Longsand Head buoy at the beginning of the flood. From this point the course is cross-tide, N of the Sunk LtF, Roughs Tower and close S of Cork Sand (R) buoy, for the last 4hrs to Harwich keeping well to the S of the Harwich Channel and entering Harwich across The Shelf, with a flooding tide for a further 4hrs into the Orwell (see Chapter 6).

PASSAGE 14   HARWICH (LANDGUARD POINT) TO NIEUWPOORT ENTRANCE

Distance   82M

Duration   18hrs, 3 tidal sets

Approach routes   See Chapter 6, Route 3, and Chapter 10, Route 2

Charts   Admiralty *125, 1183, 1406, 1491, 1610, 1872, 2052, 2693*
Imray *C1, C30, Y6, Y16*

Recommended start time outward
HW Dover +3, HW Harwich +2

Arrival time outward
HW Dover –4, HW Oostende –5, (HW Nieuwpoort is 15 minutes earlier than Oostende)

Recommended start time return
HW Dover +4, HW Oostende +3

Arrival time return
HW Dover –3, HW Harwich –4

Tidal streams

Streams start out of and into Harwich at local HW and LW (HW Dover +1 and –5½). On the Belgian coast the stream turns E-going at approximately HW Dover –1 and W-going HW Dover +5. The approach to Harwich is cross-tide so there is some flexibility with the start times, but to and from Nieuwpoort it helps to have a fair tide along the Negenvaam.

Commentary

15M shorter than Passage 10 the start is 3hrs later, arrival same time, and the return journey starts at the same time, with arrival 3hrs earlier on a rising tide.

Outward   3hrs of SW and S-going tide at the start means crossing the tide following the S outer edge of the Harwich channel buoys to Pitching Ground (R) buoy, across to Cork Sand (R) buoy passing very close to its S, across N of Roughs Tower, Sunk LtF to 1M E of Longsand Head (N card) buoy. Alter course to a heading for West Hinder LtV allowing 1M clearance to the E of the LtV. This heading will cross the Noord Hinder south TSS at right angles, with the SW-going stream pushing the vessel towards but not too close to the F3 Lanby (very useful at night). The stream turns NE-going just before leaving the TSS, and the vessel crosses Fairy Bank reaching the West Hinder on its E side. However, with these tidal sets the track crosses North Falls bank (N end of bank 8.2m minimum sounding) and, beyond the TSS, Fairy Bank (S end of bank 5.5m minimum sounding). In winds of Force 5 and over, therefore, it is safer to sail from Longsand Head buoy close up to the Galloper Lanby, then take a heading at right angles to the TSS, missing N Falls, locate Garden City (W card) wreck buoy and using this position take avoiding action to cross N of or across the N end of Fairy Bank, approaching West Hinder LtV to the N. Whichever side of West Hinder LtV is reached the next course is across and heading at right angles to the West Hinder TSS, and then following the buoyed S edge of the TSS to the Kwintebank (N card) buoy with a continuing NE-going fair stream. The SW-going stream will have started conveniently when making the final approach. The near approaches to Nieuwpoort are a maze of banks with many buoys, not to be taken lightly in rough weather. Chapter 10, Route 2 examines this approach in more detail.

Return   Leaving Nieuwpoort on a weakening NE-going stream the Kwintebank buoy is reached at slack or SW-going, the TSS is crossed at right angles at that point and the N edge of the TSS followed with the favourable tidal stream to West Hinder LtV, then a heading set for Longsand Head buoy (1M clearance E of the buoy) that also crosses the TSS at right angles. The stream turns NE-going as the lanes are entered and the vessel is pushed relatively close to Galloper Lanby from which a course corrected for tide should bring up Longsand Head buoy at the beginning of the flood. From this point the course is cross-tide, N of the Sunk LtF, Roughs Tower and close S of Cork Sand (R) buoy, for the last 4hrs to Harwich keeping well to the S of the Harwich Channel and entering Harwich across The Shelf, with a flooding tide for a further 4hrs into the Orwell (see Chapter 6).

PASSAGE 15   HARWICH (LANDGUARD POINT) TO ZEEBRUGGE ENTRANCE

Distance   84M

Duration   18½hrs, 3 tidal sets

Approach routes   See Chapter 6 Route 3, Chapter 10 Route 4

Charts   Admiralty *97, 325, 1183, 1406, 1491, 1610, 1872, 2052, 2693*
Imray *C1, C30, Y6, Y16*

Recommended start time outward
HW Dover +3, HW Harwich +2

Arrival time outward
HW Dover −4, HW Zeebrugge −5½

Recommended start time return
HW Dover +4, HW Zeebrugge +2½

Arrival time return
HW Dover −3, HW Harwich −4

Tidal streams

Streams start out of and into Harwich at local HW and LW (HW Dover +1 and −5½). On the Belgian coast the stream turns E-going at approximately HW Dover −1 and W-going HW Dover +5. The approach to Harwich is cross-tide so there is some flexibility with the start times, but it helps a little to have a fair tide between the Kwintebank buoy and Zeebrugge.

Commentary

15M shorter than Passage 11 the start is 3hrs later, arrival same time, and the return journey starts at the same time, with arrival 3hrs earlier on a rising tide.

Outward   3hrs of SW and S-going tide at the start means crossing the tide following the S outer edge of the Harwich channel buoys to Pitching Ground (R) buoy, across to Cork Sand (R) buoy passing very close to its S, across N of Roughs Tower, Sunk LtF to 1M E of Longsand Head (N card) buoy. Alter course to a heading for West Hinder LtV allowing 1M clearance to the E of the LtV. This heading will cross the Noord Hinder south TSS at right angles, with the SW-going stream pushing the vessel towards but not too close to the F3 Lanby (very useful at night). The stream turns NE-going just before leaving the TSS, and the vessel crosses Fairy Bank reaching the West Hinder LtV on its E side. However, with these tidal sets the track crosses North Falls bank (N end of bank 8.2m minimum sounding) and, beyond the TSS, Fairy Bank (S end of bank 5.5m minimum sounding). In winds of Force 5 and over, therefore, it is safer to sail from Longsand Head buoy close up to the Galloper Lanby, then take a heading at right angles to the TSS, missing N Falls, locate Garden City (W card) wreck buoy and using this position take avoiding action to cross N of or across the N end of Fairy Bank, approaching West Hinder LtV to the N. Whichever side of West Hinder LtV is reached the next course is across and heading at right angles to the West Hinder TSS, and then following the buoyed S edge of the TSS to the Kwintebank (N card) buoy with a continuing NE-going fair stream. The adverse SW-going stream will have started when making the final approach. The approaches to Zeebrugge are not difficult and Chapter 10, Route 4 examines them in more detail.

Return   Leaving Zeebrugge and pushing a weakening NE-going stream Kwintebank buoy is left well to the S so that the N edge of the TSS can be followed with the favourable tidal stream to West Hinder LtV, then a heading is set for Longsand Head buoy (1M clearance E of the buoy) that also crosses the TSS at right angles. The stream turns NE-going as the lanes are entered and the vessel is pushed relatively close to the Galloper Lanby from which a course corrected for tide should bring up the Longsand Head buoy at the beginning of the flood. From this point the course is cross-tide, N of the Sunk LtF, Roughs Tower and close S of Cork Sand (R) buoy, for the last 4hrs to Harwich keeping well to the S of the Harwich Channel and entering Harwich across The Shelf, with a flooding tide for a further 4hrs into the Orwell (see Chapter 6).

PASSAGE 16   HARWICH (LANDGUARD POINT) TO VLISSINGEN ENTRANCE

Distance   100M

Duration   22hrs, 3½ tidal sets

Approach routes   See Chapter 6 Route 3, Chapter 10
Route 4 and Chapter 11, Route 1

Charts   Admiralty *325, 1183, 1406, 1491, 1610, 1872, 2052,
2693*
Imray *C1, C30, Y6, Y16*

Recommended start time outward
HW Dover +5, HW Harwich +4

Arrival time outward
HW Dover +2½, HW Vlissingen +0½

Recommended start time return
HW Dover +3, HW Vlissingen +1

Arrival time return
HW Dover +0½, HW Harwich –0½

Tidal streams

Streams start out of and into Harwich at local HW and LW (HW Dover +1 and –5½) and
approach and departure is cross-tide. Streams in the constricted Westerschelde entrance are
extremely fast and turn ingoing at HW Dover –3 (HW Vlissingen –5) and outgoing at HW
Dover +3 (HW Vlissingen +1). The outward and return aim is to achieve the full fair stream
in the approaches to the Westerschelde. On the return passage there is the unusual opportunity of a full 9hrs of fair stream from Vlissingen.

Commentary

Outward   3hrs of weak NE, slack then SW-going tide at the start. Follow the S outer edge
of the Harwich channel buoys to Pitching Ground (R) buoy across to Cork Sand (R) buoy
passing very close to its S, across N of Roughs Tower to pass 1M NE of Sunk LtF and 1M
E of Longsand Head (N card) buoy, then on a heading for West Hinder LtV allowing 1M
clearance to the E of the LtV. This heading will eventually cross the Noord Hinder south
TSS at right angles. The SW stream will initially push the vessel S away from the Galloper
Lanby and turn NE whilst crossing the S end of North Falls bank (15–16m). Crossing the
TSS the stream pushes the vessel towards Garden City (W card) buoy and across the N end
of Fairy Bank approaching the West Hinder LtV to its E side. In winds of Force 5 and over
care should be taken on this route to cross the two banks at their deeper ends; the S of North
Falls and the N of Fairy. The next course is across and heading at right angles to the West
Hinder TSS, then following the buoyed S edge of the TSS to Kwintebank (N card) buoy, but
by now with a foul SW-going stream continuing until the ingoing stream starts approaching
the Scheur channel, then fair into the Westerschelde, where it pays to keep well to the S side
of the entrance to hold the last of the stream. The approach route from Kwintebank buoy to
Vlissingen is not difficult and Chapter 11, Route 1 examines it in more detail.

Return   There are 9hrs of fair stream from Vlissingen taking the vessel (hopefully) to at least
the West Hinder LtV and following and keeping just outside the N edge of the TSS. From
a position 1M NE of the LtV set a course for Longsand Head buoy (1M clearance E of the
buoy) that also crosses the TSS at right angles. The stream is now NE-going and the vessel
is pushed relatively close to the Galloper Lanby at the turn of the tide SW, from which a
course corrected for tide should bring up the Longsand Head buoy, continue NE of Sunk
LtF and then past Roughs Tower to close S of the Cork Sand buoy and follow the recommended yacht route, Chapter 6 Route 3, into Harwich close to HW.

Essex rivers & Harwich to Oosterschelde

Distances, times, tidal sets and most of the commentary are the same from the Nass Beacon in the Blackwater as from the Inner Crouch buoy. Chapter 7, in particular Route 1 (first half), describes the pilotage from the Nass to the Swin Spitway.

The Galloper bank forms a narrow pyramidal ridge, 6½M long by ½M wide, soundings 2.4m in the centre, down to 15m at the NE and SW extremities. The two routes N and S of the bank are little different in distance, but the S has the advantage of the more prominent Galloper Lanby in contrast with a N cardinal buoy, North Galloper, at the N end. These are routes to be avoided in strong onshore winds on the Dutch coast where the Steenbanken (3.8 to 9m soundings) and the Banjaard shoals (0.4m drying to 7m sounding) screen off the mouth of the Oosterschelde and require careful navigation via the buoyed channels (see Chapter 11).

These passages are at a slightly more acute angle to the prevailing winds than the Harwich to Oostende passage, so that the outward passages benefit more from southwesterlies and the return from northeasterlies; the reverse wind directions mean sailing much harder on the wind. A further complication is that they also require a substantial course correction to head across the TSS at right angles

Traffic is heaviest on the Harwich side and crossing the TSS just below the Noord Hinder Junction Precautionary Area, but is conveniently much lighter coastal traffic on the Dutch side, where careful pilotage is needed through the banks. Roompotsluis Lock has to be negotiated into the Oosterschelde Estuary since the completion of the storm-surge barrier.

In summary, these are emphatically fair-weather routes.

PASSAGE 17 HARWICH (LANDGUARD POINT) TO ROOMPOT VIA N GALLOPER OR GALLOPER

Distance   91M or 92M

Duration   20½hrs, 3½ tidal sets

Approach routes   See Chapter 6 Route 3 and Chapter 11 Route 2

Charts   Admiralty *110, 1183, 1406, 1491, 1610, 2052, 2693, 3371*
Imray *C1, C30, Y6, Y16*

Recommended start time outward
HW Dover +5, HW Harwich +4

Arrival time outward
HW Dover, HW Zierikzee –3

Recommended start time return
HW Dover +3, HW Zierikzee

Arrival time return
HW Dover –2, HW Harwich –3

Tidal streams

Streams start running into and out of Harwich harbour at LW and HW Harwich (HW Dover –5½ and +1). More critical to the outward or return passage is the Roompot where the flood begins at HW Zierikzee –6 (HW Dover –3) and the ebb at HW Zierikzee (HW Dover +3).

Commentary

Outward   Follow the recommended yacht route out of Harwich, Chapter 6 Route 3, to Cork Sand (R) buoy then across the slackening NE-going tide between Roughs Tower and the wreck near Threshold shoal (see Chapter 6). Midway between the South Shipwash (S card) buoy and Sunk LtF the alternative routes N and S of Galloper bank diverge and by now the tide will be strongly S-going making the N route between the S Inner Gabbard (S card) and the N Gabbard (N card) slightly more difficult to make. The alternative route turns S round the Galloper Lanby.

Both routes from Galloper bank then take a direct course towards Middelbank (RW) Lt buoy still with a S-going stream but for only 2 or 3hrs until a starboard course change is required in order to head at right angles across the Noord Hinder south TSS. Approximately when crossing commences, the tide will turn N-going, and when leaving the lanes in the region of the Birkenfels and Callisto W cardinal buoys course should be reset for the Middelbank buoy aiming to cross the N end or close N of Noord Hinder bank (13m soundings at N end). Still with a N-going tide the N ends of the Oost Hinder and Bligh banks will be clipped; all good opportunities for line of soundings navigation.

When the tide has again turned S the vessel will be approaching the string of 5 Lt buoys in the Middelbank vicinity, and at night will be on the leading line of the powerful Noorderhoofd and Westkapelle Lts, whilst W Schouwen Lt on the N side of the Oosterschelde

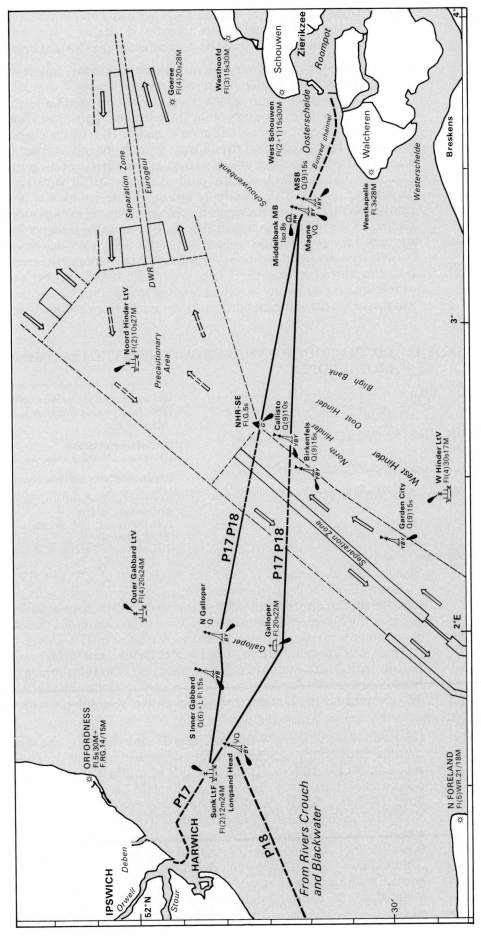

Passages 17 and 18 Essex rivers to the Roompot

Estuary will also probably be visible. There are two routes following the buoys from Middel-bank to Roompotsluis described in Chapter 11, Route 2 and the vessel will arrive with another 3hrs of rising tide to pick its destination in the Oosterschelde.

Return　Take the ebb outwards to the Middelbank buoy, following Route 2, Chapter 11, clearing the banks as the tide is turning S and slackening. From Middelbank it is advisable to take a tide-corrected course for a point midway between Callisto (W card) and NHR-SE (G) Lt buoys in order to pass N of the three banks: Bligh, Oost Hinder and Noord Hinder. Then alter course to cross the TSS at right angles. The stream will be slack and turning N so not quite as favourable as on the outward journey for this starboard course change. At this stage it is useful at twilight or in darkness to pick up one of the number of Lt buoys in this area of the TSS in order to confirm position since it will be necessary to estimate the point of departure from the TSS (electronic position-fixing equipment is invaluable in such cir-cumstances) and the course change to reach either the N Galloper (N card) buoy or Galloper Lanby. Of the two routes undoubtedly the best is via the Galloper Lanby which to the aver-age yacht has a geographical range of just under 10M; it is much more difficult to pick up the N Galloper buoy. By this time as well there will be a strong N-going stream requiring a care-ful course correction from the TSS boundary. From Galloper bank the stream will be cross-course (quartering for the N route), S-going with a correction required to make Sunk LtF, and thence via the recommended yacht route into Harwich (see Chapter 6, Route 3).

PASSAGE 18　RIVER CROUCH (INNER CROUCH BUOY) TO ROOMPOT VIA N GALLOPER OR GALLOPER

Distance　106M

Duration　23½hrs, 4 tidal sets

Approach routes　See Chapter 7 Route 3 and Chapter 11 Route 2

Charts　Admiralty *110, 1183, 1406, 1610, 1975, 3371, 3741, 3750*
Imray *C1, C30, Y6, Y17*

Recommended start time outward
HW Dover +1, HW Burnham-on-Crouch and Bradwell (Blackwater)

Arrival time outward
HW Dover, HW Zierikzee –3

Recommended start time return
HW Dover +3, HW Zierikzee

Arrival time return
HW Dover +2, HW Burnham-on-Crouch and Bradwell (Blackwater) +1

Tidal streams

Streams turn into and out of the Essex rivers at local LW and HW (HW Dover +1). At Longsand Head (N card) buoy approximately 30M from the start point, i.e. just over one tidal stream distant, they turn SW-going about 6½hrs later at HW Dover –5½, and NE-going at HW Dover +1 (near to local HW). A useful objective, therefore, in either direction on this route is to reach Longsand Head buoy at latest by HW Dover –5½ (near to local LW), and thus have a favourable tide along King's Channel, and Whitaker Channel or Blackwater. Equally critical to the outward or return passage is a fair tide through the Roompot where the flood begins at HW Zierikzee –6 (HW Dover –3) and the ebb at HW Zierikzee (HW Dover +3). Fortunately the 24hr passage enables both of these objectives to be achieved.

Commentary

Outward　Proceed along Whitaker Channel to the vicinity of the Whitaker (RW) bell buoy, continue along King's Channel N of the vicious-looking stumps of Sunk Head Tower and its attendant N cardinal buoy. It will probably be useful to do some judicious motoring at this point to cheat the start of the foul tide and round Longsand Head buoy to the N as the tide turns SW-going. At Longsand Head buoy alter course for one of the two alternative routes N and S of Galloper bank. By now the tide will be strongly S-going making the N route bet-ween the S Inner Gabbard (S card) buoy and the N Gabbard (N card) buoy slightly more dif-ficult to make. The alternative route turns S round the Galloper Lanby.

Both routes from Galloper bank then take a direct course towards Middelbank (RW) Lt buoy still with a S-going stream but for only 2 or 3hrs until a starboard course change is required in order to head at right angles across the Noord Hinder south TSS. Approximately when crossing commences, the tide will turn N-going, and when leaving the lanes in the region of the Birkenfels and Callisto W cardinal buoys course should be reset for the Middelbank buoy aiming to cross the N end or close N of Noord Hinder bank (13m soundings N end). Still with a N-going tide the N ends of the Oost Hinder and Bligh banks will be clipped; all good opportunities for line of soundings navigation.

When the tide has again turned S the vessel will be approaching the string of 5 Lt buoys in the Middelbank vicinity, and at night will be on the leading line of the powerful Noorderhoofd and Westkapelle Lts, whilst W Schouwen Lt on the N side of the Oosterschelde Estuary will also probably be visible. There are two routes following the buoys from Middelbank to Roompotsluis described in Chapter 11, Route 2 and the vessel will arrive with another 3hrs of rising tide to pick its destination in the Oosterschelde.

Return   Take the ebb outwards to the Middelbank buoy, following Route 2, Chapter 11, clearing the banks as the tide is turning S and slackening. From Middelbank it is advisable to take a tide-corrected course for a point midway between Callisto (W card) and NHR-SE (G) Lt buoys in order to pass N of the three banks: Bligh, Oost Hinder and Noord Hinder. Then alter course to cross the TSS at right angles. The stream will be slack and turning N so not quite as favourable as on the outward journey for this starboard course change. At this stage it is useful at twilight or in darkness to pick up one of the number of Lt buoys in this area of the TSS in order to confirm position since it will be necessary to estimate the point of departure from the TSS (electronic position-fixing equipment is invaluable in such circumstances) and the course change to reach either the N Galloper (N card) buoy or Galloper Lanby. Of the two routes undoubtedly the best is via the Galloper Lanby which to the average yacht has a geographical range of just under 10M; it is much more difficult to pick up the N Galloper buoy. By this time as well there will be a strong N-going stream requiring a careful course correction from the TSS boundary. From Galloper bank a tide-corrected course should be set for close N of Longsand Head buoy and during this period the stream will turn S. From Longsand Head a continuing flood will give a good push down the King's and Whitaker Channels, and again, as in the outward journey, some judicious motoring may be needed to cheat the beginning of the ebb out of the River Crouch.

Norfolk & Suffolk coast to Noord Holland

Despite the distance and time involved these five passages are the simplest and least hazardous of any. They cross only the deep water routes and the gap between the Noord Hinder and Texel TSSs, after much traffic has already turned inshore towards Antwerp and Hoek van Holland, and traffic on both coasts is lightest. There are fewer banks on each side, with the exception of those off Great Yarmouth and Lowestoft, and minor hazards are the offshore oil and gas platforms and wellheads, which are well marked and lit assisting position-fixing. One slight navigational complication, whilst not statutory: it is desirable to alter course to a right-angled heading across the DWRs. The depth sounder and the undersea contours help with estimating position particularly using the largest scale charts. Admiralty chart *1504* for example shows several large 55–60m 'holes' amidst 38–40m soundings near the junction of the the two DWRs, whilst Brown Ridge 50M W of the Noord Holland coast with average 18m soundings on its northern 15M length has a steep 45m-deep trench on its E side and a considerable S extension with 20–25m soundings and 31–38m on either side.

On the English east coast, traffic is heaviest to the south near Harwich with its ferries. It is lighter near Lowestoft and Great Yarmouth where small coasters, fishing fleets and servicing vessels for oil and gas fields are plying, and on the Dutch side there is a similar amount and kind of traffic as well as Dutch naval traffic from Den Helder, and commercial traffic to IJmuiden.

On the minus side there are fewer vessels from whom to request aid, but even so a well found yacht is far safer in deep water away from tricky shoal-water navigation. These passages also cross the prevailing winds at a more acute angle so conditions need to be carefully chosen to obtain a beam or quartering wind, avoiding long periods thrashing to windward. Extended weather forecasts need to be consulted and, as for all North Sea crossings, a lee shore on the opposite side is a major hazard to be avoided.

The two critical requirements for any of these five passages are for a favourable N or S tidal set during the course alteration across the DWRs, and for a favourable set into the destination harbour. On the diagonal passages into Harwich and Den Helder a flood tide is needed and on the more right-angled passages it helps to strike the coast N or S of the destination leaving an hour or two of fair tide towards the haven.

PASSAGE 19   HARWICH (LANDGUARD POINT) TO IJMUIDEN ENTRANCE

Distance   126M

Duration   28hrs, 4½ tidal sets

Approach routes   See Chapter 6 Route 2 and Chapter 12 IJmuiden

Charts   Admiralty *124, 1408, 1491, 2052, 2322, 2693*
Imray *C1, Y5, Y6, Y16*

Recommended start time outward
HW Dover +3, HW Harwich +2

Arrival time outward
HW Dover +6, HW IJmuiden +2

Recommended start time return
HW Dover +6, HW IJmuiden +2

Arrival time return
HW Dover –3, HW Harwich –4

Tidal streams

Near Orfordness the stream starts NE-going at about HW Dover +0½ (Harwich –0½) and SW-going at HW Dover –6 (Harwich +6). In the area of the DWR the stream starts N-going at about HW Dover +0½, and S-going at HW Dover –5½hrs. Near IJmuiden N-going starts slightly later at HW Dover +1½ and S-going also later at HW Dover –4½.

Commentary

Outward   This start time ensures 3hrs of outgoing tide to the Shipwash LtF or Orfordness, and a N-going tide for the course change across the DWR continuing northwards past the N corner of the Noord Hinder north TSS. Follow the recommended yacht route out of Harwich just outside the W and S sides of the dredged channel and cross the channel at right angles opposite the Platters (S card) buoy. From the Platters buoy there are three possible routes to the passage's departure point:

1. Via the Sledway channel passing to the E of the Cutler (G unlit) buoy to Bawd Head, rounding the NE Bawdsey (G) buoy, keeping N of the N Shipwash (N card) buoy to make a departure from the Shipwash LtF.

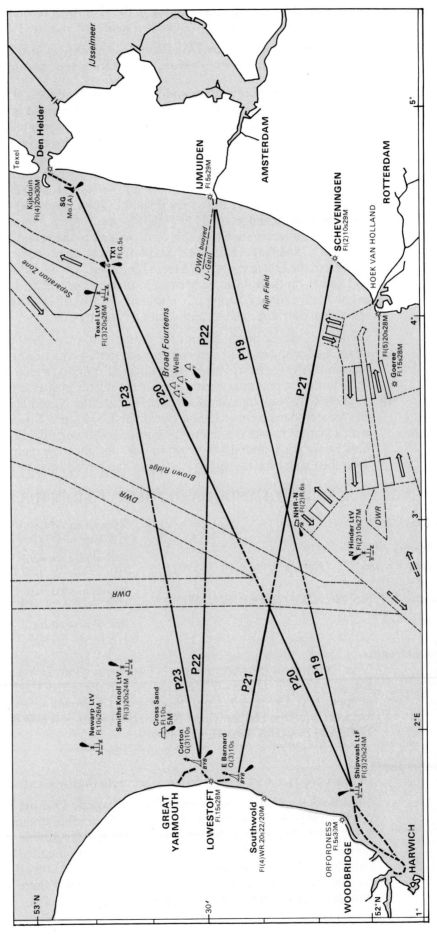

Passages 19 to 23 Norfolk and Suffolk coast to Noord Holland

2. Via the Shipway channel passing to the E of the S Bawdsey (S card) buoy and turning E close to the N Shipwash (N card) to make a departure from the Shipwash LtF.

3. Keeping approximately 1M offshore but closing in to about ½M near to Orfordness and making a departure from between Orfordness and the NE Whiting (E card unlit) buoy.

From the departure point a course direct for IJmuiden should be set and adjusted at the appropriate time to cross the DWR as close to a right angle as is practicable; this is not a large change of course since the DWR widens to fork into two branches at this point, but a good lookout must be kept in order to adjust the vessel's heading slightly when necessary depending on the traffic encountered which can be travelling on any of four possible headings. The positions of the DWR entry and exit course changes need careful estimation leaving a reasonable margin of error, say 1M. Beyond the DWR, NHR-N (R) buoy marking the NW corner of the TSS is very difficult to pick up and in any case the tidal stream should be pushing the vessel well to the north.

Tidal streams are such that this route will place the vessel conveniently close to the two lit (Morse U, 15M range) Rijn Oil Field platforms clearly visible from a distance in daylight and 15M from the Dutch coast, 30M from IJmuiden. The IJmuiden steelworks chimneys (166m) and factory buildings can be seen well offshore, or the 29M range light at night. Approach to the harbour (see Chapter 12, IJmuiden, *Approach and entrance*) should be from S of the IJmuiden-Geul approach channel with a N-going tide.

Return   A direct heading from the entrance towards the Shipwash LtF or Orfordness Lt clears the IJmuiden-Geul immediately and the S-going tide ensures picking up the Rijn Field platforms (the vessel will probably pass S of the field), which give a good departure position-fix to minimise potential error when estimating the positions of the two course changes across the DWR. A N-going tide should help to clear the Noord Hinder north TSS, a S-going tide for the starboard course change across the DWR, and 2 to 3hrs of ingoing tide to Harwich with a further 3hrs of flood into the Orwell and Stour still left. The Shipwash LtF or Orfordness Lt are good landfalls by night or day, and then the reciprocal of one of the three routes described above for the outward journey can be followed to Harwich entrance.

PASSAGE 20   HARWICH (LANDGUARD POINT) TO DEN HELDER ENTRANCE

Distance   147M

Duration   33hrs, 5½ tidal sets

Approach routes   See Chapter 6 Route 2 and Chapter 12, Den Helder

Charts   Admiralty *191, 1408, 1491, 2052, 2322, 2693*
Imray *C1, Y5, Y6, Y16*

Recommended start time outward
HW Dover, HW Harwich –1

Arrival time outward
HW Dover –4½, HW Den Helder

Recommended start time return
HW Dover +5½, HW Den Helder –2½

Arrival time return
HW Dover +1, HW Harwich

Tidal streams

Near Orfordness the stream starts NE-going at about HW Dover +0½ (Harwich –0½) and SW-going at HW Dover –6 (Harwich +6). In the area of the DWR the stream starts N-going at about HW Dover +0½, and S-going at HW Dover –5½hrs. Streams in the Schulpengat (approach to Den Helder) flood NE at HW Dover +3 (HW Den Helder –5), and ebb SW at HW Dover –4 (HW Den Helder +0½).

Commentary

In practice, over such a long distance accurate timing is extremely difficult.

Outward A start three hours earlier than in Passage 19 above is advised in order to catch the full flood into the Schulpengat. This gives a favourable stream out from Harwich to the Shipwash LtF or Orfordness but means slightly less favourable tides crossing the DWR and a ground track closer to the CTMG. Follow the recommended yacht route out of Harwich just outside the W and S sides of the dredged channel and cross the channel at right angles opposite the Platters (S card) buoy. From the Platters buoy there are three possible routes to the passage's departure point:

1. Via the Sledway channel passing to the E of the Cutler (G unlit) buoy to Bawd Head, rounding the NE Bawdsey (G) buoy, keeping N of the N Shipwash (N card) buoy to make a departure from the Shipwash LtF.

2. Via the Shipway channel passing to the E of the S Bawdsey (S card) buoy and turning E close to the N Shipwash (N card) to make a departure from the Shipwash LtF.

3. Keeping approximately 1M offshore but closing in to about ½M near to Orfordness and making a departure from between Orfordness and the NE Whiting (E card unlit) buoy.

From the departure point a course direct for the SG (RW) offing buoy at the entrance to the Schulpengat should be set and adjusted at the appropriate time to cross the DWR as close to a right angle as is practicable; this is a significant change of course at this point where the DWR widens and forks into two branches requiring a good lookout to be kept in order to adjust the vessel's heading when necessary depending on the traffic encountered which can be travelling on any of four possible headings. The positions of the DWR entry and exit course changes need careful estimation leaving a reasonable margin of error, say 1M.

The first landfall objective, particularly at night, is the Texel LtV some 2M N of the CTMG, although the wellheads of the Broad Fourteens marked by yellow Lt buoys 20M before the Texel LtV are an even more useful objective. Petten Nuclear Power Station, a square building with two 45m chimneys, 6M SSE of SG buoy is a good daylight landfall, whilst the 55m pencil-like Kijkduin Lt tower (30M range light) in the narrows opposite Noorderhaaks Island is also visible well out to sea. The Schulpengat channel is particularly well marked with R and G Lt buoys and 10M range daylight-intensity (18M at night) leading lights on Texel Island, the rear being a church spire. From Kaap Hoofd at the end of the Schulpengat/Breewijd channel the final approach route to Den Helder is E along the Marsdiep, a steep-sided channel with a number of Lt buoys near the harbour entrance which has 2 further sets of leading lights. Keep a sharp lookout for ferries coming out of the Veerhaven and heading for 't Horntje on Texel and also for naval shipping.

Return   Timing is a little more difficult than outward, and to obtain an ingoing tide from Shipwash LtF/Orfordness Lt to Harwich requires pushing the tide out of the Schulpengat for a short period. This also obtains several hours of S-going tide during the DWR course change across the wide area where the two DWRs diverge.

Follow the Marsdiep and then the Schulpengat pushing the last half of the flood to SG (RW) buoy, then a direct course to the Shipwash LtF or Orfordness Lt. To the N, Texel LtV should provide a good departure point followed to starboard by the Mobil P6-A and B platforms and buoys, and to port by the Broad Fourteens wellhead buoys on a S-going then slackening and turning tide. A change of course should be made to cross the DWR as close to a right angle as practicable as in the outward passage, and on estimating the departure point from the DWR a tide-corrected course should be set for the Shipwash LtF. Shipwash LtF or Orfordness should be passed on a strong S-going stream and Harwich reached around slack water. The Shipwash LtF or Orfordness Lt are good landfalls by night or day, and then the reciprocal of one of the three routes described above for the outward journey can be followed to Harwich entrance.

PASSAGE 21   LOWESTOFT ENTRANCE TO SCHEVENINGEN ENTRANCE

Distance   98M

Duration   22hrs, 3½ tidal sets

Approach routes   See Chapter 5 Route 2 and Chapter 12, Scheveningen

Charts   Admiralty *122, 1408, 1536, 1543, 2322*
Imray *C28, Y5*

Recommended start time outward
HW Dover −4, HW Lowestoft −2

Arrival time outward
HW Dover +5, HW Scheveningen +1

Recommended start time return
HW Dover +3, HW Scheveningen −1

Arrival time return
HW Dover, HW Lowestoft +2

Tidal streams

Outside Lowestoft entrance the S-going stream begins at about HW Dover –6 (HW Lowestoft –4) and N-going at HW Dover (HW Lowestoft +2). In the area of the DWRs the stream starts N-going at about HW Dover +0½, and S-going at HW Dover –5½hrs. Near Scheveningen N-going starts slightly later at HW Dover +1½ and S-going also later at HW Dover –4½.

Commentary

A course steered direct between the E Barnard (E card) buoy and Scheveningen entrance in either direction crosses the DWR divergence area at as close to a right-angled heading as is practicable so needs no course change until approaching Scheveningen. This passage is the same as the example in Chapter 3, Fig. 11B.

Outward   From Lowestoft entrance, head S keeping 2 cables off the end of Claremont Pier to avoid Lowestoft Bank (1m depth, 3½ cables off the entrance) and then pass close W of the Pakefield (G) buoy. Then pass close W of the S Newcome (G) buoy to the E Barnard (E card) buoy which is the passage's departure point, giving a wide berth to Barnard Sand.

Then set a direct course for Scheveningen entrance. The N-going tide in the DWR will sweep the vessel away from NHR-N (R) buoy and the Noord Hinder north TSS, and another N-going tide will similarly help approaching Scheveningen N of the Maas north TSS. A course correction will be needed crossing the tide as the entrance is approached. Scheveningen entrance (Chapter 11) is an easy one, its light powerful and the daylight marks unmistakable.

Return   A little more difficult to plan. It is critical to obtain a N set at the start, away from the Maas north TSS and above all away from the Maas Precautionary Area. This start time does this and also obtains a N set approaching and crossing the end of the Noord Hinder north TSS. There is however about one last hour of weakening foul tide along the Pakefield route into Lowestoft.

The course from the entrance is direct to the E Barnard buoy and although at night the powerful lights of Lowestoft and Southwold can be used to home in on this buoy, in daylight it is best to aim to pick up the wooded coast near to or even S of Benacre Ness in order to avoid Newcome Sand and then coast northwards to pick up the buoy. The problem at this stage, as mentioned above, is that the tide is against although weakening. The route is then the reciprocal to that described above for the outward passage.

PASSAGE 22   GREAT YARMOUTH ENTRANCE TO IJMUIDEN ENTRANCE

Distance   105M

Duration   23½hrs, 4 tidal sets

Approach routes   See Chapter 5 Route 2 and Chapter 12, IJmuiden

Charts   Admiralty *124, 1408, 1536, 1543, 2322*
Imray *C28, Y5*

Recommended start time outward
HW Dover –6, HW Gt Yarmouth (Gorleston) –4

Arrival time outward
HW Dover +5, HW IJmuiden +1

Recommended start time return
HW Dover –6, HW IJmuiden +2½

Arrival time return
HW Dover +5, HW Gt Yarmouth (Gorleston) –5½

Tidal streams

Streams off Great Yarmouth start N-going at HW Dover +6 (HW Gt Yarmouth –4½) and S-going at HW Dover –0½. (HW Gt Yarmouth +1½). In the area of the DWRs the stream starts N-going at about HW Dover +0½, and S-going at HW Dover –5½hrs. Near IJmuiden N-going starts slightly later at HW Dover +1½ and S-going also later at HW Dover –4½.

Commentary

No marks are likely to be encountered on this passage other than the Corton Channel buoys near Great Yarmouth and the Lt buoys of the IJmuiden-Geul (the approach channel). The Lowestoft and IJmuiden high lights are the major landfalls at night.

Outward   This start time with a S-going stream ensures a track S of the CTMG, particularly important to ensure approaching IJmuiden and the WNW-trending IJmuiden-Geul from the south.

From Great Yarmouth entrance the route is direct to and between the N Holm (N card) and Mid Corton (G) buoys marking the entrance to Corton Channel, and then S following the channel to round the S Corton (S card) buoy to its S and then E to the passage's departure point, the Corton (E card) buoy.

A course due E from Corton buoy will cross the first DWR at right angles. After estimating the exit point from this DWR alter course to starboard to cross the second DWR also at right angles, by which time there will be a N-going stream. Again an estimate of the departure from the DWR is required and a course set direct for IJmuiden entrance, making sure that the approach course is outside the IJmuiden-Geul. There will be a N-going tide for the last few hours closing the Dutch coast, ideal for approach to the harbour (see also Chapter 12, IJmuiden, *Approach and entrance*) from S of the IJmuiden-Geul. The IJmuiden steelworks chimneys (166m) and factory buildings can be seen well offshore, or the 29M range light at night.

Return   A direct course due W from IJmuiden entrance with the S-going tide keeps the vessel well clear of the entrance channel. The course change across the first DWR is at the beginning of another S-going stream. Another position estimate is necessary for changing course to right angles across the second DWR, and again for the final course to the Corton buoy after leaving the DWR. Without marks for confirmation of position this is not easy navigation; on the approach to Great Yarmouth it pays to take bearings of conspicuous objects (lights at night), such as the 112m chimney, and to continually take soundings to ensure picking up the Corton Channel buoys well in advance. In poor visibility it is best to wait offshore until conditions improve and lights are visible, and it is certainly inadvisable to run inshore on this stretch of coast blind without electronic aids. Cross Sand Lanby (RW) as well as Lowestoft light provide good night-time or dusk marks. From the Corton (E card) buoy the route is the reciprocal of that described above for the outward passage.

PASSAGE 23   GREAT YARMOUTH ENTRANCE TO DEN HELDER ENTRANCE

Distance   121M

Duration   27hrs, 4½ tidal sets

Approach routes   See Chapter 5 Route 2 and Chapter 12, Den Helder

Charts   Admiralty *191, 1408, 1536, 1543, 2322*
Imray *C28, Y5*

Recommended start time outward
HW Dover +5, HW Gt Yarmouth (Gorleston) –5½

Arrival time outward
HW Dover –5½, HW Den Helder –1

Recommended start time return
HW Dover –4½, HW Den Helder

Arrival time return
HW Dover –1½, HW Gt Yarmouth (Gorleston)

Tidal streams

Streams off Great Yarmouth start N-going at HW Dover +6 (HW Gt Yarmouth –4½) and S-going at HW Dover –0½, (HW Gt Yarmouth +1½). In the area of the DWRs the stream starts N-going at about HW Dover +0½, and S-going at HW Dover –5½. Streams in the Schulpengat (approach to Den Helder) flood NE at HW Dover +3 (HW Den Helder –5), and ebb SW at HW Dover –4 (HW Den Helder +0½).

Commentary

It is best not to plan the shortest course from the Corton (E card) buoy to the Schulpengat, since this cuts across the corner of the Texel TSS which surrounds the Helder Oil Field so a southerly diversion which adds very little to distance is advised. Entry and departure to and from the Schulpengat with a favourable tide is the other critical factor.

Outward   Initially on leaving Great Yarmouth for the Corton Channel the stream will be against but weakening. The tide will be pushing conveniently N on changing course across the second DWR, and the final approach to Den Helder will be with a full 6hrs of flood.

From Great Yarmouth entrance the route is direct to and between the N Holm (N card) and Mid Corton (G) buoys marking the entrance to Corton Channel, and then S following the channel to round the S Corton (S card) buoy to its S and then E to the passage's departure point, the Corton (E card) buoy.

A course due E from Corton buoy will cross the first DWR at right angles. After estimating the exit point from this DWR alter course to starboard to cross the second DWR also at right angles. Again an estimate of the departure from the DWR is required and a course set to approach 1M S of the Texel LtV. Soundings crossing Brown Ridge and the Y buoys of the Broad Fourteens to starboard, with the platforms and buoys of Mobil P6-A and B to port provide potential earlier fixes.

From 1M S of Texel LtV a course is set direct for the SG (RW) offing buoy at the entrance to the Schulpengat. If in daylight the LtV is not sighted then Petten Nuclear Power Station, a square building with two 45m chimneys, 6M SSE of SG buoy is a good landfall, whilst the 55m pencil-like Kijkduin Lt tower (30M range light) in the narrows opposite Noorderhaaks Island is also visible well out to sea, and in either case this will enable the vessel to coast N until SG buoy is sighted. The Schulpengat channel is particularly well marked with R and G Lt buoys and 10M range daylight-intensity (18M at night) leading lights on Texel Island, the rear being a church spire. From Kaap Hoofd at the end of the Schulpengat/Breewijd channel the final approach route to Den Helder is E along the Marsdiep, a steep-sided channel with a number of Lt buoys near the harbour entrance which has 2 further sets of leading lights. Keep a sharp lookout for ferries coming out of the Veerhaven and heading for 't Horntje on Texel and also for naval shipping.

Return Initially a full 6hrs of ebb will help to push the vessel well S beyond the Texel LtV. The stream will again be S-going crossing the first DWR, but there will be an adverse tide for the last hour from the Corton buoy to Great Yarmouth.

Follow the Marsdiep and then the Schulpengat with the ebb to SG (RW) buoy. From SG buoy course should then be set to cross the S end of Brown Ridge at approximately 52°35′N and continue onwards to the boundary of the first DWR where heading should be altered to starboard to cross at right angles. To the N, Texel LtV should provide a good departure point followed to starboard by the Mobil P6-A and B platforms and buoys, and to port possibly by the Broad Fourteens wellhead buoys, useful help for estimating positions for altering course on entry and exit of the first DWR as well as for crossing the second at right angles, and if necessary a further change of course estimated to home in on the Corton buoy.

Without marks for confirmation of position this is not easy navigation; on the approach to Great Yarmouth it pays to take bearings of conspicuous objects (lights at night), such as the 112m chimney, and to continually take soundings to ensure picking up the Corton Channel buoys well in advance. In poor visibility it is best to wait offshore until conditions improve and lights are visible, and it is certainly inadvisable to run inshore on this stretch of coast blind without electronic aids. Cross Sand Lanby (RW) as well as Lowestoft light provide good night-time or dusk marks. From the Corton (E card) buoy the route is the reciprocal of that described above for the outward passage.

5. Norfolk & Suffolk coast from Cromer to Orfordness

Charts   Admiralty *106, 1504, 1536, 1543, 2695*
Imray *Y5, C28, C29*

Tidal atlases   Admiralty *North Sea – southern portion*

Tidal streams   (based on HW Dover and HW Harwich)

Position	*Start times*			
	DOVER		HARWICH	
	North	*South*	*North*	*South*
Off Cromer Lt	−0115	+0430	−0145	+0400
S of Haisborough Sand	−0010	+0600	−0040	+0530
Close N of Corton buoy	HW	+0530	−0030	+0500
Off Gt Yarmouth ent.	−0020	+0600	−0050	+0530
Brush Quay in ent.★	−0030[1]	+0545[2]	−0100[1]	+0515[2]
Off Lowestoft ent.†	+0010	−0610	−0020	+0545
In entrance	−0145[1]	+0430[2]	−0215[1]	+0400[2]
2M E of Orfordness	+0020	−0550	−0010	+0605

Tidal differences and ranges   (based on HW Dover)

Place	*HW* (time)	*Springs/Neaps* (range in metres)
Great Yarmouth ent.	−0208	1.9/1.0
Lowestoft	−0133	1.9/1.1
Southwold	−0058	2.1/1.3

Notes
★ These streams turn 1½hrs after local LW and HW.
† Within 90m of entrance S-going stream only runs strongly
for 3hrs and the N-going stream can start up to 2hrs earlier.
[1] Outgoing
[2] Ingoing

Major lights

Name of light	Characteristics	Position	Structure
Offshore			
Cross Sand Lanby	LFl.10s6m5M	52°37′.0N 1°59′.2E	Wh HFP buoy, R stripes, Racon
Haisbro' LtV	Fl(4)15s12m24M Dia 60s	52°58′.6N 1°34′.6E	R hull, Racon, distress answering
Newarp LtV	Fl.10s12m26M Horn 20s	52°48′.4N 1°55′.8E	R hull, distress answering , Racon
Outer Gabbard LtV	Fl(4)20s12m24M Dia(4)60s	51°59′.3N 2°04′.6E	R hull, RC, distress answering
Shipwash LtF	Fl(3)20s12m24M Horn(3)60s	52°02′.0N 1°42′.1E	R hull
Smith's Knoll LtV	Fl(3)20s12m24M Dia(3)60s	52°43′.5N 2°18′.0E	R hull, RC, Racon, distress answering
Onshore			
Cromer	Fl(5)15s84m20M	52°55′.5N 1°19′.1E	Wh 8-sided tower
Lowestoft	Fl.15s37m28M F.R.30m18M	52°29′.2N 1°45′.5E	Wh tower, 184°-vis-211.5°
Happisburgh	Fl(3)30s41m14M	52°49′.2N 1°32′.3E	Wh tower, 3 R bands
Orfordness	Fl.5s28m30M F.GR.14m15/14M	52°05′.0N 1°34′.6E	Wh round tower, R bands 038°-R-047°-G-shore-R-210°
Southwold	Fl(4)WR.20s37m22–20M	52°19′.6N 1°41′.0E	Wh round tower 204°-R-220°-W-001°-R-032.3°

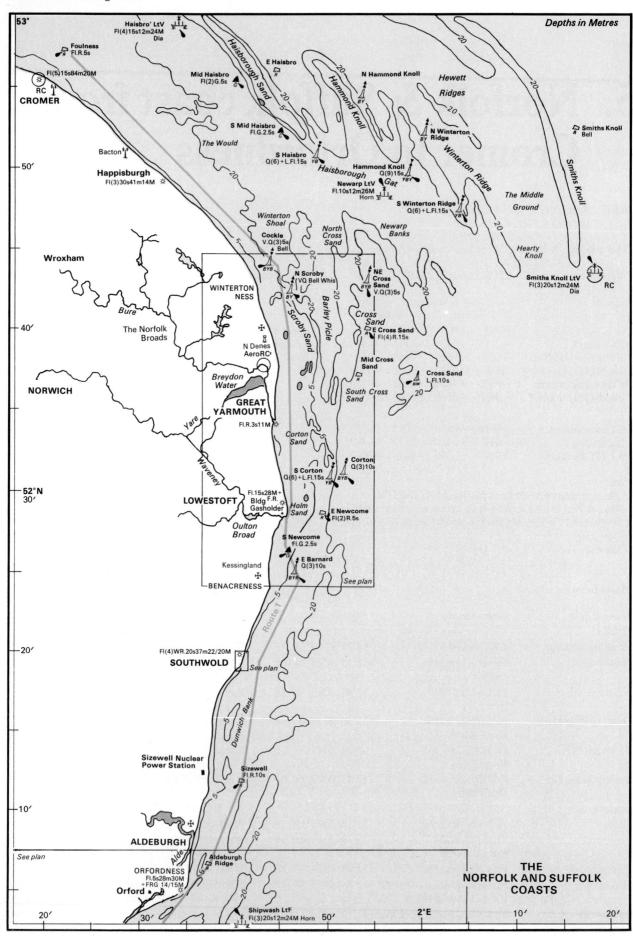

Depths in Metres

53°

Foulness
Fl.R.5s

Fl(5)15s84m20M
RC
CROMER

Haisbro' LtV
Fl(4)15s12m24M
Dia

E Haisbro

Mid Haisbro
Fl(2)G.5s

Haisborough Sand

N Hammond Knoll

*Hewett
Ridges*

N Winterton
Ridge

Smiths Knoll
Bell

Bacton

The Would

S Mid Haisbro
Fl.G.2.5s

Happisburgh
Fl(3)30s41m14M

S Haisbro
Q(6)+L.Fl.15s

Hammond Knoll
Haisborough

Hammond Knoll
Q(9)15s
Gat

Newarp LtV
Fl.10s12m26M
Horn

S Winterton Ridge
Q(6)+L.Fl.15s

Winterton Ridge

*The Middle
Ground*

Smiths Knoll
Smiths Knoll

*Winterton
Shoal*

Cockle
V.Q(3)5s
Bell

*North
Cross
Sand*

*Newarp
Banks*

*Hearty
Knoll*

Smiths Knoll LtV
Fl(3)20s12m24M
Dia
RC

Wroxham

WINTERTON
NESS

N Scroby
(VQ Bell Whis)

NE
Cross
Sand
V.Q(3)5s

Bure

*The Norfolk
Broads*

N Denes
AeroRC

Scroby Sand

Barley Picle

*Cross
Sand*

E Cross Sand
Fl(4)R.15s

Mid Cross
Sand

Cross Sand
L.Fl.10s

NORWICH

Breydon
Water

**GREAT
YARMOUTH**
Fl.R.3s11M

*South Cross
Sand*

Yare

*Corton
Sand*

Corton
Q(3)10s

Waveney

LOWESTOFT

S Corton
Q(6)+L.Fl.15s

Corton
Q(3)10s

Fl.15s28M+
F.R.
Bldg
Gasholder

*Holm
Sand*

E Newcome
Fl(2)R.5s

*Oulton
Broad*

S Newcome
Fl.G.2.5s

Kessingland

E Barnard
Q(3)10s

See plan

BENACRENESS

Route 1

SOUTHWOLD
Fl(4)WR.20s37m22/20M

See plan

Dunwich Bank

**Sizewell Nuclear
Power Station**

Sizewell
Fl.R.10s

ALDEBURGH

Alde

Aldeburgh
Ridge

See plan

ORFORDNESS
Fl.5s28m30M
+FRG 14/15M

Orford

Shipwash LtF
Fl(3)20s12m24M Horn

**THE
NORFOLK AND SUFFOLK
COASTS**

20′   30′   50′   2°E   10′   20′

Radiobeacons

Name	Freq. (kHz)	Ident.	Range (miles)	Seq.	Position
Marine radiobeacons					
Smith's Knoll Group	287.3				
Smith's Knoll LtV		SK	50	1	52°43'.5N 2°18'.0E
Goeree Lt		GR	50	2	51°55'.5N 3°40'.2E
Dudgeon LtV		LV	50	3	53°15'.5N 1°13'.5E
Outer Gabbard LtV		GA	50	4	51°59'.4N 2°04'.6E
Cromer		CM	50	5	52°55'.5N 1°19'.1E
N Hinder LtV		NR	50	6	52°00'.2N 2°51'.2E
Aero beacon					
Gt Yarmouth/N Denes	397	ND	15	Cont	52°38'.2N 1°43'.5E

Coast radio stations

Station*	Transmits & Receives (VHF Ch)	VHF Ch†	Traffic lists (times)	Storm warnings (times)	Weather messages (times)	Navigational warnings (times)
Bacton	03,**07**,16,**63,64**	07	0333 & odd H+33 (0733-2333)	0303,0903, 1503,2103	0833,2033	0233,0633,1033, 1433,1833,2233
Orfordness	16,**62,82**	62	0103,0503 & odd H+03 (0903-2303)	0303,0903, 1503,2103	0803,2003	0033,0433,0833, 1233,1633,2033

Notes
* Each station has 24hr service and watch. All times are GMT.
† Channel on which traffic lists, storm warnings, weather messages and navigational warnings are given.

Major fixed daylight marks (from N to S)

Cromer LtHo
Bacton gas terminal and nearby 2 radio masts (conspic) with 3 R Lts each
Happisburgh LtHo
Winterton Church tower (conspic) and nearby Winterton Old LtHo (disused, Wh round tower B band)
Near Caister Point a stone water tower (conspic)
Wh tower (39m) 2M N of Yarmouth entrance
2 pylons (72m) 1.2M N of Yarmouth entrance 0.4M inland
112m chimney N of Great Yarmouth entrance

Block of flats and a gasholder near Lowestoft Ness
Silo (49m), inland W of Lowestoft entrance
Kessingland Church tower
Southwold LtHo
Walberswick Church tower W of Southwold entrance
Sizewell nuclear power station (71m conspic square building)
St Peter's Church tower at Aldeburgh
Orfordness LtHo

Approach routes and tidal timing

Closing in on the 62M of coast between the Orfordness and Cromer lights requires considerable caution, particularly in strong winds from the two E quadrants or in fog. Superficially smooth, this coastline is strewn with moving sandbanks, and to most yachtsmen is merely a staging post en route to other destinations and more protected waters: inland to the Norfolk Broads, S to the Thames Estuary or N to the Wash, Humber and, especially for continental yachtsmen, Scotland. The area divides into three natural 20M long sectors from N to S.

Approach from seawards to the N sector from Cromer to Winterton Ness is not advisable. A series of dangerous shoals roughly parallel to the coast lie from inshore to 20M offshore. The outer ridges – Smith's Knoll, the Hewitt Ridges and Winterton Ridge have least depths of just over 5m, but the inner ridges – Hammond Knoll, Haisborough Sand and the Cross Sands are less than 5m and have some drying patches. All can be extremely uncomfortable and dangerous in strong winds particularly from the E quadrants. Further N another series of banks form a number of major North Sea gas fields whose pipelines come ashore at Bacton. Yachtsmen who insist on approaching or leaving this coastal area en route from or to Noord Holland are advised to use the Smith's Knoll and Newarp LtVs, 14½M apart, as the primary marks for following the channel between the main banks and the Newarp Banks and Cross

Sands, keeping 1M SW of a line drawn between these two LtVs through the Haisborough Gat into The Would, between the Haisborough Sand and the N Norfolk coast. After clipping the 14m end of Hearty Knoll the depths on this route are well over 20m, and up to 40m.

The seaward approaches to the next 20M central sector, with its two good deep-water harbours, are also fraught with dangers and a coastal approach from N or S is easier. Between Winterton Ness N of Great Yarmouth and Benacre Ness just S of Lowestoft there is a 20M-long offshore cordon of sand from the sickle-shaped Cockle Shoal covering the N approaches through the North and Middle Scrobies, the Corton and Newcome Sands and finally to the Barnard off Benacre Ness. All have least depths below 5m and many have drying patches. Inshore is a deep-water area generally of 10 to 20m but with a number of shallower banks near Lowestoft, and to which access is made via three buoyed and lit gaps: Cockle Gatway in the N (4 to 5m soundings), Corton Channel in the centre (5 to 10m soundings), and Pakefield Road leading to Lowestoft (4 to 5m soundings).

To compound the problems the offshore banks in this middle section are constantly changing; more rapidly than the Hydrographer can keep up with on the charts. On the latest copy of Admiralty chart *1536 Approaches to Great Yarmouth and Lowestoft* the inset in the top left-hand corner divides the chart area into several periods of survey, and on the chart itself there are large discontinuities of the charted contours and depth shadings at the boundaries of these areas. There is a line across the Corton Sand E of Yarmouth, S of which the last survey was in 1981–3 and N of which only a little earlier in 1979–80, yet the sand has expanded in width over the period by ½M. The message for strangers is clear: keep well within the buoyed channels and don't cross the shallower banks especially in bad weather.

The S sector from Benacre Ness southwards to Orfordness is undoubtedly the safest offshore approach route with the fewest offshore banks, and for the Dutch yachtsman approaching England for the first time it is the best objective to aim for, since he avoids crossing any of the Thames Estuary banks to the S, and in good visibility can make a clean landfall well offshore picking up Southwold Lt or the myriad lights of Sizewell nuclear power station further S or a choice of the Outer Gabbard LtV, the Shipwash LtF or Orfordness Lt at the S end.

Sheltering from bad weather inshore is not too much of a problem since both deep-water ports are easily accessible at all states of tide although the ebb runs strongly out of Yarmouth, but a lee shore at the S and N ends can be very dangerous with no easily accessible refuges. Fog is definitely to be avoided since, although Yarmouth Roads provided a protected gathering place for Sir Hyde Parker's and Nelson's fleet before the battle of Copenhagen, it is much too deep an anchorage for small yachts, and the only shallow anchorages are close to and on the shoals in the central sector.

Tidal streams follow the direction of the coast and are approximately 6hrs in each direction starting about ½–1hr earlier at the N end of the coast compared with the S and giving a slight advantage sailing S.

ROUTE 1   COASTING FROM CROMER LIGHT TO ORFORDNESS

Distance   62M

Commentary

Cromer, the N boundary of this chapter is an extremely unlikely start for any cruise, and it must be said at the outset that there are no harbours of refuge on the 50M stretch between Great Yarmouth and Blakeney, and the latter is a 2hr-entrance harbour which dries out. At the S end the situation is a little better; it is 20M between the two half-tidal shallow entrances of Southwold and Orford Haven but nearly 40M between the deep-water Lowestoft and Harwich harbours.

Starting 1½M offshore from Cromer Lt to avoid Foulness Sand the route continues parallel to the coast past the lit complex of Bacton North Sea Gas Terminal and Happisburgh Lt to the Cockle (E card) Lt buoy off Winterton Ness. Then passing ½M W of the N Scroby (N card) Lt buoy and along the Cockle Gatway buoyed (lit) channel and coasting about 1M offshore along Yarmouth, Gorleston and Corton Roads inshore of the lit buoys. The coast should be closed on the approach to Lowestoft Ness to pass close E of the obstruction marked by 2 cardinal Lt buoys lying 3 cables off the Ness, then pass 1 cable E of Ness (R) Lt buoy across the Ridge shoal (2.9 to 3.8m) then pass 1½ cables off Lowestoft entrance to avoid Low-estoft bank (1m depth, 3½ cables off the entrance), then ½M offshore past Claremont Pier and pass close W of the Pakefield (G) Lt buoy. The route then passes close to the S Newcome (G) Lt buoy and E of the E Barnard (E card) to give a wide berth to Barnard sand and then stays 1½M offshore until abreast of Southwold where a course should be set to Sizewell (R) Lt buoy in order to pass E of two minor banks: Dunwich Bank (3m at S end) and Sizewell Bank (2.7m at S end). A course should then be set for Aldeburgh Ridge (R) unlit buoy to keep E of Aldeburgh Ridge (as little as 1.2m at S end) and awkwardly close to Orfordness. At night this ridge and its hazardous buoy can be passed by using a clearing bearing on Orfordness. Finally pass Orfordness about ½M offshore to clear Whiting Bank to the S.

Tidal timing

60M is usually impossible to do in a single tide and can be broken at any of the three harbours. Approach to Great Yarmouth from the S with the last of the ebb gives a slack or ingoing tide at the entrance, useful for going up to Town Hall Quay or Breydon Water, whereas in the opposite direction on the last of the flood it is necessary to push an accelerating ebb up the river. Entering Lowestoft's Yacht Basin is not difficult in most tidal circumstances providing care is taken with the cross-tide at the entrance. Since Southwold is best entered from about half-tide rising it is easier to enter from the N, and Great Yarmouth only 16M away (or Low-estoft, 10M) can be left at slack LW with a fair tide down coast. Coming from the S around Orfordness it is better to press on to Lowestoft if Southwold is reached towards the bottom of the tide, and delay the visit for the return.

ROUTE 2 APPROACHES TO GREAT YARMOUTH AND LOWESTOFT FROM THE CORTON BUOY

Distances   6M to Great Yarmouth, 6½–7½M to Lowestoft

Commentary

Homing in from seawards on the Corton (E card) Lt buoy at night is easy, with the assistance of Cross Sand Lanby (RW) and the Great Yarmouth and Lowestoft lights. In daytime it is a little more difficult although the industrial chimneys and seaside-resort buildings of Great Yarmouth and Lowestoft provide excellent landfalls from offshore. The 112m Yarmouth chimney is the most conspicuous landmark on the coast and a bearing on this can help to locate other major daylight marks (see above) for cross bearings. Decca is an invaluable aid in such circumstances. From Corton buoy the route rounds the S Corton (S card) Lt buoy to the S, follows the Corton channel (Lt buoys) northwards and passes N of N Holm (N card) Lt buoy then direct across to Yarmouth entrance.

To Lowestoft in rough weather the route is the same as above to the N Holm buoy, but then turns SW close to NW Holm (G) Lt buoy, passes close E of the obstruction marked by 2 cardinal Lt buoys lying 3 cables off Lowestoft Ness, then passes 1 cable E of Ness (R) Lt buoy across the Ridge shoal (2.9 to 3.8m) and then into Lowestoft entrance, avoiding Low-estoft Bank (least depth 1m and 3½ cables SE of the entrance). A short cut in settled weather and a rising tide avoids the dogleg of Corton channel, taking a course due W from the S Cor-ton (S card) buoy directly across the neck in Holm Sand (depths of 2 to 3m), turning S after about 2M when 11m soundings are reached, and joining up with the last stages of the above longer route for Lowestoft Ness.

Tidal timing

The Corton buoy is usually approached cross-tide after a long North Sea crossing, so timing depends on a certain amount of luck as well as planning. Arriving at the Corton buoy around HW Dover it is slack water turning N, therefore better for going N to Yarmouth. 5½hrs after HW Dover the tide turns S, better for Lowestoft.

Harbours

All three harbours are artificial. Yarmouth and Southwold date from the Middle Ages and Lowestoft from the 1830s. Yarmouth with its 2M of double quaysides is now the dominant fishing, trading and more recently North Sea gas servicing port. Southwold, originally an important fishing port, has never recovered from the establishment of Lowestoft harbour when the Dogger Bank's 'Silver Pit' was discovered. Both Yarmouth and Lowestoft have access to the Broads and Norwich but the former is much the easier entrance.

GREAT YARMOUTH

Port radio (VHF)

Yarmouth Port Control	VHF Ch	Hrs listening
Call *Yarmouth*	12, 16	24
☎ (0493) 855151 & 663476		
Haven Bridge Gt Yarmouth	12	
Breydon Bridge	12	

Entry signals

Always contact Harbour Control by VHF before entering or leaving.

Control tower head of South Pier. F.Y Lt (day), or all Lts extinguished (night) – do not enter.

Harbour Control root of South Pier (showing N up harbour). 2 horizontal R Lts – do not leave.

Customs

Customs Office, Fishwharf, Great Yarmouth, Norfolk, NR30 3LY

☎ (by VHF) Ipswich (0473) 219481

☎ (ashore) dial 100 and ask for 'Freefone Customs Yachts'

Great Yarmouth haven entrance

Entrance

Entry against a strong mid-tidal ebb requires plenty of engine power, and since there is a further 2M to go to the visitors' berths at Town Hall Quay on the E wall just below the bridge if there is any doubt about the engine then try to go in on a rising tide or close to slack water. There are leading marks which are extremely useful at night on the right-angled turn facing the entrance, the rear one of which is Brush LtHo. This is a busy port so take great care rounding the right-angled bend inside the entrance.

Mooring and facilities

Customs clearance can be arranged and overnight fees are collected at Town Hall Quay and you must also pay here if you wish to visit the Broads. Here the tidal range and rate is considerable – the averages at the head of this chapter apply only to the harbour entrance – so you need strong warps, large fenders, and a long (telescopic?) ladder as the

Photo Great Yarmouth Port and Haven Commissioners

few sparse tiers of wooden scrambling bars are in poor shape. You may be able to find a spare berth further down the river near the many fishing boat moorings, particularly on the W bank, but the town is less accessible and there is an abundance of 'No Moorings' or 'Danger' signs, along the few lengths of empty wharf. Town Quay is very convenient for visiting the seafront or the market square in between sea and river, or a small museum nearby. There is a lot of visiting traffic, particularly from across the North Sea, so you should also leave fenders out on your offside.

It must be mentioned that the £45-million project for the building of a new harbour N of the present one with a curving 600m mole has passed through the feasibility stage, is being organised by the Port and Haven Commissioners, and lead financiers in the project have been appointed; date of completion should be around 1989.

The entrance to Great Yarmouth haven

Entrance to the Norfolk Broads

This is a far easier entry point to the Norfolk Broads than Lowestoft. Simply call Haven Bridge (often unmanned, but then Yarmouth Port Control will tell you what to do) and they will arrange a time. There are no places to tie up immediately on the other side of the bridge unless you risk lying alongside a fishing boat or go to Bure Marine, ☎ (0493) 656993, a yard on the W bank just above the bridge who also have boat repair and storage facilities. It is better to time your passage with the tide so that you go straight on through the new lifting bridge, Breydon Bridge, where there is a fine new mooring quay on the Breydon Water side, and follow the channel southwards marked by

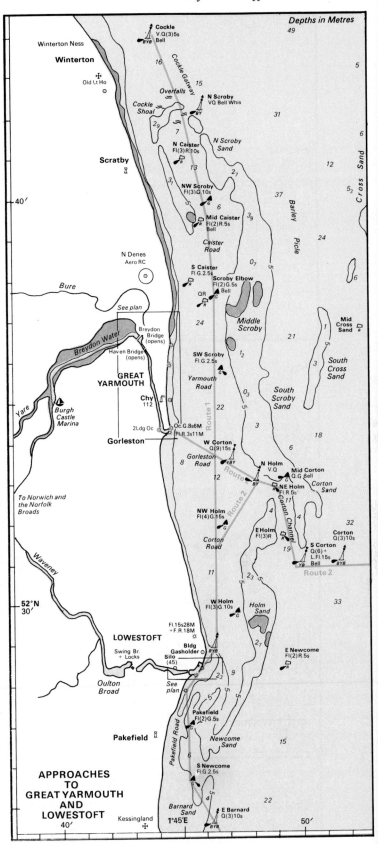

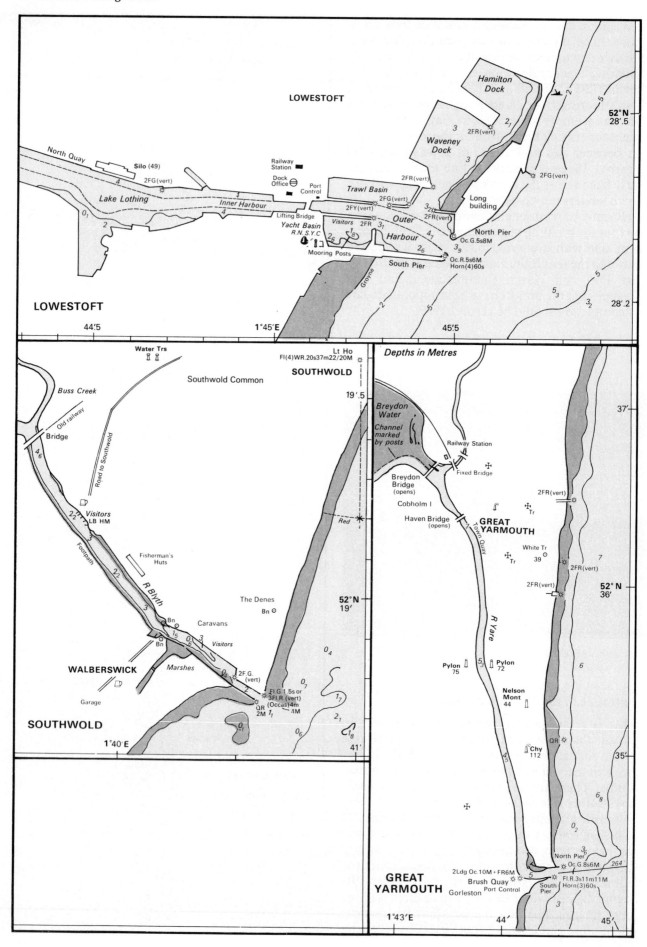

LOWESTOFT

LOWESTOFT

North Quay

Silo (49)

2FG(vert)

Lake Lothing

Inner Harbour

Railway Station

Dock Office

Port Control

Lifting Bridge

Yacht Basin
R.N.S.Y.C

Visitors

Mooring Posts

Trawl Basin

2FY(vert)

2FG(vert)

Outer
Harbour

South Pier

Groyne

Hamilton
Dock

2FR(vert)

Waveney
Dock

2FR(vert)

2FG(vert)

Long
building

2FR(vert)

North Pier
Oc.G.5s8M

Oc.R.5s6M
Horn(4)60s

52°N
28'.5

28'.2

44'.5

1°45'E

45'.5

Water Trs

Southwold Common

Lt Ho
Fl(4)WR.20s37m22/20M

SOUTHWOLD

19'.5

Buss Creek

Old railway

Bridge

Road to Southwold

Visitors
LB HM

Footpath

Fisherman's
Huts

R Blyth

Bn

Bn

Caravans

Visitors

WALBERSWICK

Marshes

Garage

SOUTHWOLD

The Denes
Bn

Red

2F.G.
(vert)

QR
2M

Fl.G.1.5s or
3Fl.R.(vert)
(Occas)4m 4M

52°N
19'

Depths in Metres

Breydon
Water

Channel
marked
by posts

Railway Station

Fixed Bridge

Breydon
Bridge
(opens)

Cobholm I

Haven Bridge
(opens)

**GREAT
YARMOUTH**

Town Quay

White Tr
Tr 39

Pylon
75

Pylon
72

Nelson
Mont
44

Chy
112

R Yare

2FR(vert)

Tr

2FR(vert)

2FR(vert)

52°N
36'

37'

QR

North Pier
Oc.G.8s6M

264

**GREAT
YARMOUTH**

2Ldg Oc.10M + FR6M
Brush Quay
Port Control
Gorleston

South
Pier
Fl.R.3s11m11M
Horn(3)60s

35'

1°40'E

41'

1°43'E

44'

45'

large port and starboard posts straight across Breydon Water, and take the port-hand fork into the River Waveney to Burgh Castle Marina, ☎ (0493) 780331, going through the small entrance on the S bank and tying up to floating pontoons. Diesel, repair and storage facilities are available here. In 1985 I found only 1.8m in the entrance at HW neaps, so even though there is only a very small range you need to pick your entry time near HW and expect to sink into soft mud during your stay. Alternatively, you may wish to visit Norwich, in which case you take the starboard-hand fork along the Yare above Burgh Castle and go on through the Reedham swing bridge (there is a large notice of next opening time on the bridge) and on through several other opening bridges to Norwich.

Breydon Water and the new Breydon Bridge

LOWESTOFT

Port radio (VHF)

Lowestoft Bascule Bridge ☎ (0502) 2286/7	*VHF Ch* **14**, 16	*Hrs listening* 24

Entry signals

Always contact Port Control on VHF before entering or leaving.

S pier head, Fl Lt below the Oc.R entrance Lts – no entry, vessels may proceed outwards only.

Customs

Customs Office, Fishwharf, Great Yarmouth, Norfolk, NR30 3LY
☎ (by VHF) Ipswich (0473) 219481
☎ (ashore) dial 100 and ask for 'Freefone Customs Yachts'

Entrance

Approach to the entrance is as in Routes 1 and 2 above. This is an easy entrance providing the strength of the cross-tide is properly gauged. The most conspicuous features near the harbour from a distance are the cranes and warehouses on the N side, and on each side of the entrance are white towers shaped like rockets or Chinese pagodas with spires on top.

Mooring and facilities

Customs clearance and mooring is in the South Basin alongside long inflated sausage-like fendoffs on the N wall. This basin is administered by the Royal Norfolk and Suffolk Yacht Club, ☎ (0502)

Lowestoft entrance

Lowestoft. The Royal Norfolk and Suffolk Yacht Club

66726, on behalf of the harbour authority, so, through no fault of the RNSYC, short stays are expensive. The club may be able to fix you up with a spare mooring to dolphins in the middle of the harbour for longer periods. The RNSYC will provide you with the harbour regulations. The South Basin is very convenient for the town centre to the N and a promenade to the S. The RNSYC is the centre of Broads One Design racing and is a stop on the Round Britain Race; facilities are excellent. The bosun of the RNSYC can direct you to the various boat repair, chandlery and storage facilities.

Entrance to the Norfolk Broads

Entering the Inner Harbour through the road lifting bridge should again be preceded by VHF contact but once through all the moorings are private, so you will probably prefer to continue on through Mutford Lock and a swing bridge into Oulton Broad where there are several marina facilities. This latter operation must be arranged 48hrs in advance with the Port Manager's Office and can be expensive.

Southwold looking up-river

SOUTHWOLD

Port radio (VHF)

Call *Southwold*	*VHF Ch*	*Hrs listening*
☎ (0502) 723502	**12, 16**	0800–1800 LT

Entry signals

Always contact the HrMr on VHF before entering.

N pier head, 3 vertical Fl.R.1.5s Lts – harbour closed. If this is exhibited and you cannot raise the HrMr on Ch 12, then call Yarmouth Coastguard on Ch 16 for advice.

Customs

☎ (VHF and ashore) Ipswich (0473) 219481

Entrance

Entry at night is not to be recommended to newcomers although the piers are lit. The entrance is tidal and depths are changing. Always take a rising tide as the ebb sluices out of this river from the lake up near Blythburgh; when I entered in mid-1985 there was about 1½m at LW springs implying safe entry in moderate conditions for 1½m draught at half-tide. Shoals extending out from the N pier and S of the S pier make it necessary to approach the entrance directly in line with the piers from about 3 cables offshore. Once inside keep more to the S pier and then cross towards the 'Knuckle' (G beacon with 2F.G (vert) Lts) on the N pier to negotiate a shallow area on the S side and thus into the River Blyth.

Mooring and facilities

Customs clearance can be arranged at the visitors' pontoons ¾M from the entrance on the N bank, which are fixed, so warps need tending, and if you are an inside boat at the pontoons you may just dry out on a soft bottom. The HrMr collects dues.

Southwold Boatyard and Chandlery, near the lifeboat house, together with the HrMr can assist in arranging repairs and storage. Southwold is ¾M walk, with plenty of shops and restaurants, an old church, the famous Adnams Brewery with many captive pubs around, the lighthouse, and the Sailors' Reading Room on the promenade, with its many old press cuttings, photographs, and models of ships, not least the famous nineteenth century 'beach yawls' upon which many subsequent lifeboat designs were based and which took part in many rescues as well as collecting much salvage money.

The council is currently trying to raise £3 million for repairs and maintenance required to the harbour, and various development schemes have so far been shelved.

6. Suffolk & Essex coast & rivers from Orfordness to the Naze

Charts   Admiralty *1183, 1491, 1593, 1594, 2052, 2693, 2695*
Imray *C1, Y6, Y16, C28, C30*

Tidal atlases   Admiralty *Thames Estuary*
North Sea – southern portion

Tidal streams   (based on HW Dover and HW Harwich)

Position	Start times			
	DOVER		HARWICH	
	Out	*In*	*Out*	*In*
2M E of Orfordness	+0020[1]	−0550[2]	−0010[1]	+0605[2]
Ore Bar	HW	+0600	−0030	+0530
Orford Quay	+0100	−0430	+0030	−0500
Deben Bar	+0015	−0610	−0015	+0545
Woodbridge	+0130	−0400	+0100	−0430
Harwich/Shotley Pt	+0100	−0500	+0030	−0530
Baltic Wf Manningtree	+0130	−0245	+0100	−0315
Ipswich	+0130	−0300	+0100	−0330
Close to Roughs Tr	HW[1]	−0615[2]	−0030[1]	+0600[2]

Notes
[1] N-going
[2] S-going

Tidal differences and ranges   (based on HW Dover)

Place	HW (time)	Springs/Neaps (range in metres)
Ore Bar	+0026	2.9/1.7
Orford Quay	+0122	
Slaughden	+0207	
Snape Bridge	+0237	
Deben Bar	+0035	3.2/2.8
Woodbridge	+0117	3.6/2.2
Harwich	+0050	3.6/2.3
Mistley	+0115	3.9/2.4
Ipswich	+0110	3.9/2.4
Walton-on-the-Naze	+0042	3.8/2.3

Major lights

Name of light	Characteristics	Position	Structure
Offshore			
Shipwash LtF	Fl(3)20s12m24M Horn(3)60s	52°02′.0N 1°42′.1E	R hull
Galloper Lanby	Fl.20s12m22M Horn Mo(A)60s	51°43′.9N 1°57′.8E	R cylinder on circular buoy, Racon
Outer Gabbard LtV	Fl(4)20s12m24M Dia(4)60s	51°59′.3N 2°04′.6E	R hull, RC, distress answering
Sunk LtF	Fl(2)20s12m24M Horn(2)60s	51°51′.0N 1°35′.0E	R hull, RC, Racon
Onshore			
Orfordness	Fl.5s28m30M F.GR.14m15/14M	52°05′.0N 1°34′.6E	Wh round tower, R bands 038°-R-047°-G-shore-R-210°

Radiobeacon

Name	Freq. (kHz)	Ident.	Range (miles)	Seq.	Position
Marine radiobeacon					
Sunk LtF	312.6	UK	10	Cont	51°51′.0N 1°35′.0E

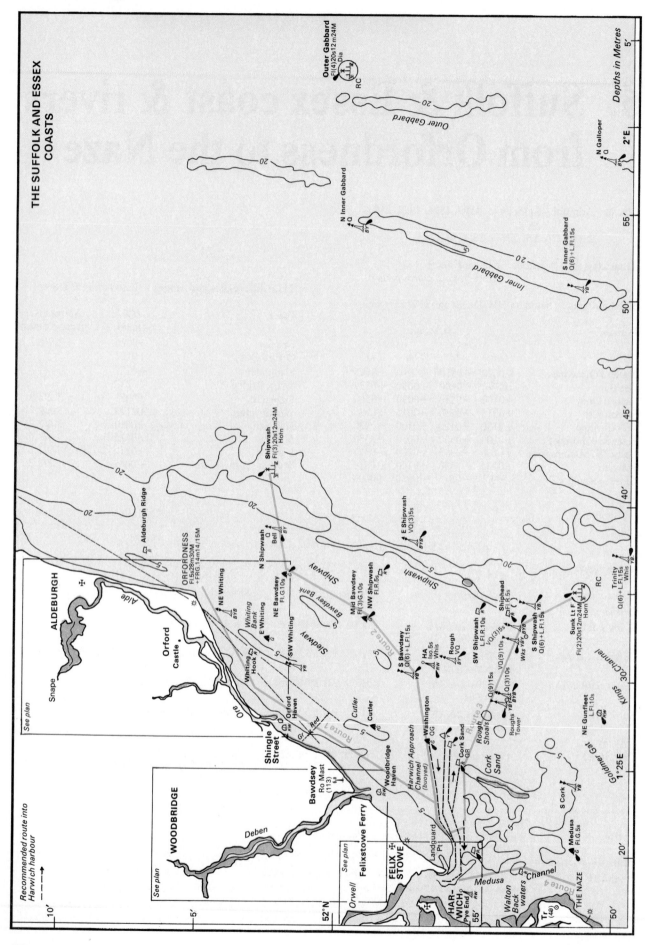

THE SUFFOLK AND ESSEX COASTS

Depths in Metres

Coast radio station

Station*	Transmits & Receives (VHF Ch)	VHF Ch†	Traffic lists (times)	Storm warnings (times)	Weather messages (times)	Navigational warnings (times)
Orfordness	16,**62**,82	62	0103,0503 & odd H+03 (0903-2303)	0303,0903, 1503,2103	0803,2003	0033,0433,0833, 1233,1633,2033

Notes

★ This station has 24hr service and watch. All times are GMT.

† Channel on which traffic lists, storm warnings, weather messages and navigational warnings are given.

Major fixed daylight marks (from N to S)

Orfordness LtHo
Orford Castle
A series of Martello towers between Orford Haven entrance and Landguard Pt
Shingle Street cottages S of Orford Haven entrance
Radio mast (113m, R Lts) N of Deben entrance
Felixstowe Church spire
The Roughs tower (double pillars and platform) approx 6M SE from Landguard Pt

Crane gantries of the container port, a conspicuous gas storage tank and a large chimney approx 1M N of Landguard Pt
Harwich Church spire on the S side of the harbour
A water tower in Dovercourt with two disused lighthouses ½M E
The Naze red cliffs with the chimney-shaped Naze Tower (49m) on top

Approach routes and tidal timing

For the continental visitor the approaches to this area begin over 20M offshore with the Outer Gabbard bank. This is one of a series of N-trending offshore undersea ridges, outliers from the Thames Estuary, rising sharply from the 20 to 30m surrounding seabed. The Outer Gabbard is 7M long, ½M wide, 5.5m sounding in centre to 18.2m at extremities, and its S extension, the Galloper, is 7M by ½M, 2.4 to 15m depth. The next one inshore is Inner Gabbard, 10M by ¾M, 4m sounding nearer the N end to 18m in the S, with its 5M-long N outlier (10.6 to 16.6m). Yet further inshore is the Shipwash, 8M long by ½M wide, drying in places and with 5 to 10m at each end, and lying from 4M offshore (N end) to 8M (S end). Lying inshore of the Shipwash from S of Orfordness to the Naze is a 'shelf' area of uneven soundings, with 3 longitudinal banks in the N – Bawdsey (1.4m least depth), Whiting (0.9m least depth and very close inshore below Orfordness), and Cutler (1.2m least depth); and a wide extremely uneven area in the S with several named areas – the Rough Shoals (4 to 10m), Cork Sand (drying ridge with an 0.9m spreading outlier, West Rocks, to the S), Stone Banks (1.9m in places), the Naze Ledge (2.4m in places) merging N in Pennyhole Bay with the drying Pye Sand. The Harwich deep-water channel provides approach access to the harbour from N of Cork Sand with a minimum of 5m and up to 10m depths, but the rivers Deben and Ore are both entered across the 5m contour and shallower bars (less than 1m soundings).

Harwich harbour is a very significant port complex with a radar surveillance area extending in a 4–5M arc offshore, a dredged and lit (buoyed) approach route, a specially recommended yacht route into and out of the entrance, and the Orwell and Stour behind, both marked and lit rivers. The Deben and Ore to the N in contrast are bar entrances and rivers which are unlit, though buoyed in places.

Apart from Orfordness the major lights are two pairs of large offshore ones. The Shipwash LtF (unmanned) 2½M NE of the Shipwash bank's N end and the Sunk LtF (also unmanned) 2½M S of the S end, each provide a landfall for two major offshore approach routes to the area (Routes 2 and 3 below). Further offshore 2 more lights mark the N end of Outer Gabbard and the S end of Galloper – the Outer Gabbard LtV and the Galloper Lanby.

As in the Thames Estuary the two problems are easterly weather and fog in an even busier shipping area. Strong winds from NE through E to S cause rolling seas which break over and near the offshore banks, create uncomfortable cruising conditions throughout the area of inshore shallows, and make the Deben and Ore bars dangerous to cross. Passage through the area in fog should be avoided if at all possible, but if inescapable and the weather is moderate it is usually possible to find an anchorage in shallow water in Pennyhole Bay or on Wadgate

Ledge well inshore away from the shipping channel, making sure of the day's tidal range and leaving reasonable keel clearance at LW.

Tidal streams offshore are not fast moving, just over 2½ knots at springs near Harwich entrance and inside Landguard Point and nearer 2 knots further offshore, except close to Orfordness and in the Ore and Deben entrances where springs average 3 knots and can be even faster when the spring rise is higher than average. Across the Ore bar they sometimes exceed 6 knots. Inside the Orwell and Stour entrances springs' rates are well below 2 and nearer to 1½ knots.

In the river entrances streams generally turn close to local HW. There is slightly more flexibility in timing the short inward passage to the top of each of the rivers with the tidal stream than the passage downstream, since with a later HW upstream depending on distance there can be a ½hr or more of extra tide upstream, and less than the 6hrs coming downstream.

Offshore the streams run roughly parallel to the coast, outgoing NE and ingoing SW, and change direction over a line extending approximately SE out to sea from the coast at right angles to the direction of tidal flow. 2M off Orfordness (and out to sea as above) the stream turns SW just before LW Harwich and NE close to HW Harwich, i.e. just over 6hrs in each direction. Near Naze Point there is no significant difference in the time of the turn of the streams from that at Orfordness.

Four approach routes are examined below.

ROUTE 1 ORFORDNESS TO LANDGUARD POINT

Distance   14M

Commentary

The lights en route are: Orfordness, the R lights on top of the Bawdsey radio mast, 2 lights in the Deben narrows which are difficult to see, Felixstowe Pier end, if indeed you can pick it out of the town lights, and the Harwich entrance buoys. From ½M off Orfordness the route is parallel to the coast to clear the Whiting bank, which is just under 1M offshore, then moving to about ¾M offshore to avoid the Orford Haven entrance bar, and 1M offshore at the Woodbridge Haven offing buoy, and making a course for Wadgate Ledge (G) Lt buoy to avoid Wadgate and Felixstowe Ledges where there are a number of shallow patches, some of which dry E of Landguard Point. Onwards to the Platters (S card) Lt buoy and a cautious right-angled crossing of the Harwich Channel, then following the recommended yacht route round the S and W edges of the channel keeping to the E of the Cliff Foot Rocks (1.6m) but outside the channel and a little less than ½M off Landguard Point, and crossing the Shelf Anchorage for commercial shipping. From April to November two unlit R can buoys mark the W side of the Shelf Anchorage, 200–300m W of the recommended yacht route, so at night to avoid these hazards it is advisable to keep close to the lit channel buoys, and watch out for ships at anchor.

Tidal timing

Tidal streams help on this route since the 14M distance gives plenty of flexibility within the 6hr tidal streams. Inwards an early rounding of Orfordness on the SW set gives a fair stream for the passage plus some slack or ingoing tide at Harwich. Outwards requires pushing the tide a little to reach Harwich entrance in time to fully use the 6hrs northwards and continue on round the Ness.

ROUTE 2 SHIPWASH LTF TO LANDGUARD POINT

Distance   18M

Commentary

There are several alternatives, depending on the weather conditions, but the shortest route is from close N of Shipwash LtF to the NE Bawdsey (G), Mid-Bawdsey (G) and S Bawdsey (S card) Lt buoys. Then the N edge of the Harwich approach channel to the Washington (G), Wadgate Ledge (G) and Platters (S card) Lt buoys, and then as in Route 1 above following the recommended yacht route.

Tidal timing

Again this is a route which is considerably helped by the tide since inwards the 18M distance gives plenty of flexibility within the 6hr SW stream for the passage plus some slack or ingoing tide at Harwich. Outwards requires pushing the tide a little to reach Harwich entrance in order to fully use the 6hrs northwards and continue beyond on a North Sea crossing.

ROUTE 3 SUNK LTF TO LANDGUARD POINT

Distance 13M

Commentary

This route is from either side of Sunk LtF to pass ½M E of the cardinal Lt buoys marking Roughs Tower, to very close S of Cork Sand (R) Lt buoy (as the Sand is to its S), and then following the recommended yacht route along the extreme S and W edges of the Harwich Channel to pass, as in Route 1 above, just under ½M W of Landguard Point and E of Cliff Foot Rocks.

Tidal timing

This is a cross-tide route, and inwards trying to catch the start of the flood near LW into Harwich for an up-river trip means crossing the last hours of the NE ebb stream, useful for avoiding Rough Shoals and Cork Sand. A later entrance means braving the SW stream. Outwards leaving Harwich at or after HW gives a NE stream away from the shoals.

ROUTE 4 THE NAZE TO LANDGUARD POINT

Distance 5M

Commentary

Keep 1M offshore of Naze Point, i.e. midway between the Naze Tower and Medusa buoy, to avoid its offlying shallows and lobster pots. Then close E of Stone Banks (R) unlit buoy to Landguard (N card) Lt buoy and finally following the recommended yacht route as in Route 1 above.

Tidal timing

The route across Pennyhole Bay is short and on the inward route visitors from the S usually round Naze Point near to Harwich LW, keeping well off Naze Sand and crossing Stone Banks near slack water taking the rise into the entrance. Outwards again a LW, or earlier than LW, slack-stream crossing is usual to catch the first of the flood down the Wallet.

The havens and rivers

Although there are 6 marinas, 1 each in the Deben and Walton Backwaters and 4 in the Orwell, this is an area of tidal mudflats and creeks, so even if you have a liferaft it is also essential to have a good, and preferably large tender – inflatable, collapsible or hard – in order to get ashore, or kedge off. In any case if the cruise is confined to marinas much of interest in the area will be missed.

As well as the many picturesque small villages off the creeks, the inland ports of Woodbridge and Ipswich are of particular historical interest. Harwich itself, which used to be the main naval base on the E coast is worth a row ashore; it has been the main ferry terminal for the Netherlands since the middle ages although the present-day monster 'blocks of flats' carrying cars and thousands of passengers are a far cry from the eighteenth-century Post Office 'packets', clinker-built gaff-rigged cutters with 20 crew carrying mail and, if one believes the claim, up to 100 passengers on a 50–60ft length, to and from Hellevoetsluis.

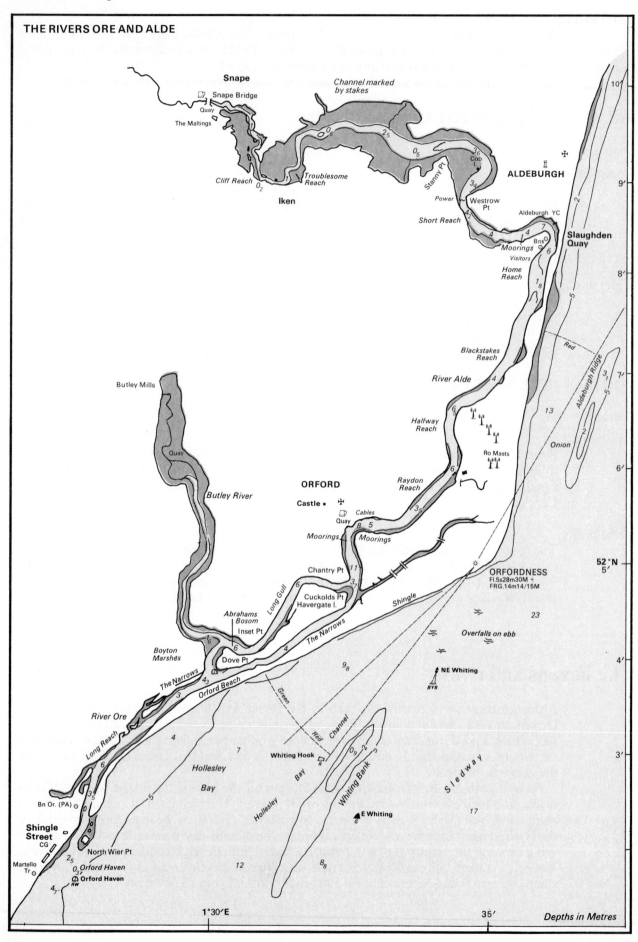

THE RIVERS ORE AND ALDE

Snape

Snape Bridge

Quay

The Maltings

Channel marked by stakes

Cliff Reach

Troublesome Reach

Iken

Stanny Pt

Cob I.

ALDEBURGH

Power

Westrow Pt

Aldeburgh YC

Short Reach

Slaughden Quay

Moorings

Visitors

Bns

Home Reach

Blackstakes Reach

Red

River Alde

Aldeburgh Ridge

Butley Mills

Halfway Reach

Ro Masts

Onion

Quay

Butley River

ORFORD

Castle

Quay

Cables

Raydon Reach

Moorings

Moorings

Chantry Pt

Long Gull

Cuckolds Pt

Havergate I.

Abrahams Bosom

Inset Pt

The Narrows

ORFORDNESS

Fl.5s28m30M

FRG.14m14/15M

Shingle

Overfalls on ebb

Boyton Marshes

Dove Pt

NE Whiting

BYB

The Narrows

Orford Beach

River Ore

Green

Red *Channel*

Long Reach

Whiting Hook

R

Hollesley Bay

Hollesley *Bay*

Whiting Bank

S l e d w a y

Bn Or. (PA)

Shingle Street

CG

E Whiting

Martello Tr

North Wier Pt

Orford Haven

Orford Haven

RW

52°N

5'

9'

8'

7'

6'

4'

3'

1°30'E

35'

Depths in Metres

THE RIVERS ORE AND ALDE

Port radio None

Entry signals None

Customs

c/o RAF Liaison Officer, USAF Bentwaters, Woodbridge,
 Suffolk
☎ (by VHF) Ipswich (0473) 219481
☎ (ashore) dial 100 and ask for 'Freefone Customs Yachts'

Entrance and river passage

Distance 14M from the offing buoy to Iken Cliff

The Ore entrance is unlit but with reasonable daylight visibility can be identified by a row of white
cottages, Shingle Street, south of the entrance,
plus a number of other cottages and a Martello
tower further south. The RW spherical offing
buoy is usually 3 cables offshore abreast the cottages. But beware, the bar moves! In 1985 entry
was made in transit between this buoy and a
beacon with an orange diamond topmark on the W
side of the entrance. In October 1986 a survey by
the Aldeburgh YC, ☎ (072885) 2562, indicated
that the transit was no longer entirely safe; entry
should be commenced by homing in, on a bearing
of 310°, on the flagstaff at the N end of the white
Coastguard cottages until the beacon comes in line
with a conspicuous chimney to its N, following
this transit until abreast the bungalow N of the
Coastguard cottages, then coasting in about 50m
off the W shore but moving to the E side of the
channel near N Weir Point to avoid a shoal near
the W bank abreast the beacon. But a caution to
exploratory visitors – wait until about 2hrs before
HW to enter; never enter on a falling tide and
never in strong onshore winds, and if you are worried about the bar contact Thames Coastguard
(VHF Ch 16) who have an auxiliary (occasional
watch) station at Shingle Street. Mr Russell
Upson, ☎ (072885) 2019, who runs the boatyard
at Slaughden Quay, is another potential source of
advice.

Orford. Note the conspicuous castle and church

The famous Maltings Concert Hall at Snape, the head of navigation
on the Alde

Timing the passage to arrive at the entrance in
order to cross the bar at half-tide rising is more difficult from Harwich and the S since there are 3hrs
of SW tide to push if you are not prepared to
anchor off. From the N it is easy. An inward passage across the bar at half-tide rising leaves only a
short time for a continued up-river cruise so the
trip up the Ore needs to be broken at Orford or
Aldeburgh in order to await a full tide for the
tricky passage to Iken.

Depths in the clearly defined channel as far as
Aldeburgh range from 3 to 7m (sometimes a little
deeper). The only danger spots are the spits at
either end of Havergate Island (the main channel
is to the S of the island) and on the edges of some
of the bends. Beyond the bend at Aldeburgh the
river shoals to around 2m near Cob Island and to
1m and less after snaking across the wide HW lake
to Bagnall's and Church Reaches. The passage to
Iken Cliff should only be attempted if the newcomer is prepared to take the ground, albeit acci-
dentally. Do not attempt it with more than 1½m
draught, and try to start, from either end, at the

earliest possible time on a rising tide, since the withy-marked channel is extremely difficult to follow near to HW. Aldeburgh YC have laid the withies voluntarily and maintain them as well as they can, and without withies only a LW, rising tide 'nudging' approach would be possible. In 1985 starboard-hand withies were occasionally marked by flags tied on with red string, and where the flags had disappeared by red string alone. Unfortunately port-hand withies which occasionally had red cans (plastic bottles etc.) as topmarks, were also tied with red string. So keep the kedge anchor available and gauge the outside and inside of the bends to 'interpret' which side to pass the many untopped withies. Since these difficulties increase nearer to Snape it is advisable to anchor, probably touching ground in the wooded Cliff Reach below Iken Cliff, and carry on by dinghy to the Maltings.

Even though it is advisable to push the stream from Iken in order to follow the winding channel to Aldeburgh with a rising tide it is not usually possible to do the whole outward passage down to and across the Orford Haven Bar without arriving rather late on a falling tide, so it is best to break the journey at Orford Quay.

For the inexperienced on the outward passage the Orford Haven Bar is better crossed at close to HW to avoid pushing the entrance tide or grounding on a falling tide. However this then means pushing the tide S towards Harwich. The N passage towards Orfordness is much more convenient with the ebb.

The 4M between the Ore and the Deben makes it feasible to pass out of one entrance and into the other, but this must be close to HW when the streams are slack, pushing the tide out of the Ore an hour before HW and crossing the Deben Bar at around HW, with a slack to S-going stream in between.

Mooring and facilities

Tides run extremely fast as far as Aldeburgh and anchorages are few. For the first 2M before the river splits around Havergate Island the holding ground is doubtful, and it is preferable to anchor in 'Abraham's Bosom', a little bay on the northern side of the west end of Havergate Island, or alternatively in the first reach of the Butley River, a creek running off west and north from the river near this point. Take care to give a wide berth to Dove Point, the shingle spit at the extreme west end of the island. Landing is not allowed on Havergate Island bird sanctuary, nor on the long shingle island from North Weir Point to Orfordness, which is the property of the Ministry of Defence. If you want to land take the dinghy up

the Butley River to Ferry Cottage whence you can walk or cycle 2 miles along the road to Orford. Alternatively you can anchor nearer Orford behind the northern end of Havergate Island at Cuckold's Point, and walk along ¾M of sea wall to the town.

At Orford Quay the HrMr is Ralph Brinkley, East View, Quay Street, ☎ (0394) 450481. There are private river moorings, which it may be possible to borrow. It is possible to lie alongside the quay, but my advice is to restrict it to a brief loading/unloading visit within an hour before or after HW to be near the top of the wall. Near the quay is the Orford Sailing Club. There are few better views on the east coast than this last approach to the town with its Norman keep (well worth a visit), church tower and old cottages. Facilities are more modest than at Aldeburgh but the HrMr will be able to direct you to local chandlery, repair, storage, slipway, scrubbing and lifting facilities. There is a restaurant (The Butley Oystery – book in advance), and a number of other ancient hostelries worth visiting.

At Aldeburgh a somewhat precarious anchorage can be had on the east bank just south of the Aldeburgh YC, but a spare mooring on the horseshoe bend is preferable after asking the HrMr, also local boatyard owner, Russell Upson, Russell Upson & Co Ltd, Slaughden Quay, ☎ (072885) 2896. There are chandlery, repair, scrubbing, storage and lifting facilities.

Anchoring is more feasible off the winding channel up to Iken; lying offshore from the mudbanks with little possibility of getting ashore, except at Iken Cliff itself, although even here it is difficult to get ashore to the sandy beach except at HW. However, there are no facilities at Iken and you have a long walk to the nearest pub, and sound carefully at the anchorage since you will probably dry out particularly at springs even though the range of tide is small. At Snape, if you risk going there, you may be able to lean against the wall or on a visiting larger vessel and the facilities are better, with pub, restaurant and tourist shops in the Maltings as well as a village nearby.

THE RIVER DEBEN

Port radio None

Entry signals None

Customs
c/o RAF Liaison Officer, USAF Bentwaters, Woodbridge, Suffolk
☎ (by VHF) Ipswich (0473) 219481
☎ (ashore) dial 100 and ask for 'Freefone Customs Yachts'

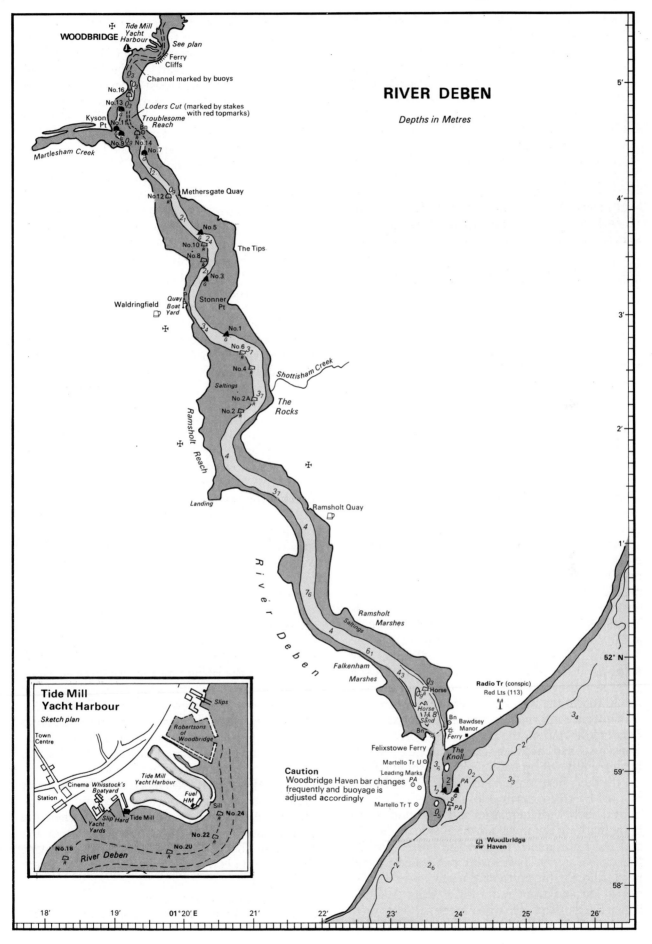

RIVER DEBEN

Depths in Metres

WOODBRIDGE

Tide Mill Yacht Harbour

See plan

Ferry Cliffs

Channel marked by buoys

No.16

No.13

Loders Cut (marked by stakes with red topmarks)

Kyson Pt

No.11

Troublesome Reach

Bn

Martlesham Creek

No.9

No.14

No.7

Methersgate Quay

No.12

No.5

No.10

2 4

The Tips

No.8

No.3

Stonner Pt

Waldringfield

Quay Boat Yard

No.1

No 6

3 7

No.4

Shottisham Creek

Saltings

No.2A

The Rocks

No.2

Ramsholt Reach

4

Landing

3 7

Ramsholt Quay

4

River Deben

7 6

Ramsholt Marshes

4

Saltings

6 1

Falkenham Marshes

4 3

Horse

Radio Tr (conspic)
Red Lts (113)

Horse Sand

Bn

Bawdsey Manor

Bn

Ferry

Felixstowe Ferry

The Knoll

Martello Tr U

Leading Marks

Caution
Woodbridge Haven bar changes
frequently and buoyage is
adjusted accordingly

PA

Martello Tr T

Woodbridge
Haven

**Tide Mill
Yacht Harbour**

Sketch plan

Slips

Robertsons
of
Woodbridge

Town
Centre

Cinema

Whisstock's Boatyard

Station

Tide Mill
Yacht Harbour

Fuel
HM

Sill

No.24

Slip Hard

Tide Mill

Yacht
Yards

No.22

No.18

No.20

River Deben

Woodbridge harbour looking down river from the Tide Mill

Entrance and river passage

Distance 9M from offing buoy to Tide Mill
Yacht Harbour

The entrance is unlit, but daylight approach from
any direction is helped by the unmistakable 113m
radio mast at Bawdsey north of the entrance,
whilst coasting from either south or north there
are a series of Martello towers to tick off before the
entrance is reached. The RW spherical offing
buoy is about 1M offshore, and in mid-1985 it was
close east of the bar buoys. This bar has moved
radically, changing the directions of entrance on
several occasions, so it is advisable to make a first
entrance on a rising tide about 2hrs or even 1hr
before HW if your draught is over 1½m, and
maybe anchoring off for a few hours from LW to
examine through binoculars the uncovered shoals
as well as the route taken by the fishing traffic. If
at all doubtful an advance telephone call to the
Tide Mill Yacht Harbour at Woodbridge may
help, ☎ (0394) 385745. In mid-1985 there was a
1.3m sounding above chart datum, but again I
would repeat the warning above – the bar moves,
seek advance information and only enter at about
2hrs before HW on the rise. In 1985 there was a
transverse gap between the S-running shingle
bank from Bawdsey and a mound of shingle leaving the bank near Martello tower 'T'. The gap was
marked by green (north side) and red (south side)
buoys, each with topmarks and a second green
buoy marking the inner northern side of the gap
and the right-angled turn to the north towards
Felixstowe Ferry. There were two leading metal
posts ashore to the north of Martello tower T, rear
board red and front red with a white triangle,
which lined up the deep water in the gap after leaving the offing buoy. Once the dogleg was turned

the channel followed the steeply sloping western
bank to the Ferry. The tide runs fast in the narrow
gap where the ferry plies, and a Q.R light marks
the W side and 2F.G (vert) on the Bawdsey side.
The main mooring area must be left well to port to
make sure that Horse Sand, around which the
moorings are disposed, is cleared. The northern
end of this shoal is marked by a red can buoy.
Depths approaching from S of the Ferry are over
3m and E of the moorings is an 8–9m 'hole'.

Timing the passage to arrive at the entrance in
order to cross the bar at half-tide rising is more difficult from Harwich and the S since there are 3hrs
of SW tide to push if you are not prepared to
anchor off. From the N the tide is fair. Since the
bar is usually crossed at half-tide rising it is often
possible to do the 9M to the Tide Mill Yacht Harbour and cross its sill, particularly as there is an
extended rise of tide, HW being ¾hr later at
Woodbridge than across the Deben Bar.

Above Horse Sand buoy (R, unlit) the channel
varying from 2 to 7.6m depths winds between
wide saltings and mudflats and is unmarked as far
as the Rocks anchorage above Prettyman's Point,
so a vessel must keep to midstream and avoid the
shallows, particularly on the convex side of bends.
Above the Rocks the width of these shallows
increases and their convex sides are marked by R
or G unlit buoys, whilst the main channel shoals to
below 1m at Kyson Point (entrance to Martlesham
Creek), above which the river effectively dries out
at LW; only to be attempted by newcomers on a
rising tide. But it is well buoyed with many moorings to follow and the approach and entrance to
the Tide Mill Yacht Harbour at Woodbridge is
best between 2hrs either side of HW when there is
at least 2m on the sill.

If you are inexperienced the bar is better crossed outwards at close to HW to avoid pushing the
entrance tide or grounding on a falling tide. It is
sometimes possible to leave Tide Mill as early as
possible on the rising tide and reach the entrance
in time to cross, but if it is late after HW and you
are in any doubt, borrow a mooring at Ramsholt or
at the Ferry and pick your departure. Leaving
near HW means pushing the tide S towards Harwich, whilst the N passage towards Orfordness is
much more convenient with the ebb.

The 4M between the Deben and Ore makes it
feasible to pass out of one entrance and into the
other, but this must be close to HW when the
streams are slack; pushing the tide out of the
Deben an hour before HW and crossing the Ore
Bar at around HW, with a slack to S-going stream
in between.

Mooring and facilities

Anchorages are plentiful with protection, depending on wind direction, behind wide bends and on the gently shelving mudbanks.

The first landing place is at Felixstowe Ferry, where you will need to borrow a mooring or anchor further up in Sea Reach and take a ½M dinghy trip. There are a café, public house, Felixstowe Sailing Club, and full chandlery and boatyard facilities, as well as plenty of fresh and smoked fish for sale at the waterside, and also a bus into Felixstowe.

The hard and quay at Ramsholt is reached 3M further along on the east bank at the beginning of the more interesting wooded reaches of the river. The quay is only accessible near the top of the tide and the whole of this stretch of river has permanent moorings but the local HrMr will help if you pick up a mooring temporarily. The only facility is the Ramsholt Arms Inn. If you cannot hire a mooring at Ramsholt then you can always anchor round the next bend at the famous Rocks anchorage with many like-minded neighbours and walk to the pub.

At Waldringfield there are moorings and all facilities – village shop, the Maybush public house and restaurant, a boatyard with chandlery, crane and all facilities, and there are scrubbing posts at the Waldringfield Sailing Club. It will be necessary to borrow a mooring by approaching the boatyard unless you are prepared to risk anchoring with a trip-line close in to the west bank.

It is possible to anchor as far upriver as about Methersgate Quay, which is private, before the river dries out, but getting ashore is difficult for the single keeler. Woodbridge Tide Mill Yacht Harbour, ☎ (0394) 385745, is the next stop entering within a 2hr range of HW, and carefully following the buoys. East of Kyson Point there is an interesting little short cut across the bend of Troublesome Reach through Loder's Cut which is marked by red port-hand can topmark withies, and which is a ½m or so deeper than the sill at the marina. The sill has about 2m, 2hrs before HW depending on the range of tide. Approaching the marina the moorings help to locate the deepest water round the bend, and you can anchor or pick up a buoy off the sill if you need to wait. Once over the sill and in the horseshoe-shaped marina turn either port or starboard on instruction from the HrMr whose office faces the sill and near which there is a clearly marked tide gauge readable from outside approaching the marina entrance. The marina and nearby boatyard has comprehensive repair and chandlery facilities. Woodbridge is a picturesque small town with full shopping facilities.

An interesting visit from Woodbridge is by ferry or dinghy to Sutton Hoo, the hill on the bank opposite the Tide Mill to visit the Saxon Ship burial excavations.

Finally for those who always push on to the bitter end there is another mile of river above Woodbridge; yet another rising tide required, taking the ground against a wall near the bridge at the top of the navigation. There are also two more boatyards (Robertson's and Melton) beyond the Tide Mill on the west bank.

Harwich as seen from offshore and approaching the Orwell and Stour

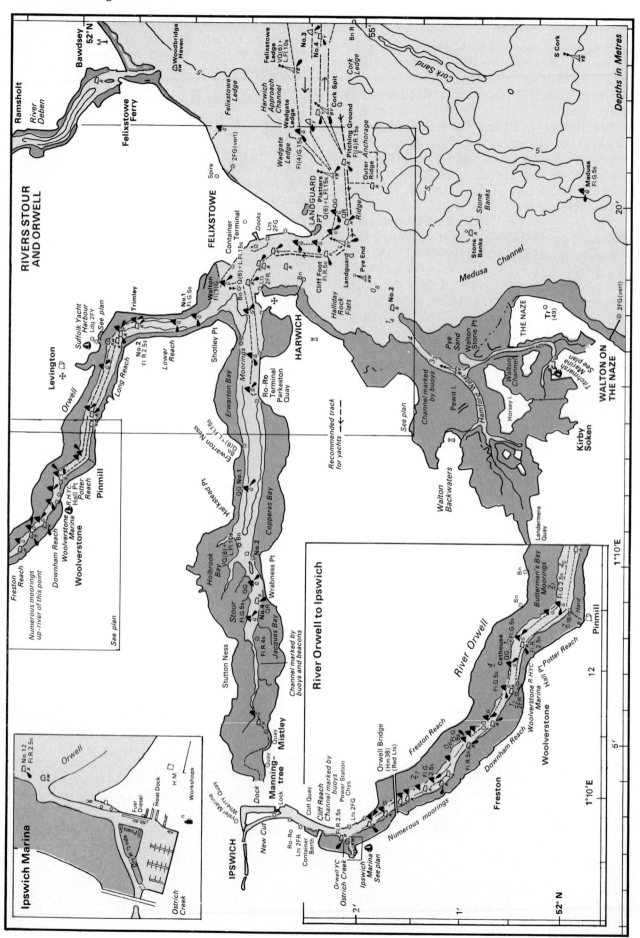

RIVERS STOUR AND ORWELL

RIVER DEBEN

Ramsholt

Bawdsey 52°N

Felixstowe Ferry

Woodbridge Haven RW

Levington

Suffolk Yacht Harbour Ldg 2FY

Trimley

No.1 Fl.G.5s

No.2 Fl.R.2.5s

Orwell

Long Reach

Lower Reach

Shotley Pt

Walton Fl(3)G

Container Terminal

Docks

Spire

2FG (vert)

FELIXSTOWE

Bn Q(6)+LFl.15s

2FR R

Cliff Foot Fl.R.5s

Lts 2FG

HARWICH

Ro-Ro Terminal Parkeston Quay

Erwarton Bay

Erwarton Ness Q(6)+LFl.15s

Harkstead Pt

Copperas Bay

Moorings

Wrabness Pt

QG

Holbrook Bay VQ(6)+LFl.10s

Bn

Stour Fl.G.5s

No.4 QR

Fl.R.4s

Jacques Bay

Stutton Ness

Mistley

Manning-tree

Quay

Landguard Pt Platters Q(6)+LFl.15s

QR

Landguard Fl(4)R.15s

Ridge R

Pye End RW

No.2 R

Halliday Rock Flats

Pye Sand

Walton Stone Pt

THE NAZE

Tr (49)

WALTON ON THE NAZE

Tichmarsh Marina See plan

Walton Channel

Horsey I.

Hamford Water

Pewit I.

Kirby Soken

Walton Backwaters

Landermere Quay

Felixstowe Ledge

Felixstowe Ledge

Harwich Approach Channel

Wadgate Ledge

Wadgate Ledge Fl(4)G.15s

No.3

No.4

Cork Ledge

Cork Spit BY Q

Pitching Ground

Outer Anchorage

Bn R

S Cork YB

Cork Sand

Stone Banks

Stone Banks

Medusa Channel

G Medusa Fl.G.5s

55'

20'

Depths in Metres

Recommended track for yachts

See plan

Channel marked by buoys

Channel marked by buoys and beacons

2FG (vert)

Pinmill

Woolverstone Marina R.HYC Hall Pt Potter Reach

Downham Reach

Freston Reach

Numerous moorings up-river of this point

See plan

River Orwell to Ipswich

Butterman's Bay Moorings

Bn

Bn

Fl.G.2.5s

Cathouse QG Fl.G.5s

Fl.R. 2.5s

Woolverstone Marina R.HYC Hall Pt Potter Reach

Pinmill

Freston

Downham Reach

Woolverstone

River Orwell

Numerous moorings

Freston Reach

Fl.G.5s

Fl.R.5s

Orwell Bridge (Hm 38) (Red Lts)

Power Station

Cliff Reach Channel marked by buoys

Cliff Quay

Lts 2FG

Ro-Ro Container Berth Lts 2FR

New Cut

Lock Gates

Oyster Marine Wherry Quay

Dock

IPSWICH

Orwell YC

Ostrich Creek

Ipswich Marina See plan

1°10'E

5'

2'

1'

52°N

12

1°10'E

Ipswich Marina

No.12 R Fl.R.2.5s

Orwell

H M

Workshops

Hoist Dock

Fuel Diesel

Posts

Dries at ½ tide

Ostrich Creek

THE RIVER ORWELL

Port radio (VHF)

	VHF Ch	Hrs listening
Harwich Harbour Operations Service Call *Harwich Harbour Control* Angel Gate, Harwich ☎ (0255) 506535/6	11, 14, 16, **71**	24
Orwell Navigation Service Call *Ipswich Port Radio* Ipswich Dock ☎ (0473) 56011	**14**, 12, 16	24
Suffolk Yacht Harbour ☎ (0473) 88465	M	0800–1730 LT

Entry signals

None for yachts at Landguard Point. It is not necessary to contact Harwich Harbour by VHF if the recommended yacht route is used, but it helps to listen in for shipping movements, particularly in fog, keeping a dual watch on VHF Channels 14 and 16, and there is no reason why advice should not be sought from Harwich Harbour Control or the Orwell Navigation Service.

Customs

Felixstowe: Customs Office, New South Quay, The Dock, Felixstowe, Suffolk, IP11 8PF

Harwich: Ray House, Foster Road, Parkeston, Harwich, CO12 4QA

Ipswich: Dock Head, The Docks, Ipswich, IP3 0DR

In all cases above, ☎ (by VHF) Ipswich (0473) 219481; ashore dial 100 and ask for 'Freefone Customs Yachts'.

Entrance and river passage

Distance 11M from Landguard Point to Ipswich Dock

Harwich, Parkeston Quay and Felixstowe make up a significant port complex with major expansion plans for both the Harwich and Felixstowe sides of the harbour, so make sure your charts are

Suffolk Yacht Harbour

up to date and be extremely wary of commercial shipping and the frequent ferries. Ipswich is also an important deep-water port so some traffic continues upstream, and it is not unusual to see deep-water container ships passing slowly between the trots.

From close W of Cliff Foot (R) Lt buoy follow the recommended yacht route to the N Shelf (R) Lt buoy and keep just outside the channel round to Grisle and Guard (both R) Lt buoys, making sure to keep well off the Guard shoal with 0.7m soundings only a cable S of the buoy. Then directly across the channel to Walton (G) Lt buoy,

The entrance to Suffolk Yacht Harbour

keeping well to the E of Shotley Spit (S card) Lt beacon and follow the starboard side of the Orwell channel. The channel shoals from 8m off Shotley Point to a dredged 5.6m at Ipswich Dock entrance, and like most of the Suffolk rivers winds between wide mudflats. However, it is very well marked with lit buoys for the commercial shipping. From Butterman's Bay onwards the river is lined with yacht moorings outside the main channel.

Tidal streams in the river are no problem over the 11M passage; the ingoing stream runs for 6½hrs from Shotley Point to Ipswich and for just under 6hrs outgoing.

Mooring and facilities

Anchorage in the river is only possible before the start of the moorings which line the river N from Long Reach just below Pinmill. It can be uncomfortable from the wash of passing ferries, ships, and pilot vessels, particularly in the entrance, where yachts can anchor in an arc round the edge of the Guard shoal sounding carefully to keep off the shoal (which means a long row to Harwich to pick up stores), but keeping just W of the commercial ship anchorage marked by the two R unlit buoys on 'The Shelf'. Inside the river proper anchorage is possible on both sides right up to the moorings; behind Shotley Point is very popular, but on the bank opposite Shotley it is best to keep well away from the 'Works in Progress' on the NW extension of Felixstowe Docks on the wide mudflats between Fagbury Point and the container terminal.

Suffolk Yacht Harbour has all the boatyard facilities as well as a club on a lightship, but is a long way from a village and shops. It is a good stag-

ing point for arrival or departure crossing the North Sea. It is entered from the S along a narrow dredged channel (1.5m at LW neaps) from a RW unlit offing buoy and between two pairs of port and starboard-hand posts. At night 2 F.Y leading lights on the N side of the marina assist entrance. There are mooring buoys if you wish to wait off the entrance.

Landing is possible at Pinmill hard with its laid up Thames spritsail barges, famous waterside inn, Chelmondiston village and boatyard, chandlery and full facilities. Vacant moorings are sometimes hard to come by in mid-season, but the HrMr from the yard will help.

Ipswich (Fox's) Marina

Continuing on through the moorings Woolverstone Marina, ☎ (0473) 84354, on the S bank is a dredged open floating pontoon marina with all facilities close by the Royal Harwich YC.

Further on and under the Orwell Bridge (do not worry – clearance is 38m above MHWS!) Ipswich Marina, ☎ (0473) 219438, is on the W bank. The approach is from No 12 (R) Lt buoy along a 300m-long dredged channel marked by port and starboard beacons, and accessible at all states of tide to yachts up to 1.8m draught. All facilities are available, the Orwell YC is nearby as well as some shops and buses into Ipswich.

After proceeding along the well buoyed Cliff Reach, Ipswich Dock can be entered through the lock and a swing bridge from between 2hrs and ½hr before local HW. There are certain tides when the timing is tight and it is always essential to call the Orwell Navigation Service (Ipswich Radio) on VHF beforehand, or at least telephone for the swing bridge to be opened. A berth can be

The bridge over the Orwell outside Ipswich

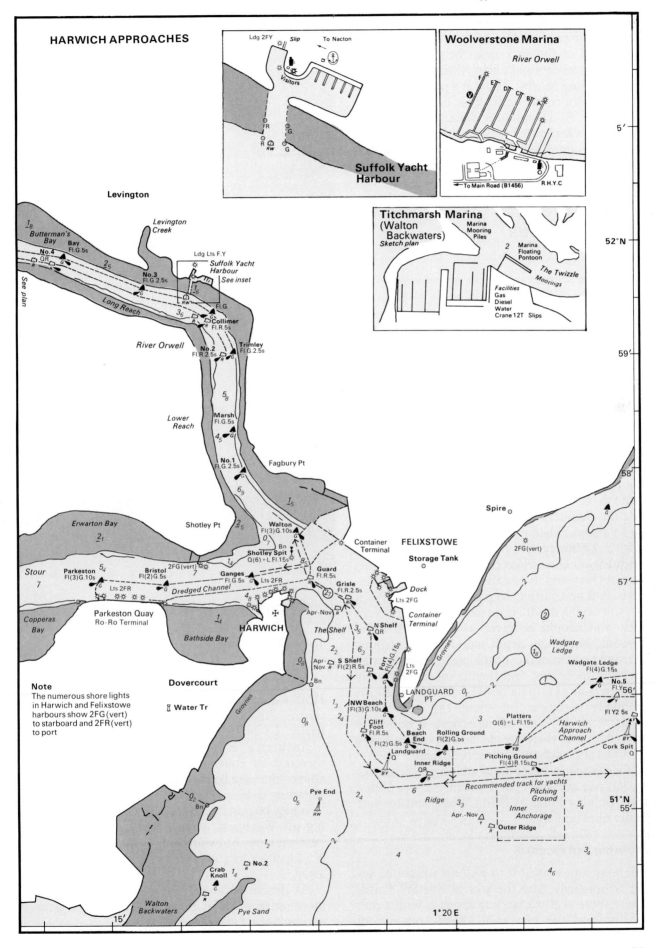

found rafted alongside the wall at Oyster World, ☏ (0473) 53999 or VHF Ch M, at Wherry Quay at the N end of the dock near the customs office. There are showers and toilets, lifting facilities and hard standing nearby. There is a pub on the quay and the town centre with all facilities including a theatre is 10–15 minutes walk away. This is another dockside area which is being developed; the island opposite with flats in the adapted malt kilns, and offices and a restaurant on Wherry quayside.

THE RIVER STOUR

Port radio None

Entry signals None

Customs
Ipswich Dock Head, The Docks, Ipswich, IP3 0DR
☏ (by VHF) Ipswich (0473) 219481
☏ (ashore) dial 100 and ask for 'Freefone Customs Yachts'

Entrance and river passage

Distance 9½M from Landguard Point to Baltic Wharf, Manningtree

This is another well marked and lit route since some commercial shipping uses Mistley Quay well beyond Parkeston Quay. From close W of Cliff Foot (R) Lt buoy follow recommended yacht route to the N Shelf (R) Lt buoy and keep just outside the channel round to Grisle and Guard (both R) Lt buoys. Then directly across the channel to close S of Shotley Spit (S card) buoy and follow N side of Stour channel keeping well away from Parkeston Quay and its North Sea ferries. The channel is wide and 5 to 8m deep up to Harkstead Point but the drying mudflats on each side are also very wide and should be given a wide berth using beacons, buoys and depth sounder. The channel narrows and closes the S bank past Wrabness with its yacht moorings, and then is more frequently marked (some lit) to Baltic Wharf. It also shallows gradually to below 1m about 1M from Manningtree, so a rising tide approach is definitely advisable, but there are no deep-water mooring facilities and anchoring afloat is difficult.

Tidal streams in the river are no problem over the 9M passage; the ingoing stream runs for 6½hrs from Shotley Point to Baltic Wharf and for just under 6hrs outgoing.

Mooring and facilities

It is sometimes possible to pick up a yacht mooring temporarily near the pier at Shotley Point, where there is also a landing slipway, with a pub nearby, and the somewhat ghostly remains of

Ipswich Dock entrance – the lock open at HW

Ipswich. Wherry Quay

HMS Ganges now partially converted to a sports centre. Anchoring inshore of the moorings is tricky. There are no facilities.

Anchoring is possible, depending on wind direction of course, on either side of the river above the dredged Parkeston Quay Channel sounding to keep off the wide drying mudflats.

It may be possible to pick up a temporary mooring off Wrabness, where landing on the sandy beach is easy, but anchoring is not to be advised. No facilities, the pub and village is a short walk away.

As described above there are no deep-water mooring facilities at Baltic Wharf, Manningtree and anchoring afloat is difficult.

THE WALTON BACKWATERS

Port radio None

Entry signals None

Customs
Walton-on-the-Naze Customs Office, The Marina, Coles
 Lane, Walton-on-the-Naze, Essex CO14 8SL
☎ (by VHF) Ipswich (0473) 219481
☎ (ashore) dial 100 and ask for 'Freefone Customs Yachts'

Entrance and river passage

Distance 5½M from Landguard Point to Titchmarsh Marina

Only to be attempted in daylight. Starting from
the recommended yacht route opposite Landguard Point it is essential to pass close to the small
Pye End unlit buoy (RW vertical stripes, spar with
round topmark) then head for No 2 (R) unlit buoy
leaving it to port. Soundings across the Sand are 1
to 1½m; not to be attempted in a strong onshore
easterly. Beyond Nos 4 (R) and 5 (G) unlit buoys
the channel deepens to 6m, and 7m after rounding
into the Walton Channel N of No 9 (N card) unlit
buoy marking its junction with Hamford Water.
Beyond Stone Point the channel gradually shallows to 3m at the bend into Twizzle Creek with 2m
off Titchmarsh Marina, and is followed by keeping to midstream and close to the continuous line
of moorings.

HW times are not significantly different in the
Walton Backwaters than at Harwich, the distances are short and tides do not run fast. It is best
to time the visit if the weather is a little doubtful to
give a reasonable depth over the Sand on a rising
tide, i.e. leaving at about half-tide. This is more
difficult for visitors from the S who may arrive at
Pye End near LW and if in doubt should go into
the Orwell. Outward it is usually best to push the
tide starting as early as feasible from Titchmarsh
Marina in order to obtain a rise over the Sand and
a continuing rise into the Orwell. Shallow-draught
visitors returning S may decide to risk running
down the Walton Channel on the last of the ebb
and scraping over the sands to catch the full flood
down the Wallet, but again if in any doubt they
should use the Orwell as the staging post and take
the rising tide passage.

Mooring and facilities

Anchorage is possible well away from any facilities
near Stone Point before the moorings in Walton
Channel start, or alternatively deep in Hamford
Water or in its branching creeks, although some of
the S ones are also crowded with moorings.

 The moorings are private and landing is difficult in the Walton Channel so it is best to turn
into the Twizzle and enter Titchmarsh Marina, ☎

Pye End buoy. The key to the channel leading into Walton Backwaters. Naze Point in background

(0255) 62185, where there are full chandlery and
boatyard facilities, but a long walk to Walton-on-
the-Naze. There is a minimum 2m of water in the
marina, but the sounding on the sill is fairly close
to chart datum so is not fully accessible ½hr either
side of dead LW, and for complete peace of mind
deep-draught boats can regard it as nearer to half-
tidal but there is a pontoon and moorings nearby
for boats to wait.

 A rising tide visit along the drying Foundry
Creek to lie alongside the Walton and Frinton YC
at the top of the tide is not to be advised for the
newcomer.

Woolverstone Marina on the River Orwell

Photo Lester McCarthy, Motor Boat & Yachting

7. The Essex coast & rivers from the Naze to Foulness

Charts Admiralty *1183, 1610, 1975, 3741, 3750*
Imray *C1, Y6, Y17*

Tidal atlases Admiralty *Thames Estuary*
North Sea – southern portion

Tidal streams (based on HW Dover)

Position	Start times	
	DOVER	
	Out	*In*
Wallet (3½M SW of Walton Pierhead)	+0100	−0530
Colne Bar buoy	+0100	−0530
Wivenhoe	+0125	−0455
Osea Island	+0125	−0425
Maldon	+0135	−0415
Burnham-on-Crouch	+0125	−0435
Hullbridge	+0140	−0420

Tidal differences and ranges (based on HW Dover)

Place	*HW* *(time)*	*Springs/Neaps* *(range in metres)*
Brightlingsea	+0105	4.6/2.6
Colchester	+0112	
W Mersea	+0107	4.6/2.6
Maldon	+0143	
Bradwell	+0111	4.8/2.9
Burnham on Crouch	+0129	5.0/3.1
Hullbridge	+0144	

Lights and radiobeacons

There are no lights of 14M or greater range nor radiobeacons in this area.

Coast radio stations

Station★	*Transmits & Receives (VHF Ch)*	*VHF Ch†*	*Traffic lists (times)*	*Storm warnings (times)*	*Weather messages (times)*	*Navigational warnings (times)*
Orfordness	16,**62**,82	62	0103,0503 & odd H+03 (0903-2303)	0303,0903, 1503,2103	0803,2003	0033,0433,0833, 1233,1633,2033
Thames	**02**,16,83	02	0103,0503 & odd H+03 (0903-2303)	0303,0903, 1503,2103	0803,2003	

Notes
★ Each station has 24hr service and watch. All times are GMT.
† Channel on which traffic lists, storm warnings, weather messages and navigational warnings are given.

Major fixed daylight marks

The Wallet and Blackwater Estuary N side (E to W)

The Naze red cliffs with the chimney-shaped Naze Tower (49m) on top
Walton-on-the-Naze Pier
Frinton: a water tower with 2 tower blocks ½M SE and the turret of the Grand Hotel on cliffs at W end of the town
5M offshore from Frinton, on Gunfleet Sand, is the Gunfleet Old Lighthouse (13m, square metal framework, unlit)
Clacton Pier
A series of Martello towers between the pier and Colne Point
The low wooded Mersea Island

Bradwell Nuclear Power Station, the most conspicuous building on the Blackwater, from Bradwell Creek

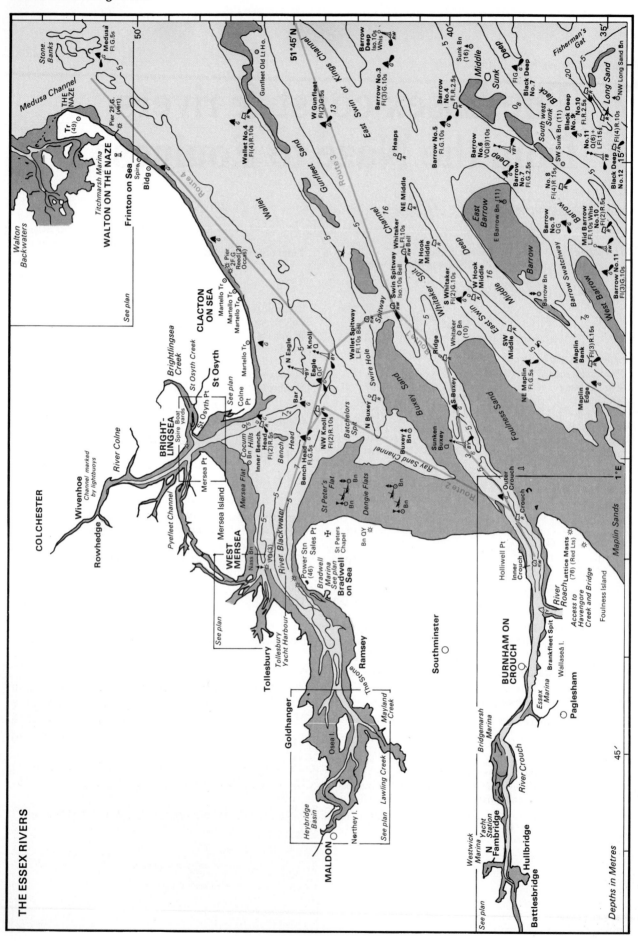

THE ESSEX RIVERS

S side of Blackwater and Ray Sand Channel (W to E)

The double buildings of Bradwell Nuclear Power Station (45m) with a lit offlying baffle wall in the river

Laid-up merchant ships often anchored off Sales Point and as far upstream as Pewit Island

The isolated St Peter's Chapel (pointed roof) on Sales Point (obscured by trees in some directions)

2 sets of 2 unlit wreck-marking beacons (E & W card) on the Dengie Flats

Buxey Beacon (8m, unlit N card).

East Swin/King's Channel, Whitaker Channel, River Crouch (E to W)

Sunk Head tower (5m, ruined concrete stump bristling with reinforcing wires, marked by N card Lt buoy)

On Gunfleet Sand is the Gunfleet Old Lighthouse (13m, square metal framework, unlit)

The Whitaker Beacon (10m, unlit, 2 B balls) 6M NE of Foulness Island

A group of 3 radio lattice masts (2 of 76m, R Lts) on the N end of Foulness Island

Approach routes and tidal timing

This area presents tricky pilotage offshore, but is generally well protected for cruising. Despite its cruising popularity but perhaps because of the minimal amount of commercial traffic it is badly lit and marked, with no major lights, no lit buoyage in the River Crouch, little in the Whitaker Channel and the Blackwater River proper, and has a shortage of radio navigational aids and port radio facilities.

There are two sandbank complexes. In the S Foulness Sand dries with a steep-to N side and a gentle slope NE ending in the Whitaker Spit. A triangular central drying area, Ray Sand and Buxey Sand with a single swatchway, the Ray Sand Channel (drying), stretches 7M gently outwards from the Dengie peninsula, and separates the Whitaker Channel from the Blackwater Estuary, together with a 13M NE extension, Gunfleet Sand which dries in patches with one major swatchway, the Spitway channel (1.2 to 1.6m soundings) and two others with 0.2 and 0.4m depths.

There are two main channels, the Whitaker and the Wallet. Whitaker Channel, branching out of King's Channel and leading into the Crouch, is shallow with a variety of uneven and unstable soundings, which are changing in places. Moving SW depths range from drying to 6m near the Swin Spitway buoy, to an 11m hole E of the S Buxey buoy, to 0.5m S of the Sunken Buxey buoy, and to 7m near shore-ends. King's Channel which extends NE from the Whitaker, is bounded by the more gently sloping shallow S edge of Gunfleet Sand but in the wide main channel typical depths are 10 to 20m.

The Wallet channel lies between Gunfleet Sand and the mainland, with 9 to 11m depths near the steep N edge of Gunfleet, and a gentle slope upwards to 2m a few cables off a steep, narrow sand and gravel beach. It is renowned for the short breaking seas such a seabed creates in strong NE and SW winds funnelled between the parallel banks of the channel. Access to the Blackwater from both the Wallet and the Spitway channel swatchway is through a narrow buoyed channel between Colne Bar and Eagle shoal to the N (0.1 to 0.3m depths) and the Knoll (dries) to the S.

From this channel the 11 to 16m main channel into the river entrance is bounded by wide drying sand and mudflats; St Peter's and Dengie Flats to the S, and Mersea Flat to the N – so compass and depth sounder are essential tools.

Commercial traffic into the area is light; small coasters carrying timber to the River Crouch wharf 1M W of Burnham on the S bank of the Crouch, very occasionally a coaster along the Roach to the quay near the mill at Rochford, varied cargoes including coal into the Colne to the wharf at Brightlingsea and the quays further up the Colne at Rowhedge, Wivenhoe and Colchester, and very little traffic to Maldon.

The tidal streams of the Thames Estuary and its tributaries are simple; rising up the rivers with progressively later HWs inland, and the turns of streams coinciding with HW and LW. The strongest streams are found in the Crouch.

Four routes into this approach area and between the river entrances are described below.

ROUTE 1 NASS BEACON TO INNER CROUCH BUOY VIA SPITWAY CHANNEL

Distance 19M

Commentary

The other routes described below branch out from or use part of this route.

First half Nass to Swin Spitway: Start from N (in the Mersea Quarters buoyed channel) or S (from the unbuoyed main Blackwater channel) of the Nass Beacon (E card). Course is taken to close S of Bench Head (G) Lt buoy, through Eagle channel (buoyed, some lit) close to Knoll (N card) Lt buoy, close to Wallet Spitway (RW) Lt buoy, through Spitway channel close to Swin Spitway (RW) Lt buoy. A word of warning – on this first half of the route in the reverse, inward direction by night or day – from the Bench Head Lt buoy the Nass Beacon is 4M away so its light, or the beacon itself, is often difficult to home in on, so at night and certainly in daytime frequent soundings should be taken to keep in the main channel.

Second half Swin Spitway to Inner Crouch: From this point the Whitaker Channel from seawards is a difficult route for newcomers and accurate compass steering and frequent sounding is essential. There is only 1 Lt buoy, the Sunken Buxey (N card), 5M from the Swin Spitway Lt buoy and 5M from the completely unlit shore-ends entrance to the River Crouch itself, also with no Lt buoys. The Whitaker Channel, however, has 4 unlit buoys which in daylight help the stranger to clear the shoal patches. Between the two Lt buoys the Buxey Sand has shown signs of extending S, so should be given a wide berth, heading for the Ridge (R) buoy first and then the S Buxey (G) buoy or at night taking soundings and bearings on the two Lt buoys to clear the shoals. The course is then down the channel between the appropriate channel buoys to shore-ends, keeping a running back-bearing at night on the Sunken Buxey.

Tidal timing

This route in either direction via the Spitway is best timed, providing your draught is modest (say up to 1.5m), by taking the last few hours of outgoing tide from the river of departure, passing through the Spitway at LW (not significantly different from Burnham LW), and the first few hours of rising tide into the destination river. If you are worried about draught then delay departure to nearer LW to push some tide in the departure channel and cross the Spitway on a rising tide with a continuing fair tide to the destination.

ROUTE 2 NASS BEACON TO INNER CROUCH BUOY VIA RAY SAND CHANNEL

Distance 13–14M

Commentary

A useful short cut, but only to be attempted near the top of a rising tide at the Burnham end. From the Blackwater to the Crouch start N or S of Nass Beacon as in first half of Route 1 above. Course to close S of NW Knoll (R) Lt buoy, then course to close SW of Outer Crouch buoy (G, unlit), then alter course down Whitaker Channel N of the Crouch buoy (R, unlit) to Inner Crouch buoy (RW, unlit). This route crosses Batchelor's Spit from NW Knoll buoy, and, with experience, the corner round St Peter's Flats can sometimes be cut closer by careful sounding on a rising tide, possibly nudging the ground, but keeping well outside the two sets of double wreck-marking beacons (unlit) on Dengie Flats, and measuring distance run down the channel from these beacons or from the Buxey Beacon (unlit) to the E on Buxey Sand. In good visibility the lattice masts (R Lts at night) on the NE end of Foulness Island can be seen on the course heading down the Ray Sand Channel.

The route in the opposite direction from the Crouch to the Blackwater is the same although it is not usually possible to cut the St Peter's Flats area too close as the tide is falling, so close S of the NW Knoll buoy is the objective.

Tidal timing

The route requires careful pilotage taking a rising tide throughout, i.e. pushing the tide out of either estuary, but aiming to cross the narrow neck of Ray Sand between about an hour before to half an hour after HW. From Burnham the timing is to reach the Outer Crouch buoy near HW and take the falling tide down Ray Sand Channel crossing the outer edge of Batchelor's Spit about 1–1½hrs later with a hard push up the Blackwater. If you make a slow passage and are worried about depth you can angle out into Swire Hole near the N Buxey buoy (R, unlit) and out near the Knoll (N card) Lt buoy before heading into the Blackwater.

From the Blackwater in moderate conditions an early start from the Nass Beacon 2hrs after LW or even earlier and sounding round/'nudging' St Peter's Flats and Batchelor's Spit can sometimes pay off, crossing Ray Sand into Whitaker Channel with another hour of ingoing tide to help up to Burnham.

ROUTE 3   LONGSAND HEAD BUOY TO INNER CROUCH BUOY OR NASS BEACON

Distance   31M

Commentary

Entering and following the N edge of King's Channel from Long Sand Head to the Swin Spitway there are 5 Lt buoys, but at rather long 4 to 6M intervals. From a position close N of Longsand Head (N card) Lt buoy pass N of the Sunk Head Tower (N card) Lt buoy, then close S of the Gunfleet Spit (S card), W Gunfleet (G) and Swin Spitway (RW) Lt buoys. Then down Whitaker Channel as in the second half of Route 1 above, remembering to sound frequently.

A variant on this route is from the Swin Spitway buoy to the Nass Beacon, taking the inward route of the first half of Route 1 above, and again sounding frequently after the Bench Head buoy.

Tidal timing

This distance takes at least 6hrs so it pays to reach the start early; on the inward route say 5hrs after HW Sheerness (HW Dover −6), or alternatively if the tide into the Crouch is lost before Shore Ends then it may be advisable to anchor well down Whitaker Channel or inside the Crouch or Roach entrance and rest for a tide. The same applies in the outward direction – start before local HW and push the tide for a little while to get the full outgoing stream in the Whitaker Channel and Swin.

ROUTE 4   THE NAZE TO NASS BEACON OR INNER CROUCH BUOY

Distance   18M

Commentary

From about 1M offshore to clear the Naze shallows a course for close N of the Knoll (N card) Lt buoy closes the shore and then angles outwards avoiding the shallows off Clacton, and the Eagle shoal. In the Wallet there are lights on the ends of Walton and Clacton piers 6M apart, and offshore three Lt buoys mark the edge of Gunfleet Sand. Then from the Knoll buoy the course is the early part of the inward route in the first half of Route 1 above, remembering to sound frequently from the Bench Head buoy.

A variant on this route to the Inner Crouch buoy – distance 22M, is to head from a position 1M off the Naze to the Wallet Spitway (RW) Lt buoy, instead of the Knoll, and then through the Spitway channel following the second half of Route 1 above, and again sounding frequently in Whitaker Channel.

Tidal timing

Distance is not critical but if on the inwards passage the departure has been from Harwich and the Orwell it pays to leave on a falling tide to cross Stone Banks at slack water or even earlier and catch the very first of the flood at the Naze, whence at least a full tide will be needed if intending to continue on up the Crouch or Blackwater beyond Shore Ends or the Nass Beacon. On the outwards trip an early start from destinations up the Crouch and Blackwater is advisable pushing the tide to pass the Inner Crouch or Nass close to HW in order to round the Naze an hour or more before LW to reach Harwich entrance with the first of the flood.

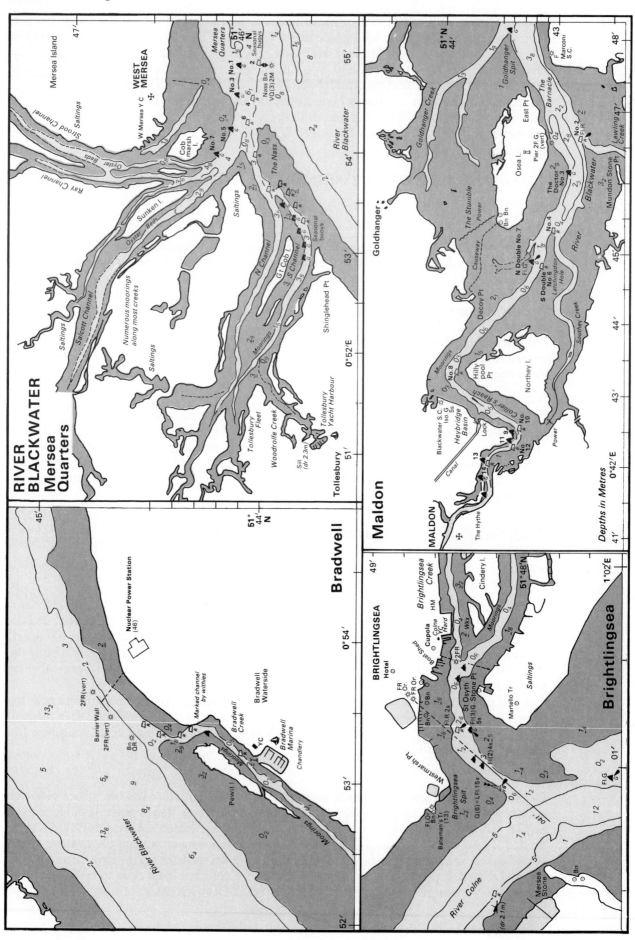

RIVER BLACKWATER Mersea Quarters

Mersea Island

Strood Channel

Saltings

Mersea Quarters

Ray Channel

Oyster Beds

Saltings

51° 46′ N

WEST MERSEA

W Mersea Y C

Seasonal buoys

No.1

No.3

No.5

Nass Bn VQ(3)2M

No.7

Cob marsh I.

The Nass

River Blackwater

55′

54′

Sunken I.

Salcott Channel

Oyster Beds

Saltings

Numerous moorings along most creeks

Saltings

N Channel

Gt Cob I.

S Channel

Moorings

Seasonal buoys

Shinglehead Pt

53′

Saltings

Tollesbury Fleet

Woodrolfe Creek

Tollesbury Yacht Harbour

Sill (dr. 2.3m)

Tollesbury

51′

0°52′E

Maldon

51° N 44′

Goldhanger Creek

Goldhanger Spit

The Barnacle

Marconi S.C.

F 43

Goldhanger

Osea I.

Pier 2F.G (vert)

The Doctor No.3

No.2 Fl.R

Lawling Creek

47′

48′

The Stumble

Power

Bn Bn

No.4

N Double No.7

S Double No.6

Causeway

Latchingdon Hole

N Double Fl.G

River Blackwater

Mundont Stone Pt

45′

Decoy Pt

Southey Creek

Moorings

No.8

Hilly pool Pt

Northey I.

Collier's Reach

44′

Blackwater S.C. Iso.G 5s

Heybridge Basin

Lock

No. 10

No. 12

Power

43′

Canal

13

6 14

11 9

MALDON

The Hythe

0°42′E

Depths in Metres

41′

Bradwell

51° 44′ N

45′

Nuclear Power Station (46)

0°54′E

Barrier Wall

2FR(vert)

Bn QR

2FR (vert)

Marked channel by withies

13₂

5

5₄

9

8₄

13₆

9₄

Bradwell Creek

Y C

Bradwell Waterside

Bradwell Marina

Chandlery

Pewit I.

Moorings

River Blackwater

Moorings

53′

52′

Brightlingsea

51° 48′ N

49′

Brightlingsea Creek

Cindery I.

HM

BRIGHTLINGSEA

Hotel

Cupola

Cole Hard

Boat Shed

Wks

Moorings

Saltings

FR Or.

2FR Or.

Bn

St Osyth Stone Pt

Fl(3)G 5s

Fl.R 2s

Martello Tr

Westmarsh Pt

Brightlingsea Spit

Q(6)+LFl.15s

Fl(2)4s

Fl.G

Fl.O Bn Bateman's T (13)

Mersea Stone

Bn

River Colne

(dr. 2.1m)

01′

1°02′E

The havens and rivers

It must be stressed, particularly to yachtsmen from the Continent, that, as in the area covered in the previous chapter there are very few marinas (currently 2 in the Crouch and a third being developed, and 2 in the Blackwater) and a proper tender – inflatable, hard or collapsible – is essential.

Burnham, Brightlingsea and Maldon are pleasant small towns, the latter being of particular historical interest, whilst the other two were to some extent created by the yachting 'industry' from the late nineteenth century onwards, which in itself provides considerable recent historical interest. The Royal Corinthian YC at Burnham for example moved from the Isle of Grain in that period, encouraged by the railway facilities from London. Burnham is still served by railway, as is Brightlingsea by the nearby Wivenhoe station, but Maldon's railway was closed after the last war. Colchester, an historic garrison town, is well worth visiting and if, as is likely, you cannot reach it afloat, then you should take a bus trip from Brightlingsea or Tollesbury. There are a few villages, some quite picturesque, worth visiting – on the Blackwater and Colne: Bradwell, West Mersea, Goldhanger, St Osyth, Wivenhoe – and on the Crouch and Roach: Paglesham, Canewdon, Althorne, Hullbridge and Fambridge.

THE RIVER COLNE

Port radio (VHF)

Colchester ☎ (0206) 575858	VHF Ch 14, 11 16	Hrs listening Mon–Fri 0900–1230, 1400–1630 LT Other times from 2hrs-HW-1hr

Entry Signals   None

Customs

Brightlingsea: Customs Office, James and Stone Shipyard, Brightlingsea, Colchester CO7 0AY

Colchester: Custom House, Hythe Quay, Colchester CO2 8JB

In both cases above, ☎ (by VHF) Ipswich (0473) 219481; ashore dial 100 and ask for 'Freefone Customs Yachts'.

Entrance and river passage

Distance   10½M from Colne Bar buoy to Colchester Hythe Quay

Between Colne Bar buoy and Pyefleet Channel entrance the Colne has 5 to 12m soundings in the main channel, but 2M further inland above Alresford Creek it effectively dries out and the passage to Colchester is advisable only on a rising tide for boats which take the ground. Pyefleet Creek dries out about 2M inland, whilst Brightlingsea Creek and its S branch to St Osyth's Creek dry out only about 1M inland at the end of Cindery Island. From the Colne Bar (G) buoy (unlit, but the group of 4 Lt buoys in the nearby Eagle channel make up for this deficiency) the Colne channel is buoyed (mainly lit) as far as the bend below Wivenhoe above which to the bridge at Hythe Quay Colchester it is a narrow winding drying creek, unlit except for the intriguing legend on the Admiralty chart 'Channel floodlit when required' i.e. near HW at night, when ships are around.

Brightlingsea – the Anchor and the Colne YC pontoon

In the Colne approach channel beware the deceptive width of water – on the W is Bench Head and on the E is Colne Bar, both with drying patches. On a rising tide a short cut route between Brightlingsea and the Blackwater to Sales Point is across Cocum Hills, a saddle between Mersea Flats and Bench Head, with a single unlit wreck-marking (E card) beacon on the drying edge of the Flats.

There is a 5 to 10 minute delay in HW between Brightlingsea and Colchester; not significant for passage timing.

Rowhedge

Wivenhoe

Mooring and facilities

Brightlingsea Creek has a lit entrance spit buoy (S card) and 2 channel-marking Lt buoys inside the creek as well as two leading lights. There is a bar close to the entrance spit buoy (0.6m sounding) which it is possible to scrape at LW, so if you are deeper draught take the rising tide an hour later.

Deep-water moorings are to be found betweeen the posts S of Cindery Island off Brightlingsea Creek. Anchoring is no longer advisable now the Wharf is used by small coasters. Landing is at the hard or the Colne YC pontoon, providing you

inform the YC. Brightlingsea has plenty of shopping facilities, a good chandlery, the Colne YC and Brightlingsea SC, and repair, scrubbing and lifting facilities. The HrMr collects dues promptly at the posts, and can advise on where to obtain services.

Somewhat precarious anchoring and landing is possible off Mersea Stone on the opposite side of the river but there is a long walk to the nearest pub in East Mersea and no shopping facilities. Anchoring but no landing is possible well up the Pyefleet Channel, taking care to avoid the moorings and the oyster beds. It is not really practical to anchor in the river above Brightlingsea because of the commercial traffic. Above Alresford Creek the river dries out and is not promising cruising country unless you are prepared to dry out.

There are private mudberths at Wivenhoe waterfront, or you may be able to borrow a mooring temporarily (drying out). Keep well away from the new ship wharf N of the village. There is a boatyard, chandlery, sailing club and good shopping facilities in the picturesque village.

At Rowhedge on the E bank it is possible but not advisable for newcomers to lean against the quayside near the pub, and there is a boat repair yard, but the creek is very narrow at this point with occasional passing traffic near HW. In navigating this stretch on a rising tide it is worth contacting the Colchester port control to see what traffic is coming, and the best place to lie to avoid manoeuvring ships (drying out against the quayside) is at Hythe Quay, Colchester.

THE RIVER BLACKWATER

Port radio (VHF)

Bradwell Marina	*VHF Ch*
☎ (0621) 76235	M

Entry Signals   None

Customs

Bradwell: Customs Post, Waterside, Bradwell-on-Sea, Essex, CM0 7QW

West Mersea: Custom House, Hythe Quay, Colchester CO2 8JB

Maldon: 112B High Street, Maldon, Essex CM9 7ET

In all cases above, ☎ (by VHF) Ipswich (0473) 219481; ashore dial 100 and ask for 'Freefone Customs Yachts'.

Entrance and river passage

Distance   10M from Sales Point to Maldon

The Blackwater proper has wide drying banks throughout and a winding channel which shoals from 10–16m soundngs off Sales Point to 6m off Stone Point, to 2m off Osea Island, with most moorings drying out above this point and a channel with least depths of 0.3m up to the fixed bridge at Maldon which is effectively available only for rising tide pilotage.

There is a ½hr delay between Bradwell HW and Maldon HW giving over 6½hrs rise and less than 6hrs fall on the passage along the river, not particularly useful over such a short distance.

The passage to Maldon requires care. At the entrance between Sales Point and Mersea Island the depth sounder is needed to keep in the deep channel, to turn round the bend of the channel (which is 4M away from the Bench Head buoy)

The moorings at Bradwell Creek – note scrubbing grid

The moorings on the opposite side of the river in Mersea Quarters

and to keep S of Nass Spit (0.2m sounding at the outer end), past the (well lit) Bradwell Nuclear Power Station to the S with its offlying baffle wall (R Lts at each end) and the Q.R beacon marking the entrance to Bradwell Creek. The course should continue in mid-channel, to a point close S of the Thirslet Spit (G) Lt buoy, sounding frequently if you are a newcomer, since the banks are very wide and shallow, the buoyage sparse and overshooting the buoy to its N means running aground. Then a course to Osea Island Pier (lit) opposite No 2 (R) Lt buoy marking the entrance to Lawling and Mayland creeks. The route to the northern bend round Northey Island is across wide mudflats, with several channel-marking buoys. At night this stretch should be taken towards the top of a rising tide, since there is only 1 lit channel-marking buoy and an Iso.G.5s Lt on the Northey bend near the River Blackwater Sailing Club. The rest of the winding channel to Maldon is buoyed but unlit.

Mooring and facilities

Deep-water anchoring is possible, depending on weather direction, on the edges of the mudflats on either side of the river from the Nass to just beyond Osea Island, the S side of which is the most popular anchorage, probably followed by a wide area at the entrance near Bradwell.

On the N side of the river entrance at the W end of Mersea Island a number of very narrow creeks packed with deep-water moorings fan out inland from Mersea Quarters N of Nass Spit. Follow the G buoys (unlit) and then the moorings. With 1, 2 and 3m soundings care should be taken not to take the ground on the mudflat edges of the channels or to go too far inland where the moorings dry out. Thorn Fleet is best for the visitor offering access to West Mersea with slipways, scrubbing posts, lifting and repair facilities, two clubs (West Mersea YC and Dabchicks SC), but it will be necessary to borrow a mooring or a space between posts, since anchoring out in the Quarters is extremely inconvenient.

Tollesbury Yacht Harbour, ☎ (0621) 869202, is at the end of Woodrolfe Creek leading off the southernmost creek out of the Quarters, Tollesbury Creek's South Channel. The R can buoys (keep to the N) should be followed out of Mersea Quarters to the one behind Nass Spit marking the South Channel entrance, then the G buoys (keep to the S) S of Cob Island to Woodrolfe Creek. The creek and sill into the marina are available for 1.5m-draught vessels for about 2hrs either side of HW, when there can be less than 2m over the sill at neaps but the average is 2.3m. Pick up a visitor's mooring off the creek entrance if you need to wait.

There are full repair facilities, chandleries, two clubs (Tollesbury Cruising Club and Tollesbury SC), and a 15 minute walk to the village shops.

S of the stone embankment below the power station, Bradwell Creek leading to Bradwell Marina (Ch M and ☎ (0621) 76235) is a tricky entrance for newcomers. SW from the very clearly visible power station baffle wall (2F.R vert Lts at each end) is a beacon (Q.R) marking the W side of the creek entrance, which must be passed on the E side following the channel between the R (unlit) port-hand buoys (ex beer barrels – you can expect a warm welcome at BQYC!) and the starboard-hand withies, although the latter are frequently in poor shape, broken and below the surface near HW. The channel heads direct for the N end of Pewit Island, just before which (the shallowest water is here) it turns a dogleg SE towards a G (unlit) buoy amongst the first of the moorings in the creek. This buoy is rounded (keeping it to starboard going inwards) and the main moorings closely followed on the E side of the creek (please take care, my boat is often on one of these!) to a R buoy (unlit, keep to its W side, there is a shallow bar just before this point) and round between the withies marking each side of the clearly visible narrow marina entrance channel. For up to 1.5m draught the creek and marina can be entered about 4hrs either side of HW. The marina has full repair, chandlery and lifting facilities and Bradwell Cruising Club. If you pick up a mooring temporarily, many of which dry out, particularly above the marina, you should check its availability with Bradwell Quay YC. This club is close to the top of the slipway in the creek below the marina; it hires out scrubbing posts, is open at weekends, and welcomes visiting yachtsmen. There is a village store close by. The Essex Sailing Centre is also near the BQYC.

The next deep-water moorings are off the exposed Stone Point and Stansgate Abbey on the S bank just below Osea Island. There is a pub and, a short walk away, a shop. Goldhanger Creek marked by a single G (unlit) entrance buoy and with drying moorings near the top is for experienced locals not newcomers. Lawling Creek and Mayland Creek marked by a R can buoy are lined with drying moorings, but the former has a good repair yard with slipways and lifting facilities, and two clubs; Maylandsea Bay YC and Harlow YC.

The last deep-water moorings can be obtained by locking through into the Heybridge Basin (lock-house ☎ (0621) 53506) of the Chelmer and Blackwater Canal, just below Maldon town and available for about an hour before HW. Berths are alongside (or rafted) the canal bank. Chandlery and limited repair facilities. Blackwater SC is a short walk from the lock. You can anchor offshore near the channel whilst waiting for the lock or rowing ashore to enquire.

There are drying moorings in several stretches of river up to Maldon. Finally a drying berth at the head of the navigation can be obtained by leaning against Hythe Quay or more likely one of its Thames barges, or finding a mud-berth in one of the several boatyards. There are full repair and lifting facilities, a large shopping centre, Maldon YC and Maldon Little Ship Club.

The lock into Heybridge Basin

The quay at Maldon's Fullbridge

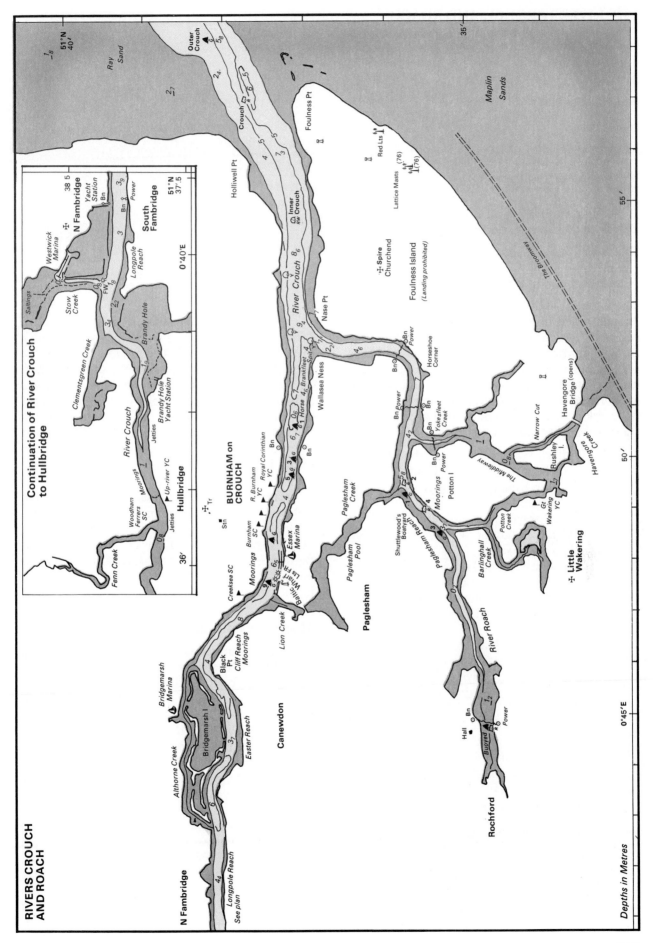

RIVERS CROUCH AND ROACH

N Fambridge

Continuation of River Crouch to Hullbridge

N Fambridge

South Fambridge

Hullbridge

BURNHAM on CROUCH

Canewdon

Paglesham

Rochford

Little Wakering

Depths in Metres

THE RIVERS CROUCH AND ROACH

Port radio (VHF)

	VHF Ch	Hrs listening
Essex Marina, Crouch ☎ (03706) 531	M	0900–1700 LT

Entry Signals   None

Customs

Burnham-on-Crouch: Customs Office, The Quay, Burnham-on-Crouch, Essex
☎ (by VHF) Ipswich (0473) 219481
☎ (ashore) dial 100 and ask for 'Freefone Customs Yachts'

Entrance and river passage

Distance   Inner Crouch buoy to Battlesbridge 14M
Inner Crouch to Paglesham Reach 5M

The Crouch shallows from 7m at the entrance to 3m near Fambridge, 1.2m in Brandy Hole Reach, and dries completely from just E of Hullbridge to the head of the navigation at Battlesbridge. There is no buoyed channel, but the banks in the deep-water channel are relatively steep-to, and even at night rising tide navigation, providing the visibility is good, is no problem. There are six tricky spots moving W from the entrance.

1. The mudflat, 1½ cables wide, on the S bank at Foulness Island together with Brankfleet Spit (Y unlit buoy) on the opposite side of the Roach entrance.
2. The Horse shoal in mid-river (0.6m least depth, G unlit buoy at W end) with a deep channel on either side.
3. Burnham Fairway on the S side of the river opposite the town bypasses the crowded moorings with two G buoys leading into the E end of the Fairway and one at the W end (unlit).
4. The spit on the N bank off Creeksea (0.8m depth on outer edge, G unlit buoy to E).
5. Bridgemarsh Island (drying) but flooded at half-tide with a rim of discontinuous banks around it.
6. From W of N Fambridge Yacht Station are wide drying areas which again flood deceptively and are unmarked except for moorings, which mainly dry out and become more numerous towards Hullbridge, and where the best tactic is to take the rising tide as early as possible and follow the moorings of the larger vessels. Above Hullbridge it is a case of following an empty, winding tidal channel through the meadows to the mounds of mud on each bank at Battlesbridge.

Burnham-on-Crouch – Royal Burnham YC to the right

The River Roach entrance is tricky with 1.7 to 3.7m soundings with a wide W sweep under Brankfleet Spit (Branklet to the locals) and away from the shallow, drying E bank, and from then on has a 4 to 9m deep, winding and narrowing channel to Paglesham Creek but with tricky drying edges particularly at the entrance to Yokesfleet. The maze of drying creeks beyond Yokesfleet and Paglesham, some leading to the lifting Havengore Bridge and to Rochford, are a subject beyond the scope of this pilot.

From Burnham there is a 10 minute delay in HW reaching Paglesham on the Roach and Fambridge on the Crouch, 15 minutes in reaching Hullbridge, and as much as 25 minutes for Battlesbridge; useful for the full 14M passage but not so helpful downstream particularly if continuing out into Whitaker Channel.

Mooring and facilities

Despite the limited navigational facilities, this is the only east coast river with a Harbour Authority (Crouch Harbour Authority, ☎ (0621) 783602) which charges harbour dues. Do not worry if you are liable they will find you!

Good anchorages are available throughout the river system, but the most popular are at the two river entrances (on the S bank of the Crouch and S of Brankfleet Spit) and at Cliff Reach on the N bank 2M above Burnham. The Roach creeks, where they do not dry out are also popular.

The first available yard-laid moorings on the Roach are in Paglesham Reach off Shuttlewood's Boatyard, ☎ (03706) 226, where there are repair, storage and lifting facilities as well as a pub, but it is a long walk to village shops. Further into the Roach drying creeks is Wakering YC at the S end of Potton Island where the Middleway (from

Yokesfleet) and Potton Creek (from Paglesham, through a swing bridge) join and with many drying moorings, and still further the Havengore Bridge with the shallow-draught route into the Thames Estuary.

The Burnham moorings are a yachtsman's paradise. There is a multitude of yards with every facility, a modest-sized shopping High Street with almost a restaurant for every shop, and at least six yacht clubs – Royal Corinthian YC, Royal Burnham YC, Crouch YC, Burnham SC, Creeksea SC, and The London Hospital SC. Moorings can be borrowed on arrangement with the yards. The Royal Corinthian YC, Royal Burnham YC and Prior's boatyard all have floating pontoons to provide unloading facilities. A new marina is planned behind the N bank close W of the built-up area.

Essex Marina, Wallasea Bay ☎ (03706) 531, on the S bank is remote from any village, but there is a shop, full marina and boatyard facilities and Wallasea Bay YC.

At the E end of Bridgemarsh Island is Althorne Creek with 0.5 to 0.7m soundings to drying just beyond the boatyard which has several small walled 'docks' into the mainland bank which dry out, with moorings in the creek some of which have single floating pontoons with a boat on each side. Anchorage in the entrance is just possible, and entry is best when the banks are uncovered on a rising tide, keeping somewhat closer to the island side through and from the entrance.

North Fambridge Yacht Station, ☎ (0621) 740370, and Westwick Marina, ☎ (0621) 741268, is an appreciable combined business. The former lets a large number of deep-water moorings in the river off Fambridge, and has good landing facilities. The latter is a pontoon marina its entrance to be avoided at dead LW but available at most other states of the tide via Stow Creek with a starboard entrance beacon (F Lt) and starboard withies up to the right-angled E offshoot forming the dredged marina. The yard has most facilities, there are two clubs, Westwick YC and North Fambridge YC, and there is a pub and shop in the village.

The river dries after the few deep-water moorings in Brandy Hole Reach. There are four clubs, Brandy Hole YC, Up River YC, Hullbridge YC, and Woodham Ferrers YC, and at Brandy Hole a boatyard with modest facilities. Hullbridge and Woodham Ferrers both have appreciable shopping facilities. Above Hullbridge the river narrows and it is only feasible for a newcomer to visit Battlesbridge on a rising spring tide, tying alongside the wall near the Barge Inn and leaving soon afterwards or taking the ground on a long mud slope down from the wall.

North Fambridge Yacht Station moorings

The head of navigation on the Crouch at Battlesbridge

Tollesbury Marina
Photo Lester McCarthy, Motor Boat & Yachting

8. The Thames Estuary from Foulness to North Foreland

Charts Admiralty *1183, 1185, 1186, 1605, 1607, 1610, 1834, 1835, 1975, 2151, 2484, 2571, 2572, 3337, 3683*
Imray *C1, C2, Y6, Y7, Y18*

Tidal atlases Admiralty *Thames Estuary*
North Sea – southern portion

Tidal streams (based on HW Sheerness and London Bridge)

Position	LONDON BRIDGE		SHEERNESS	
	Out	*In*	*Out*	*In*
Longsand Head buoy (rotatory, approx)			−0030	+0530
N Edinburgh Chan S ent.			+0100	−0530
Southend Small Ship Anchorage	−0045	+0515	+0030	−0600
London Bridge	+0015	−0520		

Start times

River Medway

Outgoing and ingoing tides as far up as Rochester start approximately just after HW and LW Sheerness.

The Swale

1hr after LW Sheerness streams start to run inwards at both ends meeting somewhere near (depending on wind and varying tidal rise) Fowley Island N of Conyer Creek. At HW Sheerness the stream from Fowley Island turns E-going so whole stream in Swale is running from Medway to Thames Estuary. An hour later, from HW +0100, streams in W entrance from Long Point become W-going into the Medway and this separation point moves progressively E (+0200 Kingsferry Bridge, +0300 Milton Creek) reaching Fowley Island at about HW +0330 and continuing from that point until slack water just after LW Sheerness.

SW Sunk beacon

Tidal differences and ranges (based on HW Dover)

Place	HW (time)	Springs/Neaps (range in metres)
Whitaker Beacon	+0105	
Southend	+0130	5.2/3.4
Tilbury	+0203	6.0/3.9
Greenhithe	+0130	5.1/3.3
London Bridge	+0252	6.6/4.2
Sheerness	+0136	5.1/3.3
Queenborough	+0140	
Strood Pier (Rochester)	+0154	
Harty Ferry	+0136	
Whitstable	+0135	4.9/3.0
Margate	+0100	4.3/2.5

Major lights

Name of light	Characteristics	Position	Structure
Offshore			
Tongue LtF	Fl(2)10s12m24M Horn 20s	51°30′.6N 1°23′.0E	R hull
Onshore			
North Foreland	Fl(5)WR.20s57m21–16M	51°22′.5N 1°26′.8E	Wh 8-sided Tr, RC Shore-W-150°-R-200°-W-011°
Isle of Grain	QWRG.20m13–7M	51°26′.6N 0°43′.5E	R Tr, 220°-R-234°-G-241°-W-013°
CEGB Power Stn	Aero Oc.R.242m Aero F.R.205m Aero F.R.167m Aero F.R.117m	51°26′.7N 0°42′.8E	Main chimney
Kingsnorth Power Stn	Aero F.R.198m15M	51°25′.1N 0°36′.3E	Chimney

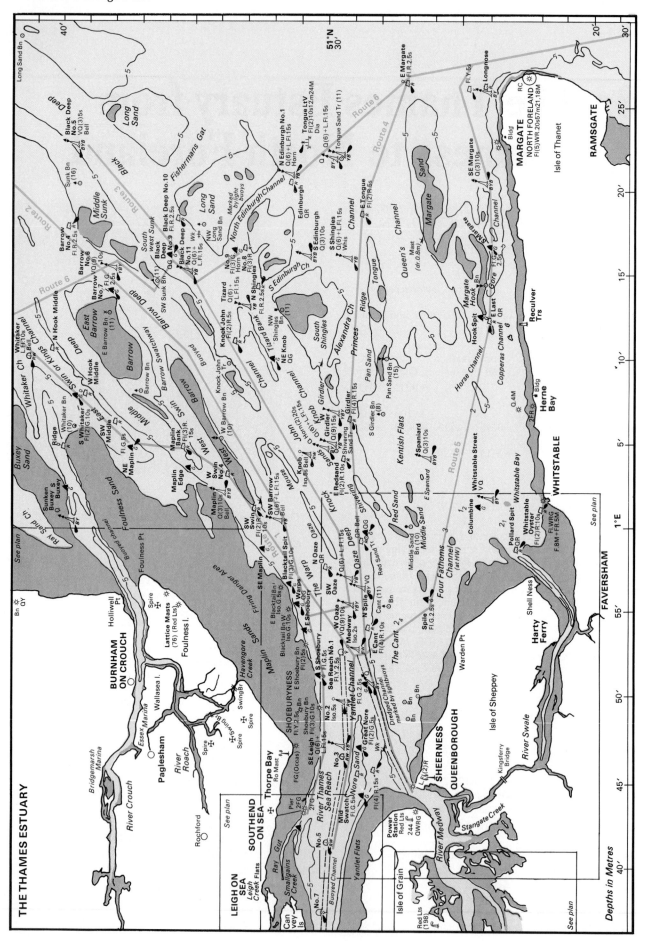

THE THAMES ESTUARY

Depths in Metres

Shornmead	Fl(2)WRG.10s12m17–13M	51°27′.0N 0°26′.6E	R metal framework Tr, Shore-G-080° 080°-R-085°-W-088°-G-141°-W-205°-R-213°
Northfleet Lower	Oc.WR.5s15m17/14M	51°26′.9N 0°20′.4E	R framework Tr 164°-W-271°-R-S shore Gravesend reach
Northfleet Upper	Oc.WRG.10s30m16–12M	51°26′.9N 0°20′.2E	On roof of building 126°-R-149°-W-159°-G-269°-W-279°
Crayford Ness	Fl.5s19m14M	51°28′.9N 0°12′.8E	Grey framework radar Tr

Radiobeacons

Name	Freq. (kHz)	Ident.	Range (miles)	Seq.	Position
Marine radiobeacon					
N Foreland Lt	301.1	NF	50	Cont	51°22′.5N 1°26′.8E
Aero beacon					
Southend	362.5	SND	20	Cont	51°34′.6N 0°42′.1E

Coast radio station

Station	Transmits* (kHz) (VHF Ch)	Receives* (kHz) (VHF Ch)‡	Freq VHF Ch†	Traffic lists (times)	Storm warnings (times)	Weather messages (times)	Navigational warnings (times)
N Foreland (GNF)	2182,**1848**, 2698,2733	2182‡, 2002(Ch11), 2016(Ch2), 2548(Ch7), 2569(Ch10)	1848	0103,0503 & odd H+03 (0903-2303)	0303,0903, 1503,2103	0803,2003	0033,0433,0833, 1233,1633,2033
	05,16,**26**,**65**, 66	05,16,26,65, 66	26	0103,0503 & odd H+03 (0903-2303)	0303,0903, 1503,2103	0803,2003	0033,0433,0833, 1233,1633,2033

Notes
* 24hr service
† Frequency (kHz) and VHF Channel on which traffic lists, storm warnings, weather messages and navigational warnings are given.
‡ 24hr watch
All times are GMT

Major fixed daylight marks

N shore of the Thames Estuary (E to W)

The Whitaker Beacon (10m, unlit, 2 B balls) 6M NE of Foulness Island

A group of 3 radio lattice masts (2 of 76m, R Lts) on the N end of Foulness Island

Havengore Bridge when lifted and Great Wakering Church spire can occasionally be seen in good visibility

A gasholder just N of Shoeburyness and a radio mast (31m) just W of the point

Southend Pier with hotel (conspic) at its root and green spire to the E

S shore of the Thames Estuary (E to W)

N Foreland LtHo with nearby radio mast (42m)

Building (conspic) SW of Margate stone pier (Margate promenade pier is being demolished)

Birchington Church spire

The Reculvers (twin towers)

Block of flats (conspic) near root of Herne Bay pier with clock tower to the E and water tower and windmill (conspic) inland to the SE

Whitstable Church tower (square)

Leysdown-on-Sea holiday camp with St Clements Church (conspic) on Isle of Sheppey

Clay cliffs stretching W from Warden Point

Garrison Point on Isle of Sheppey, with fort on top

Grain Power Station chimney (244m, R Lts) on Isle of Grain, the most prominent landmark in the Estuary

Thames Estuary channels (E to W)

King's Channel, E & W Swin & the Warp

Sunk Head Tower (5m, ruined concrete stump bristling with reinforcing wires, marked by N card Lt buoy)

Gunfleet Old Lighthouse (13m, square framework, unlit) on Gunfleet Sand

The Whitaker Beacon (10m, 2 B balls) 6M NE of Foulness Island

A group of 3 radio masts (2 of 76m, R Lts) on the N end of Foulness Island

Blacktail E & W Lt beacons (10m, B framework towers)

Maplin survey platform (10m, square framework tower, lit) close E of E Shoebury beacon (upward-pointing triangle topmark, unlit)

Barrow Deep

E Barrow beacon (11m, x topmark, unlit)

Barrow beacon and SW Sunk beacons (11m, square frameworks, triangular tops, unlit)

Knock John Tower (double towers with platform on top, unlit)

W Barrow beacon (10m, E card topmark, unlit)

Black Deep, Knock John Channel, Knob Channel, Oaze Deep

Little Sunk (18m), Long Sand (18m), Sunk (16m), SW Sunk (11m), NW Longsand (11m) beacons (square frameworks, triangular tops, unlit)

Knock John Tower (double towers with platform on top, unlit)

NW Shingles beacon (11m, can topmark, unlit)

Shivering Sand Tower (group of 6 concrete pile structures with huts on top, unlit, marked by 2 cardinal Lt buoys)

Red Sand Tower (group of 7 concrete structures similar to Shivering Sand Tower, unlit, marked by R and G Lt buoys)

Princes Channel

Tongue Sand Tower (11m, double towers with platform and structure on top, unlit, marked by 2 cardinal Lt buoys)

Margate Hook beacon (11m, S card topmark, unlit)

'Overland Passage', South Channel, Gore Channel, Horse Channel, Four Fathoms Channel

Pan Sand beacon (15m, S card topmark, unlit)

Middle Sand beacon (10m, circular topmark, unlit)

Sea Reach

N of Medway dredged channel is the wreck of *Richard Montgomery* (explosive carrier) with masts and part superstructure above water, marked by 4 Y Lt buoys.

Approach routes and tidal timing

Like the Dover Strait this is a well marked, well lit area which, despite its sandbanks is easy to navigate in good conditions given a sound knowledge of tides, but in very strong winds, especially those funnelling in from the N round to E, and in wind-over-tide conditions, it can be dangerous, building short choppy seas, and the banks should be given a wide berth.

There is a modest amount of freight and ferry traffic to Tilbury and the London River, and to Sheerness and the Medway, as well as fishing vessels, but not to the extent of the Westerschelde or the Nieuwe Waterweg. Nevertheless care has to be taken to keep on the edges of the shipping channels and to cross them at right angles, whilst passage in fog should be avoided if at all possible. If fog does come down after the start of a passage it is usually possible to find sheltered anchorages in shallow water away from traffic in some of the shallow swatchways and on the edges of the banks, making sure of the day's tidal range and leaving reasonable keel clearance at LW; e.g. on Maplin Edge NE of Blacktail Spit, round the edges of the Barrow Swatchway, in many of the swatchways across Sunk and Long Sand, and inside Margate Sand.

S of Whitaker Spit the Estuary has two wide shallow edges. The N edge, Foulness and Maplin Sands, dries out completely with a steep S edge into the Warp channel (10 to 15m soundings) and a gentle N slope extending to Whitaker Spit. The S edge, Kentish Flats, is a wide shallow shelf of typically 2 to 4m soundings with a narrow drying edge in Kent and Sheppey and a number of drying 'hillocks', the Cant, Spile, Red Sand, and Middle Sand in the W, and the Pan Sand and Margate Sand in the E. These two shore-banks merge W towards the London River entrance into the drying Southend, Leigh and Chapman Sands on the N bank, and the Yantlet Flats and Blyth Sands on the S.

Between the Estuary shore-banks are three major sandbars: the Barrows, Sunk and Long Sands, which in turn separate four major shipping channels; the W and E Swins and Middle Deep, Barrow Deep, the Oaze and Black Deeps, and Princes Channel. The sandbars dry in patches with occasional swatchways across; e.g. Barrow Swatchway 6–11m sounding, Little Sunk (S of the beacon) 3.2m, SW Sunk (S of the beacon) 2.3m, Knock John (NE of the tower) 0.6m, Fisherman's Gat 6.2m, and strangest of all, N Edinburgh Channel where the tidal stream has broken in a southerly arc through Long Sand reaching depths of over 15m shallowing to 7m across the southern bar. Typical depths in the deeps are 15 to 20m but the S ends of Middle Deep and E Swin are shallow, 2 to 5m, with an ominous scattering of wrecks.

This is an oversimplified picture. In the N, West Swin cuts through the Barrows and joins Barrow Deep but is poorly marked and though deep is of no use as a short cut to anywhere. Black Deep, at its S end, is compressed by Tizard Bank (0.4m least depth) into the narrow but deep Knock John Channel, and S of the bank Knob Channel curves W out of N Edinburgh Channel. In the S, Princes Channel crosses two 5–9m patches S of the Girdler and on Shivering Sand, whilst Queen's Channel is a sham, petering out into the Kentish Flats. The Gore, Horse and Four Fathoms Channels closer to the Kent and Sheppey coasts form a shallow-water yacht route from North Foreland across the Kentish Flats. All of these channels and banks focus on the single Yantlet dredged channel leading into the London River between Canvey Island and the Isle of Grain. All are well marked with IALA Lt buoys, towers with Lt buoys around them, and many, usually unlit, beacons.

Finally a major warning – the shallower swatchways, and the banks and edges of swatchways are unstable and moving, and frequent reduced soundings or drying patches are reported. Up-to-date charts are needed, and a reasonable error margin for under-keel clearance should be left.

Compared with the Dutch Delta tidal streams in the Estuary are simple except in the Swale and N Edinburgh Channel (see above under *Tidal streams*) and in the extreme outer approaches and near N Foreland, generally turning at local HW and LW. The Admiralty tidal atlas is a must; as well as 13 hourly charts it also contains co-tidal (charted lines of times of HW) and co-range (charted lines of tidal range) lines.

The tide moves into and out of the Estuary in a 'wave' which remains parallel to a line from Clacton to Margate, but which bulges slightly E. The range of tide increases progressively from this line inwards to Southend and Sheerness, and the rate of the tidal stream also increases with this funneling effect reaching 2½–3kn average (do not forget averages can be exceeded!) spring rate near the Nore, and greater rates within the London River proper. Near Southend the tide is slack at HW Sheerness and runs out until about 5½hrs after HW Sheerness at Long Sand Head, at the outer ends of the channels. In the opposite direction there is an advantage with about 7hrs of ingoing tide from Long Sand to Southend, useful over a 40M distance.

There are a variety of yacht passage routes within the Estuary and its tributaries. Below are summarised 5 access and 1 crossing routes.

ROUTE 1   THE SWINS

Distance   41M Longsand Head buoy to Southend Pier

Commentary

This route is well buoyed and lit with very light traffic. From N of Longsand Head (N card) Lt buoy across the end of Black Deep N of Sunk Head Tower (and N card Lt buoy) into King's Channel (10 to over 20m soundings) with widely spaced Lt buoys (see Chapter 6), it passes N of Middle Sand along the E Swin buoyed (some lit) channel – Middle Deep to the S is not buoyed – and shallows to just over 3m, crossing a number of obstructions (over 3m soundings) between Maplin Edge (G, unlit) buoy and Maplin Bank (R) Lt buoy. Then into W Swin (buoyed on each side, some lit), and finally along the Warp channel, well buoyed (mainly lit) on Maplin Edge.

Tidal timing

The inward route has the advantage with 7hrs of flood, although still not usually quite enough for the 41M distance, and if the departure is to be from Harwich across the tide then it is advisable to start early enough to reach Longsand Head buoy with an hour or more of weakening ebb to push obtaining the maximum of flood to Southend. On the return with only 5½hrs of flood it pays to start at half flood pushing 3hrs of tide and getting a full ebb to Longsand Head buoy before the cross-tidal leg to Harwich.

ROUTE 2   BARROW DEEP

Distance   41M Longsand Head buoy to Southend Pier

Commentary

Also well buoyed and lit. Heavier traffic than the Swins and thus to be avoided when possible. From N of Longsand Head (N card) Lt buoy across the end of Black Deep N of Sunk Head Tower (and N card Lt buoy) into King's Channel (10 to over 20m soundings), the Barrow Deep (mainly 16 to 17m), across Mouse shoals (over 4m soundings) and along the Warp channel, well buoyed (mainly lit) on Maplin Edge.

Tidal timing

As in Route 1 above the inward route has the advantage with 7hrs of flood, although still not usually quite enough for the 41M distance, and if the departure is to be from Harwich across the tide then it is advisable to start early enough to reach Longsand Head buoy with an hour or more of weakening ebb to push obtaining the maximum of flood to Southend. On the return with only 5½hrs of flood it pays to start at half flood pushing 3hrs of tide and getting a full ebb to Longsand Head buoy before the cross-tidal leg to Harwich.

ROUTE 3   BLACK DEEP AND OAZE DEEP

Distance   42M Longsand Head buoy to Southend Pier

Commentary

Heavier traffic than the Swins and to be avoided when possible. Well buoyed and lit, and with 10 to over 20m soundings throughout. From Longsand Head (N card) Lt buoy down Black Deep, Knock John Channel and Oaze Deep to W Oaze (W card) Lt buoy, then a right-angled heading across the Warp turning to follow the buoyed, mainly lit Maplin Edge.

Tidal timing

As in Route 1 above the inward route has the advantage with 7hrs of flood, although still not usually quite enough for the 41M distance, and if the departure is to be from Harwich across the tide then it is advisable to start early enough to reach Longsand Head buoy with an hour or more of weakening ebb to push obtaining the maximum of flood to Southend. On the return with only 5½hrs of flood it pays to start at half flood pushing 3hrs of tide and getting a full ebb to Longsand Head buoy before the cross-tidal leg to Harwich.

ROUTE 4   PRINCES CHANNEL

Distance   32M N Foreland to Southend Pier

Commentary

From 1M off N Foreland to close E of E Margate (R) Lt buoy passing E of Diffusers (Y) Lt buoy, then cross Princes Channel at right angles from E of the E Tongue (R) Lt buoy and follow the starboard N edge of the buoyed (lit) channel crossing the shallow patch (over 5m) between the Girdler and Pan Sand, then S of Shivering Sand Towers and its Lt buoys (over 6m soundings) then down Oaze Deep to W Oaze (W card) Lt buoy, then a right-angled heading across the Warp turning to follow the buoyed, mainly lit Maplin Edge.

Shallow-draught yachts (up to 1m) heading for the rivers Crouch and Roach and crossing the Maplins from the E Shoebury beacon and its nearby light platform to Havengore Creek and Bridge often find this a convenient route from Ramsgate or Calais, and on the falling tide in the opposite direction having passed through Havengore near HW.

Tidal timing

In contrast with the N channels, here the 32M outward passage which aims to round N Foreland and travel S has 6hrs of fair tide to N Foreland and a further 3½hrs stream sweeping S from N Foreland. If, as in Passage 4 in Chapter 4, Calais is the objective then it pays to start well in advance of HW and push some tide in order to achieve the full down-channel tidal stream round the outer side of the Goodwins. The inward passage has to push a strong foul tide which runs from 2 to 3hrs before LW Sheerness in order to round N Foreland and take the full 6hrs flood (see also Chapter 9 *Approach routes and tidal timing*).

ROUTE 5   OVERLAND ROUTE

Distance   30M from N Foreland to Southend Pier

Commentary

No significant traffic. However it is definitely to be avoided in rough weather, particularly with wind-against-tide, since the Four Fathoms and Horse Channels have occasional minimum soundings of 1.8m, and even the Admiralty's legend 'at High Water' after the name 'Four Fathoms Channel' on the chart, is a small exaggeration.

From 1M off N Foreland round N of Diffusers (Y) Lt buoy, then W through South Channel and Gore Channel (3–7m generally with some 14m 'holes') S of two Lt buoys and Margate Hook beacon to the Hook Spit (G, unlit) and E Last (R, lit) buoys at the entrance to Gore Channel, with the Reculver towers 1.2M to the S (a particularly good landmark travelling in the outwards direction). Then compass course via Horse Channel to close S of the Spaniard (E card) Lt buoy, and through Four Fathoms Channel to close S of Spile (G) Lt buoy. This course clears an obstruction to the S and some drying patches to N and S of the course. From Spile buoy, head for SE Leigh (S card) Lt buoy, crossing the Cant (2–3m soundings) close to an unnamed, unlit beacon en route (so take care or detour at night) as well as the Cant beacon (11m, unlit) to the E; also take care to cross both the Medway buoyed channel and Yantlet Channel on right-angled headings. Then a right-angled heading across the Warp and along the N buoyed (lit) edge to Southend Pier.

Tidal timing

Again as in Route 4 and in contrast with the N channels, here the 30M outward passage which aims to round N Foreland and travel S has 6hrs of fair tide to N Foreland (weakish over Kentish Flats) and a further 3½hrs stream sweeping S from N Foreland. If, as in Passage 4 in Chapter 4, Calais is the objective then it pays to start well in advance of HW and push some tide (again remembering the usefully weak streams across Kentish Flats) in order to achieve the full down-channel tidal stream round the outer side of the Goodwins. The inward passage has to push a strong foul tide which runs from 2 to 3hrs before LW Sheerness in order to round N Foreland and take the full 6hrs flood (see also Chapter 9 *Approach Routes and tidal timing*).

ROUTE 6   CROSS-ESTUARY

Distance   25M Whitaker bell buoy to N Foreland

Commentary

Tidal timing is critical since some depths are marginal. From close N of the Whitaker (RW, lit) bell buoy proceed across NE Middle, N of the NE Middle (R, unlit) buoy (3.7m, further S it is as shallow as 0.6m), then N of E Barrow Sand to Barrow No 7 (G) Lt buoy, sounding round the sand and keeping well off the wreck awash. Then follow the W side of Barrow Deep and cross the channel at right angles when opposite to Barrow No 8 (R) Lt buoy and then between this buoy and the SW Sunk Beacon and directly across Sunk Sand (2.3m at this point), continuing at right angles across Black Deep towards the NW Long Sand Beacon. 8 cables NW of the beacon in 20m soundings near the edge of the channel (keep clear of the remains of *Radio Caroline*, marked by an unlit R buoy, which are close to the edge of the channel and at this point) alter to a S course sounding carefully across a 3m corner of Long Sand and picking up the No 9 (G) and No 8 (R) lit entrance buoys to N Edinburgh Channel. Follow the edge of the channel (all buoys lit), then cross the outer entrance of the Princes and Queen's Channels to the E Margate (R) Lt buoy passing about halfway between Tongue Sand Tower and the Tongue LtV, and finally down to 1M off N Foreland taking care to pass E of Diffusers (Y) Lt buoy.

There are a number of more difficult variations on this route, which are relatively easier now that accurate Decca fixes are available to many yachtsmen. A few examples are: S of E Barrow Sand (depths have been fluctuating here), Fisherman's Gat and the S Edinburgh Channel, whilst yachts from the Harwich area via an inner route can choose between other swatchways across Sunk and Long Sands.

Tidal timing

From N to S the 25M passage is best started near the Whitaker bell buoy at 1hr before HW Sheerness meaning plenty of water in the critical early stages and obtaining some fair or slack tide down to N Edinburgh Channel, although there will be an inevitable push probably down Barrow Deep, then an outgoing tide through N Edinburgh and beyond reaching N Foreland at around 4 to 5hrs after HW Sheerness, with 5 to 6hrs of continuing S-going tide at N Foreland giving a good push if necessary to Calais.

From S to N in order to get the last of the ingoing stream at N Edinburgh and an outgoing stream beyond requires starting from N Foreland when the tide is just turning favourably

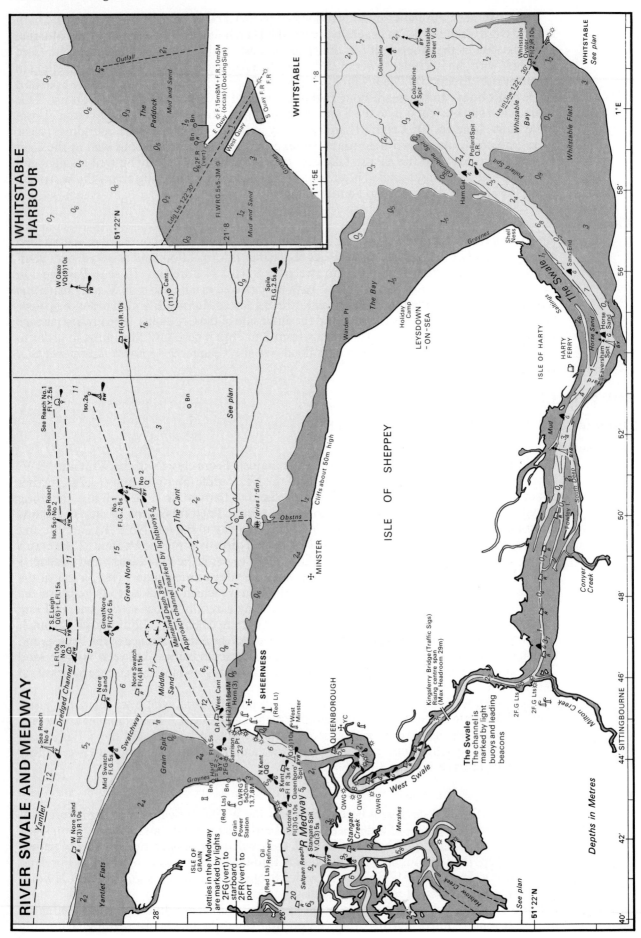

WHITSTABLE HARBOUR

WHITSTABLE

WHITSTABLE
See plan

1°E

RIVER SWALE AND MEDWAY

ISLE OF SHEPPEY

LEYSDOWN
-ON-SEA

ISLE OF HARTY

HARTY
FERRY

The Swale
The channel is
marked by light
buoys and leading
beacons

Kingsferry Bridge (Traffic Sigs)
Rising centre span
(Max Headroom 29m)

SITTINGBOURNE

Depths in Metres

Jetties in the Medway
are marked by lights
2FG (vert) to
starboard
2FR (vert) to
port

ISLE OF GRAIN

SHEERNESS

MINSTER

QUEENBOROUGH

northwards, 3hrs before HW Sheerness, which means reaching the Whitaker bell buoy with 2 to 3hrs of outgoing tide left, helpful for continuing on to Harwich, but difficult for doubling back into the Crouch or Blackwater.

The rivers and havens

With a dual-watch VHF radio it is advisable when in the Estuary to listen on Channel 12, Thames Navigation Service (Gravesend Radio) as well as 16 to be informed of ship movements, and on 14 (Woolwich Radio) and 16 in the river.

This is another area requiring a good tender since going aground is all too easy, and walking dryshod ashore direct from the boat is only feasible at the two marinas in the Medway, at Gravesend canal basin, and at St Katharine's Dock.

Our Dutch friends will enjoy re-visiting this area, the scene of Admiral De Ruyter's attack on the Medway and Gravesend approaches in 1667, and the cutting out of the British flagship *The Royal Charles* at Chatham taking her back to the Netherlands. In the case of the Dane who visits he can console himself that his forebears did an even better job of it by capturing most of the country from bases which included Canvey Island and Sheppey. We have learnt much since then and the French and Germans have never quite made it; witness the Martello towers of Essex and Suffolk and the anti-aircraft towers on the Estuary sandbanks!

The London River itself not only gives access to a berth in St Katharine's Dock to visit London itself, but also to the whole of England's inland waterways (see end of this chapter, p. 128).

Havengore Bridge: Although this guide primarily covers deep-water cruising, in the unlikely and inadvisable event that you should wish to run for shelter across Maplin Sands on a rising tide it is worth warning that the present rolling lift bridge (built in 1919) is not operated in winds of Force 5 and over, and a new bridge is scheduled for completion by autumn 1989. The offshore entrance to Havengore Creek has been marked by Wakering YC with R (unlit) port-hand buoys, and a RW buoy on the Broomway, the shallowest part of the channel, which is effectively only available to vessels of 1¼m draught close to HWS, and is very difficult if not impossible at neaps. However, Gravesend Radio on Ch 12 gives a navigational bulletin on the hour and 30 minutes past the hour, and when the tide gauge at Southend Pier shows 5m there is 1½m of water over the Broomway. Foulness Island is Ministry of Defence property with a firing range over the Maplins so it is always essential to contact the bridgekeeper by telephone in advance (a link call?). He is available 2hrs either side of HW at weekends on ☎ (03708) 2271 ext. 436, and at other times the Range Planning Officer must be contacted on the same number ext. 211/212.

Southend on Sea: In late 1986 when going to press a company owned jointly by Brent Walker and Southend County Borough Council was conducting a feasibility study into the possibility of building a marina. This would radically improve cruising facilities in the Estuary, but there are a large number of interests to consult and estuary effects to be examined before it can go ahead.

THE RIVERS MEDWAY AND SWALE

Port radio (VHF)

Whitstable ☎ (0227) 274086	*VHF Ch* 09, 12, 16	*Hrs listening* Mon–Fri 0800–1700, LT. Other times from 3hrs–HW–1hr
Kingsferry Bridge, West Swale	10	24
Medway Call *Medway Radio* ☎ (0795) 663025	09, 11, 16 22, 73, **74**	24
Hoo Yacht Harbour ☎ (0634) 250311	M	
Gillingham Marina ☎ (0634) 54386	M	0900-1700

Entry Signals None in the Swale. In the Medway a powerful light, Fl.7s, on Garrison Point when tankers over 5000 tons and cargo vessels over 10,000 tons are approaching the entrance. When shown to seaward this light signifies such a vessel is outward bound, when shown into river the vessel is inward bound. Contact Medway Radio for advice if necessary.

Sailing on the River Medway

Entrance and river passage

Distances   Medway, Garrison Point to Rochester Bridge 12M
Swale, Columbine Spit buoy to Queenborough Spit buoy 15M

The Medway enters the Thames over 25M from N Foreland, at a point where the Estuary is only 5M wide. All the harbours along the Estuary coast dry out, and in view of the extra distance and crowding at Ramsgate it is very unfortunate that the Margate Marina project has been postponed indefinitely; there is little doubt that, after St Katharine Yacht Haven this project could have been Britain's premier marina, with all the local inhabitants learning French, Dutch and German!

The earliest protected route to the Medway is via the shallow Swale channel behind Sheppey which starts 18M from N Foreland. The Swale is a cruising ground in its own right rather than a short cut and this route entails a total 33M passage to the Medway from N Foreland, whilst the channel almost dries out (0.4m least depth) in its middle reaches, requiring careful rising-tide pilotage.

Garrison Pt at the entrance to the Medway

Customs

Sheerness: Customs Office, Garrison Point, Sheerness, Kent ME12 1RS
Faversham: 7B Market Place, Faversham, Kent ME13 7AG
Chatham: Anchorage House, High Street, Chatham, Kent, ME4 4NW

In all cases above, ☎ (by VHF) Dover (0304) 202441; ashore dial 100 and ask for 'Freefone Customs Yachts'

In the E, the Swale is entered across a shallow 2 to 3m bay, which can be extremely rough in strong northeasterlies but which deepens to 7 to 8m in the protection of the E channel curving below Sheppey. In the central reaches it shelves to 0.4m winding across wide mudflats, and deepens again to 3m and more well before Kingsferry Bridge and to 6m near the W entrance, with some even deeper holes.

The E Swale entrance is best approached by finding Whitstable Street (N card) Lt buoy, and then (in daylight) heading W to close S of Columbine Spit buoy (G, unlit) and then following a course straight down the channel between Ham Gat (G, unlit) and Pollard Spit (R, lit) buoys to Sand End (G) Lt buoy over 2M from Pollard Spit. At night it is necessary to use the echo sounder between the Whitstable Street and Pollard Spit Lt buoys, which are 2M apart, to avoid straying too far S into the shallow Whitstable Bay and onto the drying Pollard Spit. From Sand End buoy the channel winds between wide drying banks with numerous drying creeks and is marked by unlit (mainly R) buoys at reasonably frequent intervals but with a large gap between Elmley Reach and Kingsferry Bridge. The bridge is lit, has 3.3m height closed and a rising centre span to 29m (MHWS); a VHF call in advance is advisable or hoist a flag or bucket to signify your intention to pass through. From then on the channel is buoyed/lit but look out for the wide banks and take care on the convex sides of bends, and at the W Swale entrance keep close to the dolphins on the E bank and keep to the E of Queenborough

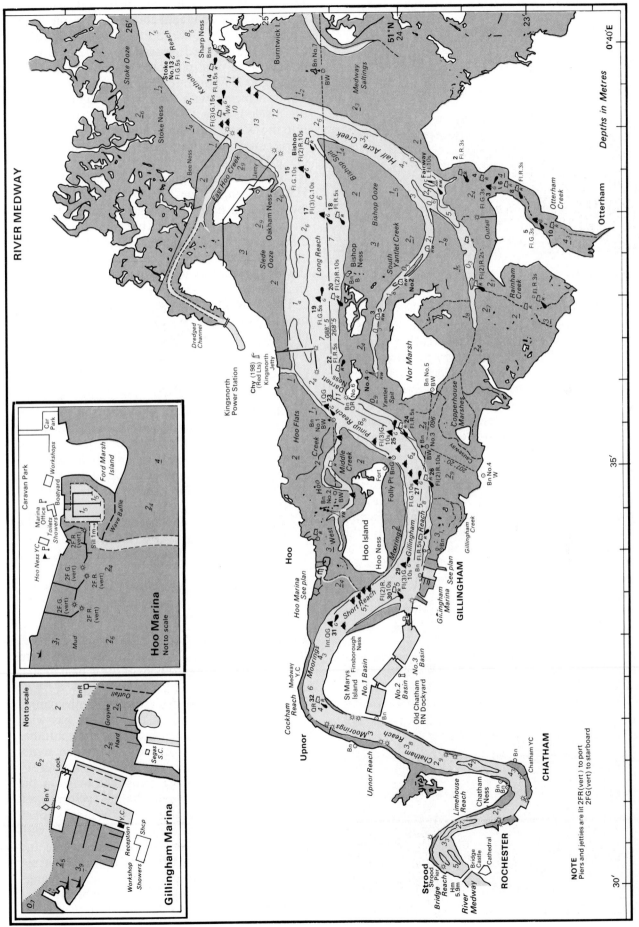

RIVER MEDWAY

Depths in Metres

0°40'E

Stoke Ooze
Stoke Ness
Bee Ness
Stoke No.13 Reach
Fl.G.5s
Sharp Ness
Bns
Kethole Reach
Fl(3)G.15s
14
Wk
10
13
12
Burntwick I.
Bn No.7
BW
Medway Saltings

East Hoo Creek
Jetty
Bishop
Fl(2)R.10s
Bishop Spit
Half Acre Creek
Fairway
L.Fl.10s
RW
51°N
24
23'

Oakham Ness
Fl.G.10s
15
Slede Ooze
17
Fl(3)G.10s
6
18
Fl.R.5s
Bishop Ooze
South Yantlet Creek
Fl.G.3s
3
Otterham Creek
Otterham
Fl.G.3s

Long Reach
20
Fl(2)R.10s
Bishop Ness
B
No.2
Fl(2)R.2s
Rainham Creek
Fl.R.3s

Dredged Channel
Kingsnorth Power Station
Chy (198) (Red Lts)
Kingsnorth Jetty
19
Fl.G.5s
7
088.5
268.5
22
Fl.R.5s
Darnett Ness
23
QG
Bn No.1
Hoo Flats
Hoo Creek
Middle Creek
Pinup Reach
No.4
Bn OR
Yantlet Spit
24
Fl.R.5s
No.4
Nor Marsh
Bn No.5
BW
Copperhouse Marshes
Causeway
207 20
Bn No.4
W

Hoo
West
Bn No.2
BW
Hoo Island
Hoo Ness
Fort
Folly Pt Bn G
25
Fl(3)G.10s
26
Fl(2)R.10s
27
Fl.G.10s
5
BW No.3 096
Gillingham Reach
Fl.R.5s
Gillingham Creek
GILLINGHAM
Gillingham Marina See plan

Hoo Marina
See plan
29
Fl(3)G.10s
Short Reach
30
Fl(2)R.10s
31
Int.QG
Moorings
St Marys Island
Finsborough Ness
No.1 Basin
No.2 Basin
No.3 Basin
Old Chatham RN Dockyard

Cockham Reach
Medway Y.C.
32
OR
Upnor
Upnor Reach
Chatham Reach
Moorings
Limehouse Reach
Chatham Ness
Chatham YC
Bn
CHATHAM

Strood
Strood Pier
Bridge Reach
Hm 5.9m
Bridge Castle
Cathedral
ROCHESTER
River Medway

35'
30'

Hoo Marina
Not to scale
Caravan Park
Car Park
Ford Marsh Island
Hoo Ness YC
Marina Office
Boatyard
Workshops
Toilets
Showers
Wave Baffle
Sill 1m
2F.R (vert)
2F.G (vert)
2F.R (vert)
Mud

Gillingham Marina
Not to scale
Bn R
Outfall
Groyne
Hard
Lock
Segas S.C.
Bn Y
Y.C.
Workshop
Showers
Reception
Shop

Kingsferry Bridge on the Swale

Spit (E card) Lt buoy. It is essential to note the direction of buoyage at each end of the Swale: S of Kingsferry Bridge in the middle reaches down to the E entrance near Harty Ferry the direction is westwards with the eastern flood, i.e. R buoys should be kept to the S side of the vessel and G to the N; and N of Kingsferry the direction is the reverse with the flood from the W entrance with G buoys on the mainland side of the channel and R on the Sheppey side.

The times of the Medway tides are not significantly different from Sheerness and the streams turn just after HW and LW but the Swale is a little more complicated (see above under *Tidal streams*). The streams to and from each end meet on the flood and separate on the ebb so that starting at or soon after LW from either end on an initially rising tide means pushing the tide when the tidal 'watershed' or junction is passed. On a normal W passage from N Foreland it is sometimes possible to continue to take the tide to reach Fowley Island

Queenborough

area by HW and the falling tide from then on, but of course the tide is foul continuing up the Medway so a stop at Queenborough is usually called for. On the outward passage from the Medway a strange local tidal phenomenon can be used: by starting at HW the outward passage from W to E can be achieved with a fair tide throughout, since the Swale is only 13M long and the stream runs into the W end until some time after HW whilst it starts running out from Fowley Island in the middle at HW, i.e. there is a period when the stream is running E throughout the Swale. Starting at HW gives ample time (up to 3½hrs) to reach Fowley Island on an E-going stream catching the outgoing stream at the E end, but remembering that throughout the passage the tidal level is falling so pilotage between the buoys must be careful and the engine must be used if necessary to cross the shallows as early as possible.

The River Medway proper is navigable for fixmasted vessels as far as Rochester Bridge, but pilotage is trickier than on the London River. There are three approach routes.

1. The buoyed and lit Medway approach channel and the lights, power station chimney and fort on the Isle of Grain, and Garrison Point are unmistakable marks, but keep to the channel's extreme starboard edge, or pass through the ship anchorages on each side taking care to avoid the wreck of the *Richard Montgomery* with its surrounding buoys on the W side of the channel.
2. This enters the channel from the N via the Nore Swatchway (two Lt buoys), sound across Sheerness Middle Sand (2.3 to 4.2m in its middle area).
3. This enters the channel from the E across the Cant, sound to keep generally N of the 2m contour and Cheyney Spit.

All three routes pass between Garrison Point (well lit), which can be passed close to, and Grain Hard (G) Lt buoy marking the edge of the wide E drying area.

The River Medway has a winding main channel which shallows from 25m to 3m, and has wide areas of drying mud and sand flats, and a number of islands and creeks on each side which sometimes provide routes for precarious (but interesting!) rising tide pilotage across the loops of the main channel. The main channel is very well buoyed and lit, although none of the creeks entering from N or S are effectively buoyed at all, with the major exception of the Swale of course and Hoo Creek which leads to some drying moorings at Hoo, N of Hoo Island, and which has an entrance beacon and 4 unlit buoys on the bends.

Whitstable Harbour

Mooring and facilities in the Swale

Sheltered anchorage is possible throughout the Swale if you carefully select the place appropriate for the conditions, but landing can be difficult; the E end is best since there is little shipping.

Whitstable Harbour, ☎ (0227) 274086, (dries out) is approached across a drying area and should only be attempted by newcomers after half-tide rising. However, it can be a tempting objective so some description is called for. An ancient stone causeway, supposedly part of a submerged town, projects 1M northwards from the coast E of the harbour. At night approach can be made from Whitstable Street (N card) Lt buoy aiming for about 1 cable W of the main entrance light until the G sector of the light is picked up about ½M off, then head direct for the light, and from close N of it follow the two leading lights (F.R) in. Take care not to confuse the town lights and harbour lights with the leading lights, and keep an eye on the Whitstable Oyster buoy (R, Fl(2)R) 3 cables NW from the entrance to be sure you are on the right heading. In daytime a course can be taken from Columbine Spit (G, unlit) to Whitstable Oyster buoy and into the harbour, picking up the entrance well in advance from its distinctive square silo-type tower with a pointed roof on the W Quay side. The harbour is used primarily by small fishing vessels. It is possible to lie up in the mud alongside the wall or a fishing vessel at the W end; there are ladders, Whitstable YC, several boatyards, and the quaint old town is close by. Call the HrMr on VHF in advance. There are also Entry Signals: 1 B ball on signal staff by day or F.R Lt below main harbour Wh Lt by night

means port closed (contact HrMr on VHF). By day docking Lt 3hrs either side of HW: Fl.3s means port open, and F means port closed (contact HrMr on VHF).

Faversham and Oare Creeks both dry out and are outside the scope of this guide, but are used by quite large vessels and have good boatyard repair facilities particularly at their junction.

At Harty Ferry there are moorings on the S side of the bend and a landing hard, and on the N side is a popular protected anchorage on the W side of Horse Sand with access to another hard and an inn on Sheppey.

Conyer Creek, also with good boatyard facilities, dries out and its winding, withy-marked (occasionally!) approach is for shallow-draught connoiseurs.

Milton Creek dries out, is industrial and to be avoided, but has moorings at the entrance.

At Queenborough the river is lined with moorings from close to the dolphins at Queenborough Point round into Loden Hope. There are two clubs, Sheppey YC and Queenborough YC which has visitors' moorings near the hard. There is a yard and chandlery.

Mooring and facilities in the River Medway

Moorings in the Swale tend to be free floating, but throughout the Medway many of the moorings are stem and stern, so please make sure which type you are picking up and be careful not to pick up a mooring line between buoys on your prop shaft!

Stangate Creek and its partner Sharfleet Creek are popular yacht anchorages, but keep away from the floating pipeline moorings at the S end of the former. Halstow Wharf is drying country of the old Thames barges' fame.

Hoo Yacht Harbour

Half Acre Creek leading to the drying Rainham and Otterham Creeks has some commercial shipping so is not too popular an anchorage.

Hoo Creek is marked by Medway Port Authority buoys but leads only to some drying moorings at Hoo and is a somewhat inconvenient route to the marina which is best approached from the W side of Hoo Island.

Hoo Yacht Harbour (☎ (0634) 250311) in August 1986 was putting the finishing touches to a basin with 120 floating pontoon berths, to the E of its already excellent drying marina with 125 berths. Most of the pontoons had been floated and many were occupied. The floating harbour is dredged to ½m below chart datum and its sill into the creek to 1m above chart datum, it is protected from the SW by a wave baffle, and is available for up to 1m draught 4hrs either side of HW and to 2m draught for 2hrs either side of HW, providing the creek on the W side of Hoo Island is followed. The latter was eventually to be marked with additional withies and port and starboard entrance markers (either buoys or posts), and it commences near Hoo Ness trot, a line of ship mooring buoys just W of Hoo Island. In August 1986 there were offlying mooring buoys at which to wait and a green buoy marking the entrance to the creek winding alongside Hoo Island and into the floating marina entrance marked by a line of port-hand (and a single starboard-hand) withies. For about 1½hrs either side of HW (well certainly before, maybe after is a little risky!) a route can be taken across the drying flats to the marina. There are boatyard facilities as well as the Hoo Ness YC and the Hundred of Hoo SC nearby. But please ring or call on VHF in advance if you wish to visit.

In summer 1986 Gillingham Marina (☎ (0634) 54386) was hot on the heels of its competitor across the river but with twice as many berths, 450 in total, about half to be in the nearly completed deep-water section, and the rest already in use in the drying section to the W. The deep-water lock, allowing access to the floating marina, was due for completion at the end of 1986 ready for the 1987 season, and there was a gap to the W of this allowing access to the projected floating marina but in its temporary drying state, with a small number of pontoon drying berths. Access will be 2hrs either side of HW, and possibly more. The marina entrance is N of two unmistakable large gasholders and has port and starboard topmarked posts leading into each section. There is every facility and a marina club with showers and toilets as well as the nearby Medway Cruising Club. Once again please ring them or call on VHF in advance if you wish to visit.

In the final reaches to Rochester Bridge, there are many rows of yacht moorings on each side, but

Gillingham Marina

all are private and if you wish to make a temporary stop it is essential to pick up a mooring near several other empty moorings to minimise potential inconvenience for the tenant, and check its availability ashore. Anchorage is not really advisable since there is a lot of traffic and few convenient spots. There are also several deep-water jetties, where only the most temporary stops can be made, so watch out carefully for the armed gamekeeper! There is a galaxy of well known yacht clubs and in some cases nearby repair facilities: Medway YC (near Upnor, the largest number of moorings and a very helpful club), Wilsonian SC, Upnor SC, Royal Engineers YC, Chatham YC and probably others who will never forgive me for overlooking them!

LEIGH AND CANVEY ISLAND

Port radio None

Entry signals None

Customs

Southend-on-Sea (London Port): Southend Airport, Southend-on-Sea, Essex SS2 6YB
☎ (by VHF) Southend (0702) 547141 ext 26
London 01-626 1515 ext 5861/5864
Thames Haven (London Port): SFP Administration Building, Shell Haven, Stanford-le-Hope, Essex SS17 9LT
☎ (by VHF) London 01-626 1515 ext 5861/5864
In both cases above, ashore dial 100 and ask for 'Freefone Customs Yachts'.

Entrance and facilities

30M from N Foreland on the N bank, Southend Flat and Leigh Sand provide drying out moorings for many yachts, and a long shallow (3m shelving to 0.6m) buoyed (unlit) channel, Ray Gut, provides a useful deep-water anchorage. The Gut provides a route into Smallgains, Hadleigh Ray and Benfleet Creeks round the E end of Canvey Island, all with drying moorings. A partially buoyed but completely drying branch, Leigh Creek leads N into Leigh-on-Sea. All of these offer very difficult pilotage for visitors unless they have local assistance or are prepared to explore on a rising tide or are not worried about drying out before reaching their objectives. There are several boatyards with a very wide range of facilities at Leigh-on-Sea, on Two Tree Island and on Canvey Island near Smallgains Creek, and of course yacht clubs – probably more per square mile than anywhere else in Europe! I hope I have listed them all – Benfleet YC, Essex YC, Leigh-on-Sea SC, Alexandra YC, Thorpe Bay YC, Thames Estuary YC, Island YC, Chapman Sands SC.

It is not possible to navigate round Canvey Island and the only other feasible anchorage (possibly borrowing a mooring) is in Holehaven Creek at the W end where there is a 0.8m bar, to be avoided near LW, and a 2.1m 'hole' N of the PLA pier, but it is crowded with moorings. The famous Lobster Smack Inn is just over the sea wall.

The times of the Canvey Island and Medway tides are not significantly different from Sheerness and the streams turn just after HW and LW.

THE LONDON RIVER

Port radio (VHF)

	VHF Ch	Hrs listening
Tilbury Docks, Entrance Lock	04	
Call *Tilbury Dock*		
Port of London: Thames Navigation Service ☎ (0474) 60311 & 57724		
Gravesend Radio (E of Crayford Ness)	12, 14, 16, 18, 20	24
Woolwich Radio (W of Crayford Ness, with substation for Thames Barrier)	14, 16, 22	24

Broadcasts
Gravesend Ch 12 H+00, H+30 seaward of Crayford Ness
Woolwich Ch 14 H+15, H+45 westward of Crayford Ness

Barking Creek Control Centre	14	4hrs-HW-4hrs (N Woolwich)
Thames Barrier Navigation Centre	14, 16, 22	24
Call *Woolwich Radio* ☎ 01-855 0315		
King George V Dock Lock	05	
Call *KG Control*		
India and Millwall Docks Lock	04	
Call *West India Dock Control* ☎ 01-987 7260		
St Katharine Yacht Haven	06, 12, 14	See note
Call *St Katharines* ☎ 01-488 2400		

Note Hours: 2hrs-HW-1½hrs; summer 0600–2030, winter 0800–1800 LT.

Entry Signals

None, but it is essential to contact Woolwich Radio, Thames Barrier Navigation Centre, to obtain permission to pass through the Thames Tidal Barrier; the Traffic Controller covers the whole area from Margaret Ness to Blackwall Point. It is also advisable to contact St Katharine's Dock (or any other dock lock you require to enter if you have a few spare hundred pounds for the opening fee!) in advance to request passage through the lock.

Customs

Gravesend (London Port): Customs House, Gravesend, Kent DA12 1BW

Tilbury Dock (London Port): Tilbury Dock, Tilbury, Essex RM18 7EJ

London Port: Customs House, Lower Thames Street, London EC3R 6EE

Royal Docks (London Port) including St Katharine Yacht Haven: No 8 Office, Pier Head, King George V Dock, London E16 2PL

In all cases above, ☎ (by VHF) London 01-626 1515 ext 5861/5864; ashore dial 100 and ask for 'Freefone Customs Yachts'

Gravesend Canal basin sea lock

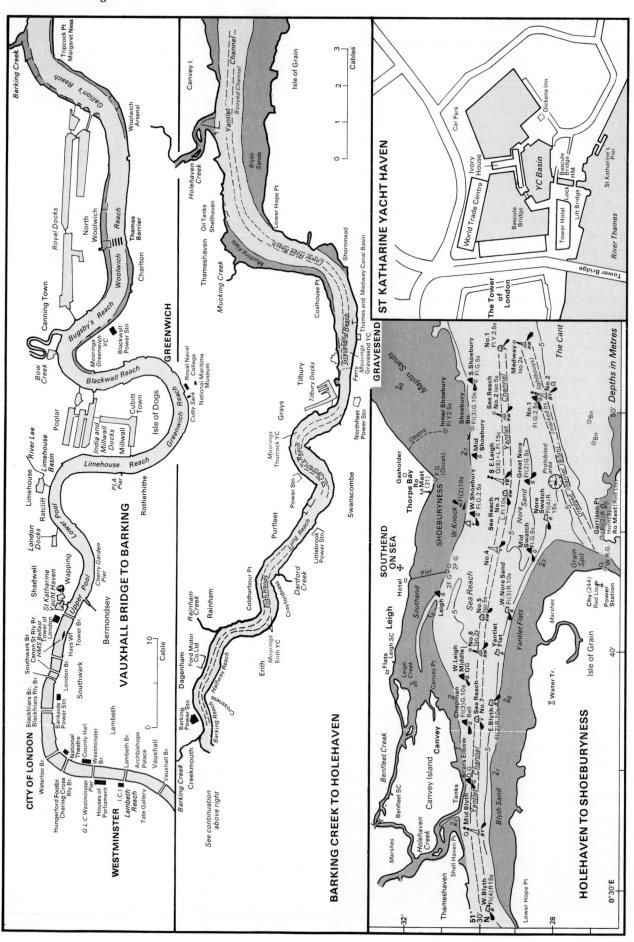

Entrance and river passage

Distance 38M from Southend Pier to Tower Bridge

Again, given a sound tidal knowledge, pilotage is not difficult for small vessels. From Lower Hope Reach to Tower Bridge soundings in the main channel shelve from 10m to about 6m but in the early reaches, fortunately buoyed (lit), there are some wide drying areas on the banks, e.g. Mucking Flats and Higham Creek, followed by a series of increasingly wide meanders which in general, but not always, are shallow on the convex bank and steep on the concave. The river is marked by light buoys to Gravesend and then by light beacons on the convex point of every bend as far as Margaret Ness, and with a multitude of wharfs and slipways (mainly lit) along each bank. The main navigational hazards are the unlit ship mooring buoys closer inshore and the Thames Tidal Barrier at Silvertown, Woolwich, which is well lit, has traffic lights and requires advance radio contact to pass through. Spans A on the south side, and H, J and K on the north are permanently closed to traffic (they dry out), spans B to the south and G to the north have a minimum depth of only 1.2m, which leaves 4 spans from south to north – C, D, E, F – which are available. Traffic is allowed in one direction only between spans; red crosses on each side pillar mean span closed to traffic from that direction and green arrows mean span open.

Tidal streams run extremely fast and must be worked by a sailing yacht of normal power. Like Routes 1, 2, and 3 above the 38M ingoing passage from Southend Pier is the easiest with 7½hrs of rising tide, since HW London Bridge is 1hr and 20 minutes later than off Southend, so the ingoing route starts at LW Southend/Sheerness and reaches St Katharine's Dock at HW London Bridge in time to lock through into the marina. The outgoing passage is tricky since with only 5hrs of tide from just after HW London Bridge it is usually advisable to anchor or more likely borrow a mooring at Gravesend, Thurrock or Holehaven, or alternatively motor at good speed to the entrance of the Medway and take the rising tide inwards.

Mooring and facilities

The Public Relations Department, Port of London Authority, Europe House, World Trade Centre, St Katharine's Way, London E1 9AA (☎ 01-481 8484) should be contacted for advisory literature, regulations etc. *London's Waterway Guide* by C. Cove-Smith is the most comprehensive handbook available. The Cruising Association (☎ 01-481 0881) publish *A Visiting Yachtsman's Guide*

to the London River.

Sailing is possible throughout the river, but strongly discouraged in Woolwich Reach and through the Thames Barrier where, if at all possible, engine should be used. Beware of gustiness and wind funneling as the banks become more built-up inland. There are very few possible anchorages to be recommended and all are exposed to wash from passing vessels and should be regarded as emergency only: E of Southend Pier end, Greenhithe near the causeway, E of Margaret Ness on the S bank, and below Greenwich Pier abreast the Naval College. If you pick up a mooring anywhere in the river then make sure you know whether it is a free floating or stem and stern mooring, since entangling your prop shaft on a mooring line between buoys is definitely to be avoided in the strong tidal streams.

The first moorings are on the S bank at Gravesend and the Gravesend Sailing Club, ☎ (0474) 533974, has two visitors' moorings just off the entrance to the Thames and Medway Canal and is extremely helpful. But take care, they allow 5 tons maximum and one may dry out. Some of the other moorings also dry out, so use the depth sounder. For a longer stay you may be able to arrange to lie at the E end of the canal basin (lockkeeper ☎ (0474) 352392/337488), lock available

Greenwich Yacht Club's Tideway Sailing Centre peeping over the S bank in Bugsby's Reach

near to HW, 1hr's notice required in daytime, but 24hrs' notice for the period 2100–0700.

Between Gravesend and the London docks there are three places with clubs and river moorings: Gray's Thurrock on a northerly bend opposite Broadness Point (Thurrock YC, free floating moorings); Erith on a southerly bend W of Crayford Ness (Erith YC has a clubhouse in an old

The entrance to St Katharine's Dock from the Thames

The Outer Basin at St Katharine's Dock

ferry ship, free floating moorings); and then 18M further on at Bugsby's Reach, Greenwich on the S bank (Greenwich YC, stem and stern moorings – take care, they dry out). At all of these there is landing access, and it is just possible to anchor with care, making sure not to dry out. By the N bank of the river near the Isle of Dogs and St Katharine's Dock there are other moorings which are stem and stern and may be possible for a night's stop providing permission can be obtained.

Finally the jewel of the river, St Katharine Yacht Haven (☎ 01-488 2400, VHF Ch 06, 12, 14) has a tidal lock available from 2hrs before to 1½hrs after HW, opened only in daytime 0600 to 2030 in summer and 0800 to 1800 in winter. There is a floating pontoon for waiting yachts at St Katharine's Pier just off the entrance. Please note that the marina does not use VHF Ch M. At Ivory House, as well as a marina club, showers and toilets, the Haven houses the headquarters of the Cruising Association (☎ 01-481 0881) Britain's premier cruising organisation with around 5000 members, a helpful full-time general secretary, social facilities, and a library of charts and constantly updated yacht cruising information unequalled anywhere in the world.

England's inland waterways can be accessed from the Thames in four ways, but in all cases you will need to unstep your mast to continue. It may be possible to do this at St Katharine's Dock and store it there. There are several entrance points for which you will need further information on licence fees, rules, etc. Inland from Tower Bridge, three authorities are in control:

Lower river to Erith: Port of London Authority, Thames Navigation Service, Royal Terrace, Pier Road, Gravesend, Kent DA12 2BG, ☎ (0474) 67684

Upper river to Teddington lock: Port of London Authority, Tower Pier Extension, Tower Hill, London EC3, ☎ 01-481 0720, Kew Pier ☎ 01-940 0634

Beyond Teddington lock: Thames Water Authority, Nugent House, Vastern Road, Reading Berks, ☎ (0734) 593333

The British inland canals and rivers are controlled by the British Waterways Board, Amenities Services Division, Clarendon Road, Watford, Herts, ☎ (0923) 31363. There are four points of access from the River Thames:

Bow Creek Tidal Locks, entrance to the River Lee, ☎ 01-987 5661

Limehouse Ship Lock (on request), entrance to Limehouse Cut/River Lee, ☎ 01-790 3444

Brentford Creek, leading to Thames Locks into the Grand Union Canal (tidal, 2hrs either side of HW), ☎ 01-560 1120 or 01-560 8942

Oxford Canal locks at Oxford (contact the British Waterways Board, see above)

9. Dover Strait

Charts   Admiralty *323, 1352, 1610, 1698, 1827, 1828, 1892, 2449*
Imray *C8, C12, C30*

Tidal atlases   Admiralty *Dover Strait*
North Sea – southern portion

Tidal streams   (based on HW Dover)

Position	Start times	
	DOVER	
	N or E	*S or W*
Folkestone Breakwater	−0155	+0320
S Foreland/Deal	−0150	+0415
Kellet Gut	−0130	+0440
Off Ramsgate Harbour	−0100/HW	+0500/+0600
N Foreland/Foreness	−0450 to −0120 weak/irregular	
	−0120 to +0045 NW-going	
	+0045 to +0440 irreg./turbulent	
	+0440 to −0450 SE-going	
	N or E	*S or W*
1M NW Cap Gris Nez	−0200	+0345
½M NW Calais ent.	−0215	+0355
Calais entrance	−0305	+0255

Tidal differences and ranges   (based on HW Dover)

Place	HW (time)	Springs/Neaps (range in metres)
Ramsgate	+0020	4.5/2.6
Dover	–	5.9/3.3
Calais	+0048	6.2/3.8

Reculver towers on the N coast of Kent before rounding N Foreland

Major lights

Name of light	Characteristics	Position	Structure
Offshore			
Dunkerque Lanby	Fl.3s10m25M	51°03′.1N 1°51′.9E	R cylinder on circular buoy, Racon
E Goodwin LtF	Fl.15s12m26M Horn 30s	51°13′.0N 1°36′.3E	R hull, Racon
Falls LtF	Fl(2)10s12m24M Dia Mo(N)60s	51°18′.1N 1°48′.5E	R hull, RC, Racon
F3 Lanby	Fl.10s12m22M Horn 10s	51°23′.8N 2°00′.6E	R cylinder on circular buoy, Racon
N Goodwin LtF	Fl(3)20s12m26M Horn(3)60s	51°20′.3N 1°34′.3E	R hull
S Goodwin LtF	Fl(2)30s12m25M Horn(2)60s	51°07′.9N 1°28′.6E	R hull
Sandettié LtV	Fl.5s12m25M	51°09′.4N 1°47′.2E	R hull, Racon
Varne LtF	Fl.R.20s12m22M Horn 30s	51°01′.2N 1°24′.0E	R hull, Racon
Onshore			
Calais (town, N side)	Fl(4)15s59m22M	50°57′.7N 1°51′.2E	Wh 8-sided tower, B top, RC
Calais (Jetée Est)	Fl(2)R.6s12m17M Reed(2)40s	50°58′.4N 1°50′.5E	Grey tower, R top
Calais (W side)	Dir.F.G.5m14M	50°58′.0N 1°50′.8E	Grey column, G top, 291°-intens-297°
Calais (Gare Maritime)	Dir.F.R.14m14M	50°58′.0N 1°51′.4E	Grey pylon, R top, 115.5°-intens-121.5°
Cap Gris Nez	Fl.5s72m29M Siren 60s	50°52′.2N 1°35′.0E	Wh tower, B top, RC, 005°-vis-232°
Dover (Admiralty Pier)	Fl.7.5s21m20M Dia 10s	51°06′.6N 1°19′.8E	Wh tower

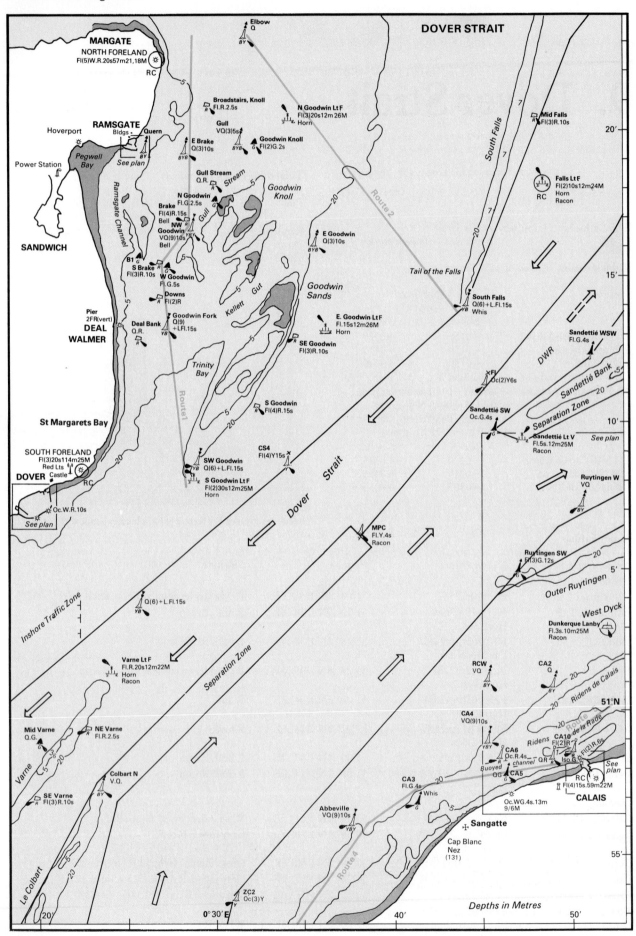

MARGATE
NORTH FORELAND
Fl(5)W.R.20s57m21,18M
RC

DOVER STRAIT

Elbow
Q
BY

Broadstairs, Knoll
Fl.R.2.5s

Hoverport
RAMSGATE
Bldgs
Quern
BY

Gull
VQ(3)5s

N Goodwin Lt F
Fl(3)20s12m 26M
Horn

Mid Falls
Fl(3)R.10s

20'

E Brake
Q(3)10s
BYB

Goodwin Knoll
Fl(2)G.2s
BYB

Pegwell
Bay

Power Station

Gull Stream
Q.R.

Gull Stream

Falls Lt F
Fl(2)10s12m24M
Horn
Racon
RC

N Goodwin
Fl.G.2.5s

Goodwin
Knoll

Brake
Fl(4)R.15s
Bell

E Goodwin
Q(3)10s
BYB

South Falls

7

7

20

20

SANDWICH

NW
Goodwin
VQ(9)10s
Bell

Ramsgate Channel

B1
S Brake
Fl(3)R.10s

W Goodwin
Fl.G.5s

Downs
Fl(2)R

Goodwin Sands

Tail of the Falls

South Falls
Q(6)+L.Fl.15s
Whis
YB

15'

Pier
2FR(vert)
DEAL
WALMER

Deal Bank
Q.R.

Goodwin Fork
Q(9)
+LFl.15s
YB

Kellett Gut

E. Goodwin Lt F
Fl.15s12m26M
Horn

Sandettié WSW
Fl.G.4s

DWR

Sandettié Bank

Separation Zone

20

5

SE Goodwin
Fl(3)R.10s

Trinity
Bay

XFl
Oc(2)Y6s

Sandettié SW
Oc.G.4s

Sandettié Lt V
Fl.5s.12m25M
Racon

10'

See plan

St Margarets Bay

S Goodwin
Fl(4)R.15s

Ruytingen W
VQ
BY

SOUTH FORELAND
Fl(3)20s114m25M
Red Lts
DOVER
Castle
RC

SW Goodwin
Q(6)+L.Fl.15s
YB

CS4
Fl(4)Y15s

Dover Strait

MPC
Fl.Y.4s
Racon

Ruytingen SW
Fl(3)G.12s

5'

Oc.W.R.10s

See plan

S Goodwin Lt F
Fl(2)30s12m25M
Horn

Outer Ruytingen

20

West Dyck

Dunkerque Lanby
Fl.3s.10m25M
Racon

Inshore Traffic Zone

Q(6)+L.Fl.15s
YB

RCW
VQ
BY

CA2
Q
BY

Ridens de Calais

20

Varne Lt F
Fl.R.20s12m22M
Horn
Racon

Separation Zone

CA4
VQ(9)10s

51°N

20

20

Route

Mid Varne
Q.G.

NE Varne
Fl.R.2.5s

Ridens de la Rade

CA10
Fl(2)R
YBY
CA6
Oc.R.4s
channel
QR
Iso.G
Fl(2)R.6s

Varne

Colbart N
V.Q.
BY

Buoyed
QG
CA5
G

See
plan

SE Varne
Fl(3)R.10s

CA3
Fl.G.4s
Whis

RC

Fl(4)15s.59m22M
CALAIS

Abbeville
VQ(9)10s
YBY

Oc.WG.4s.13m
9/6M

Le Colbart

✠ **Sangatte**

Cap Blanc
Nez
(131)

55'

20

ZC2
Oc(3)Y

0° 30' E

40'

Depths in Metres

50'

Dover (S breakwater, W head)	Oc.R.30s21m18M	51°06′.8N 1°19′.9E	Wh tower
Dover (S breakwater, knuckle)	Oc.WR.10s15m15/13M	51°07′.0N 1°20′.6E	Wh tower, 059°-R-239°-W-059°
Dungeness	Fl.10s40m27M F.RG.37m11M Horn(3)60s	50°54′.8N 0°58′.7E	B round tower, Wh bands, floodlit, RC 057°-R-073°-G-078°, 196°-R-216°
Folkestone (break-water head)	Fl(2)10s14m22M Dia(4)60s Bell(4)	51°04′.5N 1°11′.8E	Tower Close SW Occas
North Foreland	Fl(5)WR.20s57m21-16M	51°22′.5N 1°26′.8E	Wh 8-sided tower, RC, Radio Shore-W-150°-R-200°-W-011°
South Foreland	Fl(3)20s114m25M	51°08′.4N 1°22′.4E	Wh square castellated tower, RG 194°-vis-058°, 194°-obscd-222° within 7M

Radiobeacons

Name	Freq. (kHz)	Ident.	Range (miles)	Seq.	Position
Marine radiobeacons					
Falls Group	305.7				
Falls LtF		FS	50	1	51°18′.1N 1°48′.5E
W Hinder LtV		WH	20	3	51°23′.1N 2°26′.4E
Oostende		OE	30	4	51°14′.2N 2°55′.9E
Calais		CS	20	5	50°57′.7N 1°51′.3E
S Foreland Lt		SD	30	6	51°08′.4N 1°22′.4E
N Foreland Lt	301.1	NF	50	Cont	51°22′.5N 1°26′.8E
Dungeness Lt	310.3	DU	30	6	50°54′.8N 0°58′.7E
Cap Gris Nez Lt	310.3	GN	30	1,4	50°52′.1N 1°35′.1E
Aero beacon					
Calais/Dunkerque	275	MK	15	Cont	50°59′.9N 2°03′.3E

Coast radio stations

Station	Transmits* (kHz) (VHF Ch)	Receives* (kHz) (VHF Ch)‡	Freq VHF Ch†	Traffic lists (times)	Storm warnings (times)	Weather messages (times)	Navigational warnings (times)
N Foreland (GNF)	2182,**1848**, 2698,2733	2182‡, 2002(Ch11), 2016(Ch2), 2548(Ch7), 2569(Ch10)	1848	0103,0503 & odd H+03 (0903-2303)	0303,0903, 1503,2103	0803,2003	0033,0433,0833, 1233,1633,2033
	05,16,26,65 66	05,16,26,65 66	26	0103,0503 & odd H+03 (0903-2303)	0303,0903, 1503,2103	0803,2003	0033,0433,0833, 1233,1633,2033
Calais	**01,16,87**	01,87	87	None	On receipt	0633,1133	None

Notes

* 24hr service

† Frequency (kHz) and VHF Channel on which traffic lists, storm warnings, weather messages and navigational warnings are given.

‡ 24hr watch

All times are GMT. Calais weather broadcasts are given 1hr earlier when DST is in force.

Major fixed daylight marks

Kent coast (N to S)

N Foreland LtHo (with nearby radio mast, 42m)

Ramsgate: Granville Hotel tower (conspic) with large build-ings (conspic) 60m W, and SW, Ramsgate entrance piers, dome 1M W of entrance

Richborough Power Station chimney (135m) and 3 cooling towers (103m)

Deal: Sandown ruined castle, gasholder (conspic), Deal Pier, water tower (conspic) 1M SW of Deal

Dover Patrol Memorial (obelisk on cliffs)

S Foreland LtHo with a windmill close NE

3 radio masts (conspic, R Lts) ¾M NE of Dover Castle (con-spic)

Dover harbour walls

Radio mast (366m, R Lts) 3M WSW of Dover Castle

French coast (W to E)

Cap Gris Nez LtHo and radar surveillance station (tower plus
　　low building)
Tardinghen square belfry (42m)
Cap Blanc Nez with Dover Patrol Monument (obelisk 131m)
Sangatte square belfry (27m)
Calais: entrance jetties, main Lt tower 0.7M SE of entrance
　　(59m), Hotel de Ville tower 1.2M SSE of entrance (clock
　　tower 82m unlit)
Walde Lt structure 1M offshore (18m B pylon on hut)

North Foreland lighthouse

Approach routes and tidal timing

The Dover Strait proper between South Foreland and Cap Blanc Nez, just over 18M wide, is the favoured crossing place for large numbers of yachts from all northern European countries. Unencumbered with banks it is tide-scoured deep water with soundings generally over 20m and reaching as much as 70m or more.

The SW approaches to the Strait are similarly deep and trouble-free with the exception of two very pronounced undersea ridges, both rising from surrounding 25m soundings: The Varne, in the middle of the SW-going traffic lane, marked by the LtV at its N end and reaching as little as 3.2m sounding at its peak; and The Ridge/Le Colbart further S forming part of the TSZ and reaching only 2.7m at its N end which is marked at its deeper extremity by a N cardinal buoy. Yachts approaching the Strait from the Solent or from the Seine/Cherbourg Peninsula are well advised to take inshore routes outside the TSS boundaries, and cross the Strait from Dover or Cap Gris Nez thus avoiding closing either of these banks.

The NE approaches to the Strait in contrast are cluttered with 6 longitudinal banks fanning outwards into the North Sea which together with the TSS limit the approaches by yachts, and which should be avoided except in the lightest weather (Force 3 and less) and even then only cross their deeper extremities.

The Goodwin Sands complex off the Kent coast dries over a number of wide areas, and there are a number of N-trending unmarked gutways through; Kellett Gut has soundings up to 18m or more. The E side of the Goodwins is reasonably marked by LtFs and Lt buoys, as also is Gull Stream round the W side of the northern section, Goodwin Knoll, with its scattering of historic wrecks, but the W side of the southern Goodwins is dangerously unmarked and should be given a wide berth. The Brake/Cross Ledge bank to the W of Gull Stream is an outlier of the Goodwins, also dries out in places, and screens off the shallow (2.5m and less in places) inshore Ramsgate Channel.

E from the Goodwins rises that extraordinary, near vertically sided ridge, South Falls, ½M wide and 15M long with least depths of 7 to 8m rising from a surrounding 40m seabed, marked by a S cardinal buoy at its S end which is on a corner of a N turn of the outer boundary of the TSS.

The Sandettié Bank yet further SE is wider (up to 2M) shorter (13M) and a little less pronounced rising from over 30m, but has a series of 'hillocks' with depths as little as 4m. It forms a most unusual traffic separation zone between two traffic lanes both going in the same direction, approximately northeastwards, and separates the DWR to the N from the rest of the traffic to the S; it is marked at its SW end by the LtV and a Lt buoy.

The Outer Ruytingen and West Dyck banks further S are both outside the TSS and are wide, broken-up areas of shallows similar to the Sandettié though longer and merging into the Belgian offshore banks, with soundings sometimes as little as 1.7m rising from over 20m surrounding seabeds; both are marked by Lt buoys at their W ends whilst the latter also has the Dunkerque Lanby.

The Dunkerque Lanby marks the 1M gap between West Dyck and the sixth of the banks – the Ridens de Calais and its southern sister Ridens de la Rade which screens the Calais approach. Ridens de Calais has least soundings of 9 to 10m, but Ridens de la Rade reaches 1m near Calais and gradually approaches drying out as it joins the coast near the Walde Lt.

There is no problem at all with night sailing; this is probably the best-lit approach area in the world with no less than 15 onshore and offshore lights of over 20M range and more than adequate RDF coverage. In fog, which should be avoided if at all possible, it can be a nightmare because of the traffic from all directions, although the high degree of radar awareness of the traffic as well as shore-radar surveillance from each side is some compensation for this. In fog the only feasible shallow anchorages away from traffic, and which should be used as soon as fog danger arises at the beginning of a crossing, are: in the N the Brake bank, Ramsgate Channel, some of the W sides/bays of the Goodwins, and less effectively well inshore on the Downs which is steeply shelving; and in the S on or inside the protective banks screening Calais but keeping away from the entrance and its W approach with its ferry traffic. The whole of the TSS of course is prohibited for anchoring.

The banks and the associated disposition of the TSS and the lit buoys and LtF's limit the approaches of yachts from and into the North Sea to 5: 2 from the N, 1 from the English S coast in the W, 1 along the French coast from the SW, and 1 along the French coast from the E.

Tidal streams are extremely fast both in the narrow neck of the Dover Strait and on the Kent coast and Gull Stream, so every effort should be made to work them and to avoid foul streams. The streams run NE from approximately 1½hrs before, and SW from 4½hrs after HW Dover; approximately 6hrs NE and 6½hrs SW.

The problem area is between North Foreland and Foreness Point where for (approximately) 1hr each side of HW Dover the stream sweeps N and W round the point and for 1hr each side of LW, E and S. For the intervening 4hrs of rising tide the stream is weak and irregular and on the 4hrs of falling tide irregular and turbulent. Coming out of the Thames Estuary is easy, since catching the S-going tide at North Foreland 1hr before LW Dover follows 4hrs of falling tide out of the Estuary and is at the start of another 6hrs of S to SW-going tide round the Goodwins and down the Dover Strait; a 14 to 15hr trip from the Essex rivers or Medway/Leigh areas can achieve 10hrs of fair tide if timed properly. Going into the Estuary is much more difficult since achieving a fair tide at N Foreland, i.e. 1hr before HW Dover, requires pushing the tide from the Ramsgate area, and leaves only 2hrs of ingoing tide into the Estuary, although this is not so bad for yachts crossing the Estuary through the Edinburgh Channels and pushing straight on to the Essex rivers with the outgoing tide. To obtain a full 6hrs of ingoing Estuary tide requires motoring hard into it to round N Foreland 5hrs before HW Dover.

ROUTE 1   THE GULL STREAM & THE RAMSGATE CHANNEL INNER ROUTE

Distances   15M from off N Foreland to S Goodwin LtF via Gull Stream
16M via Ramsgate Channel

Commentary

In SW weather a sheltered northern route is close round North Foreland across Broadstairs Knolls and through the Gull Stream, or via Ramsgate and the Ramsgate Channel, and this is equally good in the same weather travelling in the opposite direction. Continuing S when the S Goodwin LtF is rounded into the teeth of the weather there is always the possibility of bolting into Dover for shelter. There are Lt buoys all the way.

Going N in very strong NE to E winds this inshore route can again provide shelter, this time from the Goodwins, and it is often advisable to take the Ramsgate Channel and make a short stay in Ramsgate to wait for better conditions to round N Foreland. However the 2 G buoys at either end of the Ramsgate Channel are unlit.

Tidal timing

The distance is short but is usually part of a longer passage, and travelling S full benefit of S-going tide is obtained by rounding N Foreland 5hrs after HW Dover. Travelling in the opposite direction N from S Goodwin/S Foreland leaving 1hr before HW Dover is ample time to take a fair tide to Ramsgate and possibly to push on across the Estuary but turning into the Estuary means pushing 5 or 6hrs of adverse tide. To get a fair tide into the Estuary for 6hrs means taking a break at Ramsgate and leaving 5hrs before HW Dover and plugging a strong foul tide for the short distance N to N Foreland taking a full flood into the Estuary, and at springs it is probably advisable to start even earlier at 4hrs after HW Dover rounding N Foreland against a weak stream and once round into a weakening stream against.

ROUTE 2   THE NORTH GOODWIN OUTER ROUTE

Distance   11M Elbow buoy to E Goodwin LtF

Commentary

In strong northeasterlies Route 1, close to N Foreland and across Broadstairs Knolls (with a seabed like its name), can be extremely uncomfortable and a route from just outside the Elbow (N card) buoy and E of the Goodwins is often better, particularly going S to Calais. E Goodwin LtF can be passed well to the E depending on the strength of tide down the Strait. There are 3 LtF/Vs in range and several Lt buoys.

Tidal timing

Like Route 1 the distance is short but is usually part of a longer passage, and travelling S full benefit of S-going tide is obtained by rounding N Foreland 5hrs after HW Dover. Travelling in the opposite direction N from E Goodwin LtF, having left Calais at HW, means arriving at the E Goodwin 4hrs after HW Dover with the tide turning against. There are two options at this stage: one, at neaps to use the engine and push on against 3hrs or so of adverse tide to N Foreland taking a full 6hrs of flood into the Estuary, although this is no good for yachts crossing the Estuary; or two, motoring up to Goodwin Knoll and then across tide into Ramsgate, taking a break there and leaving 5hrs before HW Dover and plugging a strong foul tide for the short distance N to N Foreland taking a full flood into the Estuary, and at springs it is probably advisable to start even earlier at 4hrs after HW Dover rounding N Foreland against a weak stream and once round into a weakening stream against. Similarly yachts crossing the Estuary can also wait at Ramsgate and leave at HW Dover −3hrs, push a weak stream for an hour and obtain a favourable slant across the Estuary.

ROUTE 3   DUNGENESS TO DOVER

Distance   17M from off Dungeness to Dover W entrance

Commentary

Approach and departure to Dover and S Foreland from the W and Dungeness is across deep water with no problems, except the need to choose weather and tide. It is outside three Lt buoys across Hythe Bay.

Tidal timing

Eastwards, starting 3½hrs before HW Dover can give up to 9hrs of slack to fair tide, time to reach Dover easily and also Ramsgate and take some fair tide out across the Estuary, but not into it. Westwards, returning down Channel the streams are much less favourable; leaving Dover 4½hrs after HW obtains a little more than 5hrs slack to W-going tide and starting from Ramsgate an hour earlier (i.e. 5½hrs before HW Dover) gives about the same period of fair tide.

ROUTE 4 BOULOGNE TO CALAIS

Distance 21M

Commentary

From Boulogne entrance the main problem in rough weather is keeping inshore of the Bassure de Basse ridge (least depths 4.6m) marked by Lt buoys (S card at S end and N card at N end), and a course about 1½M offshore takes a vessel inside the N cardinal, then clear Cap Gris Nez by about 1M, pass the W cardinal Lt buoy W of Cap Blanc Nez and then outside CA3 and CA5 (G) Lt buoys and follow the remaining Calais channel-marking buoys (rising tide is E-going) to the entrance. As it will be near LW any decision to carry on beyond Calais means going N of CA6 (R) Lt buoy and giving a wide berth to the Ridens bank.

Tidal timing

Northwards, a start from Boulogne 1½hrs before HW Dover gives 7hrs of slack to fair tide but means waiting in the Avant Port for the Calais lock-opening. Returning southwards, leaving Calais 4½hrs after HW Dover also means waiting in the Avant Port after locking through but gives 6hrs of slack to fair tide.

ROUTE 5 DUNKERQUE TO CALAIS

Distance 21M Dunkerque Est to Calais

Commentary

Only one SE approach, an inshore one, is recommended. Forget the initially wide channel between the Outer Ruytingen and West Dyck banks running eastwards 5M off the French coast; it is unmarked and in 12M disintegrates into a series of channels between a number of banks, the Inner Ruytingen, Oost Dyck, Le Dyck, and the Binnen and Buiten Ratel, with only the very occasional buoy making navigation to Oostende or beyond extremely tricky, although with Decca equipment and accurate navigation even this is now more feasible.

The easiest and best route is inshore S of the West Dyck bank, and from Calais making eastwards and leaving near HW to use the E-going tide it is usually possible to cross the Ridens de la Rade, whilst the Ridens de Calais can be ignored except in the very worst weather. In the opposite direction similarly the Ridens de la Rade will be crossed near HW. The lit channel from Dunkerque Lanby into Dunkerque Ouest (see Chapter 10) should be avoided since it is well used by commercial traffic.

Tidal timing

Westwards, leaving Dunkerque 5hrs after HW Dover gives about 5hrs of W-going tide and a crossing of the Ridens probably on the top half of a rising tide. Travelling E and constrained to leave the Calais lock at only an hour or less before HW Dover also gives an easy crossing of the Ridens and around 6hrs of fair tide. See Chapter 10 for further details of this route.

The harbours

Ramsgate, Dover and Calais are busy cross-channel ports. Permission to enter and leave all three harbours should always be sought on VHF, and with a dual-watch VHF radio it is advisable to listen on Ch 14 as well as 16 when in the vicinity of Ramsgate to obtain information on ship movements and particularly ferries out of the port. Use Ch 74 near Dover, and Ch 12 near Calais for port operations; Ch 16 warns of broadcasts on Ch 11 by Dover Strait, Channel Navigation Information Service.

All three have inner docks entered through tidal locks, have pontoon mooring and are crowded, and Dover has few facilities. If you crave for complete peace and can take the ground you can lean against the wall or sink in the soft mud of Sandwich Quay just over 4M from the Ramsgate harbour entrance, crossing the mudflats to the SW towards Richborough

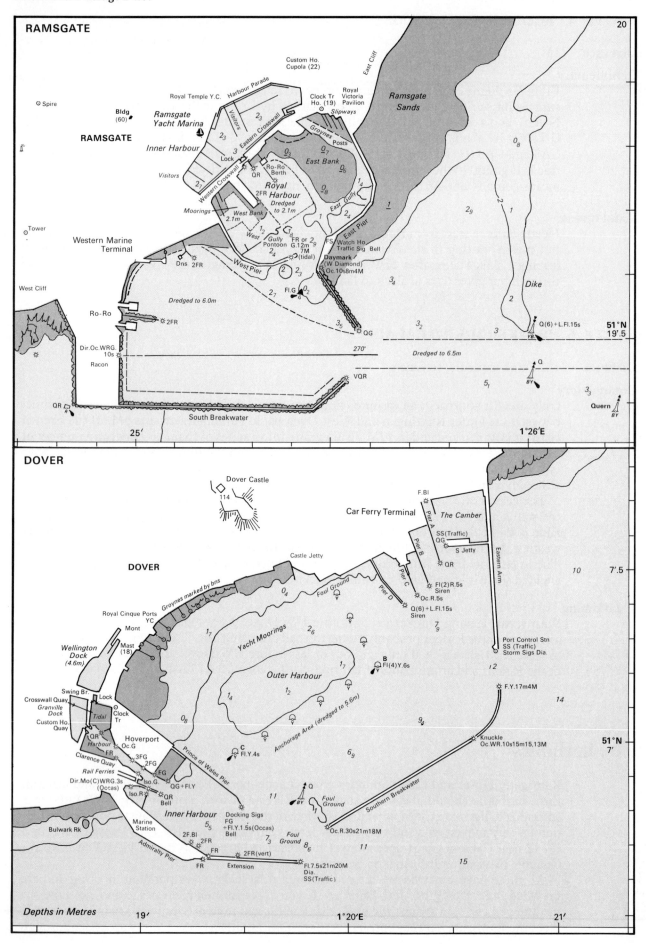

RAMSGATE

20

Custom Ho.
Cupola (22)

East Cliff

*Ramsgate
Sands*

Spire

Bldg
(60)

Royal Temple Y.C.   Harbour Parade

Clock Tr   Royal
Ho. (19)   Victoria
Pavilion

RAMSGATE

*Ramsgate
Yacht Marina*

Visitors

Slipways

2₃

2₃

0₈

Inner Harbour

2₃

Eastern Crosswall

3

Groynes
Posts

East Bank

Lock

0₃

0₇

0₆

Visitors

2₇

QR

Ro-Ro
Berth

1₄

Western Crosswall

2FR

*Royal
Harbour*
Dredged
to 2.1m

0₈

East Gully

1

2₉

2₇

2₁

Tower

Moorings

West Bank
2.1m

1₂

1

East Pier

*Western Marine
Terminal*

West

Gully
Pontoon

2₄

1₈

FR or
G.12m
7M
(tidal)

2₉

FS   Watch Ho
Traffic Sig Bell

Dike

Dns 2FR

West Pier

2₄

2

2₃

Daymark
(W Diamond)
Oc.10s8m4M

3₄

2

West Cliff

Dredged to 6.0m

2₇

Fl.G
G

0₂

3₅

3₂

3

Q(6)+L.Fl.15s
YB

51°N
19'.5

Ro-Ro

2FR

QG

Dir.Oc.WRG.
10s
Racon

270°

Dredged to 6.5m

Q
BY

5₁

3₃

QR
R

VQR

South Breakwater

Quern
BY

25'

1°26'E

DOVER

Dover Castle

114

F.Bl

Car Ferry Terminal

The Camber

Pier A

DOVER

Castle Jetty

SS (Traffic)
QG
S Jetty

10

7'.5

Groynes marked by bns

0₄

Pier B

QR

Foul Ground

Pier C

Fl(2)R.5s
Siren

Royal Cinque Ports
YC

1₇

Yacht Moorings

2₆

Pier D

Oc.R.5s

7₉

Eastern Arm

Mont

Q(6)+L.Fl.15s
Siren

*Wellington
Dock
(4.6m)*

Mast
(18)

1₇

Port Control Stn
SS (Traffic)
Storm Sigs Dia.

Swing Br.
Lock

1₄

Outer Harbour

B
Y

1₂

Fl(4)Y.6s

1₂

F.Y.17m4M

14

Crosswall Quay
*Granville
Dock*
Custom Ho.
Quay

Clock
Tr

Y

9₁

Knuckle
Oc.WR.10s15m15,13M

51°N
7'

Tidal

0₈

QR

Harbour
Oc.G

Hoverport

C
Y

Fl.Y.4s

Anchorage Area (dredged to 5.6m)

6₉

FR

3FG
2FG

Prince of Wales Pier

Y

Southern Breakwater

11

Clarence Quay
Rail Ferries

FG
Iso.G.
QG+Fl.Y

Q
BY

*Foul
Ground*

Dir.Mo(C)WRG.3s
(Occas)

Iso.R

QR
Bell

1₇

Bulwark Rk

Inner Harbour

Marine
Station

5₅

Docking Sigs
FG
+Fl.Y.1.5s(Occas)
Bell

7₃

*Foul
Ground*

8₆

11

Oc.R.30s21m18M

2F.Bl 2FR

FR

FR

2FR(vert)

Fl.7.5s21m20M
Dia.
SS (Traffic)

15

Admiralty Pier   Extension

Depths in Metres   19'

1°20'E   21'

Power Station at about HW Ramsgate −0130. From the entrance first pass between three pairs of port (R) and starboard (G) unlit channel buoys, then between a series of port (R) and starboard (G) unlit marker posts starting just N of a Fl.R.10s beacon, and snake round Shell Ness Point into the River Stour where the middle of the winding channel should be followed.

RAMSGATE

Port radio (VHF)

Ramsgate Harbour	VHF Ch	Hrs listening
☎ (0843) 592277/8	14, 16	24

Entry Signals

Always contact by VHF first if possible.

W Pier LtHo, G Lt – less than 3m between piers; R Lt – more than 3m between piers.

E Pier Signal Stn. Day 1 B flag – entry to harbour only, no exit; 2 B Balls – exit from harbour only, no entry. Night Fl Lt – entry only, no exit; Fl.R Lt – exit only, no entry.

Yacht Marina entrance, daylight signals as above, at lock signal station. Night 2 F.G vert Lts – entry only, no exit; 1 F.R over 2 F.G vert – exit only, no exit.

Customs

Ramsgate: Custom House, Harbour Parade, Ramsgate, Kent CT11 8LR

☎ (by VHF) Dover (0304) 202441

☎ (ashore) dial 100 and ask for 'Freefone Customs Yachts'

Entrance

With the completion of the new Sally Line Harbour at time of going to press a preliminary *Notice to Mariners* announced a number of amendments to Lt buoys and Lt beacons due to be completed by end November 1986, promising prompt issue of a revised version of Admiralty chart *1827*. These are shown in this book's Ramsgate approach plan; and include 6 additional dredged-channel-marking buoys, Lt beacons near the end of each of the outer breakwaters, a G 'Harbour' Lt buoy marking the end of the remains of the shoal off the old East (inner) Pier, a Lt beacon at the root of the new South Breakwater, and 2 R can Lt buoys outside of and close to the SW corner of the new harbour.

The marina in the inner harbour at Ramsgate

The dredged (6.5m) buoyed (lit) approach channel for the Sally Line Dunkerque ferries extends 1.1M due E from the outer entrance to the two outer buoys. Yachts should avoid the channel wherever possible and keep to the extreme outer edges when unavoidable. An approach from the N should enter the channel just E of the Dike (S card) Lt buoy to clear the shoal of that name. An approach from the Ramsgate Road, where there is an anchorage and a route from B2 (G, unlit) buoy at the end of Ramsgate Channel, should enter just round the South (outer) Breakwater. An approach from E should keep clear to N or S of the channel edges until reaching the Dike buoy (N side) or the N cardinal Lt buoy just opposite on the S side of the channel. However in closing this latter channel buoy, beware of the Quern bank (1.8m at its N end but 0.3m at its S end) which is marked by yet another, but this time unlit, N cardinal buoy 1½ cables ESE of the channel buoy.

There is a double entrance to this harbour, and all traffic including the ferries must use the outer entrance between the two new breakwater Lt beacons, so clearly if you do not wish to enter with a ferry a VHF call is essential. Yachts and fishing traffic then make a wide sweep to starboard rounding the new 'Harbour' (G) Lt buoy entering the inner piers parallel with the E Pier landing stage but closer to W Pier to avoid the remnants of a shoal (0.1m) extending about 80m SW of the E Pier entrance light.

Looking out across the entrance between the new breakwaters at Ramsgate

Mooring and facilities

In Royal Harbour mooring for a temporary period is to a pontoon on the W side with *portakabin* toilets on the wall nearby. For a longer stay the lock is available to enter the inner harbour marina from approximately 2hrs before to 1hr after HW. There are currently plans afoot to relocate the entrance to the marina to the clock tower area. There are all facilities round the inner harbour; showers, toilets, lifting, chandleries and repair shops, but no club. The Royal Temple Yacht Club, however, on the cliffs above the harbour is very helpful to visitors.

The town is a very popular holiday resort with all the associated amenities, but if you are of a quieter frame of mind there are some good views and walks on the cliffs.

DOVER

Port radio (VHF)

Dover Strait: Channel Navigation Information Service (CNIS).

Shore stations	VHF Ch	Hrs listening
Dover Coastguard	11, 16, **69**, 80	24
Gris Nez Traffic	11, 16, **69**, 79	24

Broadcasts
Dover Coastguard, Ch 11 H+40; Gris Nez Traffic, Ch 11 H+10. Additionally at H+55, H+25 respectively when visibility is less than 2M.

Dover Harbour
Call *Dover Port Control*   12, 16, **74**   24
☎ (0304) 206560/206585

Entry Signals

Contact Dover Port Control by VHF if possible to request permission to enter or leave.

Vessels without VHF should flash the following local signals in morse code (Aldis lamp, or other suitable) to the Admiralty Pier Signal Station (W entrance) or Port Control (E entrance), S V – I wish to enter port; S W – I wish to leave port. Reply from the entrance stations will be 'OK' or 'WAIT'.

Traffic signals, applying to yachts as well as to commercial shipping, shown from W entrance (Admiralty Pier Extension head), and from E entrance (Port Control on E Arm head), directed seaward or towards the harbour and obscured in opposite direction: 3 F.R Lts (vert) – vessels shall not proceed; G/W/G Lts (vert) – vessel may proceed only after specific orders to do so, one way traffic only.

Wellington Dock entrance, traffic signals as above plus Fl.Y Lt warning of opening of the swing bridge.

Customs

Dover (west): Watch House, South Pier, Dover CT17 9DL
☎ (by VHF) Dover (0304) 202441
☎ (ashore) dial 100 and ask for 'Freefone Customs Yachts'

Entrance

For advance information write to either The Managing Director, or The Public Relations Manager, Dover Harbour Board, Harbour House, Dover, Kent CT17 9BU; or telephone the Duty Marine Officer, ☎ (0304) 206560 ext 4522.

Passing vessels are warned to keep at least 1M off the entrances. Your intentions will be obvious, therefore, if you approach one of the entrances close to.

The W entrance is used for pleasure vessels but beware of the hovercraft negotiating this entrance. The E entrance is used for the ferries, and for pleasure craft only occasionally, in emergencies etc. In very strong southwesterlies the W entrance can be difficult to enter and Port Control should be contacted for an alternative well in advance since the tide runs very fast along the S Breakwater.

It is always advisable to contact Port Control on VHF from 1M off on approach, but if VHF is not available a pilot launch usually runs out as a vessel approaches the entrance to give advice at close quarters, so do not drop your contraband overboard and turn and run, they are coming to help!

Having passed through the W entrance an anchorage can be obtained in the Outer Harbour N of the edge of the dredged anchorage area marked by a line of Y spherical, lit and unlit buoys but well offshore from the Royal Cinque Ports YC. Use the depth sounder to pick the best depth, it is shallow in places, keep away from private mooring trots, and do not leave your vessel unattended, particularly in bad weather. There can be an uncomfortable swell.

Wellington Dock, Dover   *Photo* Port of Dover

Entrance to Wellington Dock through the lock and swing bridge is available from about 1hr before to HW, and from about 2½hrs before HW a temporary berth alongside the pontoon next to the dock office at Crosswall Quay can be obtained. It is essential to contact the Dock in advance however.

Tides in and around the harbour are complicated and shown on the Admiralty harbour chart:

Outer Harbour: The streams tend to rotate clockwise round the Outer Harbour, except for a long period (3½hrs) of slack water from about 3½hrs after to 5½hrs before HW, so do not be surprised to lie at anchor facing SW throughout the day.

Western entrance: Off the W entrance the stream starts running E about 2hrs before HW and reverses W 4½hrs after HW, but comes out of the entrance for about 10hrs of the tidal cycle (entering the entrance between 3½ to 1½hrs before HW).

Eastern entrance: Off the E entrance the stream sets across from N to S for 10hrs of the tidal cycle, being slack or N-going only from 2hrs before to ½hr after HW. It sets into the E entrance from 5hrs after to 2½hrs before HW, and, somewhat intermittently, sets mainly out of the entrance for much of the rest of the cycle.

Mooring and facilities

Contrary to popular myth yachtsmen are not unwelcome in Dover and the Port Control are very helpful. The local authority have merely thought fit not to develop yachting facilities beyond a certain limited stage.

There are pontoons in Wellington Dock and visitors are usually directed to those adjacent to Union Street. There are all the usual boatyard facilities, including a chandlery nearby, but no showers or toilets. The Royal Cinque Ports YC, which has the only social facilities, is a short walk along the front and is extremely helpful. Landing by dinghy on the steep shingle beach can be difficult in swell.

If you have time on your hands and are fit and healthy a climb up to Dover's Norman Castle is worthwhile, although a lot of exercise is involved in visiting the dispersed shopping facilities nearer to the W end of the harbour. Also worth seeing is a well preserved Roman town house with wall paintings and an underfloor heating system, and on the cliffs a Roman Pharos, lighthouse, 40 feet high.

CALAIS

Port radio (VHF)

Dover Strait: Channel Navigation Information Service (CNIS).

Shore Stations	VHF Ch	Hrs listening
Dover Coastguard	11, 16, 69, 80	24
Gris Nez Traffic	11, 16, 69, 79	24

Broadcasts
Dover Coastguard, Ch 11 H+40; Gris Nez Traffic, Ch 11 H+10. Additionally at H+55, H+25 respectively when visibility is less than 2M.

Calais Ecluse Carnot (lock)	12, 16	
Calais	12, 16	24

☎ (21) 96 31 20/96 69 59

Entry Signals

Contact with Calais harbour by VHF in English is advised.

Gare Maritime signal station shows full international port traffic signals (see Appendix XIII), not to be complied with by yachts. Additional signals are: 3 W Lts – priority for small vessels; for ferries (all movement prohibited), ball or R Lt – ferry leaving, cone point down or G Lt – ferry entering; for tankers (all movement prohibited), 2 R Lts – tanker entering, R/G Lts – tanker leaving.

Lock/road bridge into Bassin de l'Ouest, Y Lt – standby for lock/bridge opening; G or R Lt – pass through or passage prohibited respectively.

Customs

Rue Lamy ☎ 26 34 75 40

Calais. The Hotel de Ville clock tower

Calais. The start and finish of many a passage

Photo Chambre de Commerce et d'Industrie de Calais

Entrance

In moderate to good weather the Ridens de la Rade shoal can be crossed between 3hrs either side of HW close to CA10 (R) Lt buoy opposite the harbour entrance. At other times particularly in onshore winds when rolling seas can build up, and if in any doubt at all, follow the entrance channel S of the R buoys out to CA4 (W card) Lt buoy, and if aiming to go W this buoy can be cut inside just beyond CA6 (R) Lt buoy keeping in 10m and deeper soundings throughout.

The new mole round the N side of the harbour is now complete with its entrance light, and the inner one is demolished, but it is still advisable to keep to the W side of the harbour to avoid the ferries, and head for the Avant Port and the high LtHo taking care to give a wide berth to the knuckle of wall at the root of the Jetée Ouest. There are a few uncomfortable mooring buoys outside the Bassin de l'Ouest, whilst Bassin du Petit Paradis (a misnomer if ever there was one when you see the debris strewn across its gleaming mud!) dries out, as do parts near the walls. So try to time your entry to be about an hour before HW to go straight through the swing bridge and lock

into the Bassin de l'Ouest. In May 1986 the bridge was permanently open and being reconstructed, and there were varied lengths of lock-opening time from 2hrs before to 1hr after HW depending on day of week and holiday periods.

Mooring and facilities

The Chambre de Commerce et d'Industrie de Calais runs Port de Plaisance de Calais, Bassin de l'Ouest ☎ 26 34 55 23, a pontoon marina on the N side of the basin, with every yachting and boat repair facility nearby including a scrubbing grid and a comfortable bar, but without a sailmaker I discovered, to the extreme discomfort of my right fingers and thumb – Dunkerque is apparently the nearest. It is the base for the YC du Nord de la France, and is a very popular rallying rendezvous for S and E coast English yachtsmen who are welcomed with open arms by the Calais municipality.

Calais has an interesting history but there is little left of it. If you are of macabre frame of mind you can visit the War Museum in the park opposite the Hotel de Ville; it is housed in what was the German wartime headquarters bunker.

Calais. Port de Plaisance

Calais. Entrance lock to the canal system

It is possible to store one's mast at Port de Plaisance and continue on into the French waterways system via the Ecluse Carnot into the Bassin Carnot. The lock gates are open from 3hrs before to 1hr after HW, but it is advisable to contact the HrMr or the lock-keeper to arrange passage and find out the rules.

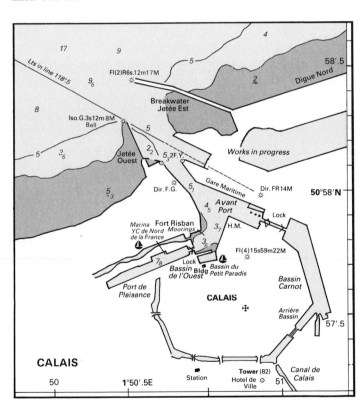

10. The French & Belgian coast from Gravelines to Zeebrugge

Charts Admiralty *97, 125, 323, 325, 1350, 1872, 2449*
Imray *C30*
Dutch Hydrographic *1801*

Tidal atlases Admiralty *North Sea – southern portion*

Tidal streams (based on HW Dover)

Position	Start times DOVER	
	East	*West*
½M NW Calais ent.	-0215	+0345
N of Gravelines ent.	-0140	+0350
Dunkerque Port Est ent.	-0115	+0500
Dunkerque to Zeebrugge offshore	-0120	+0440
Off Oostende	-0045	+0445
Off Blankenberge	-0135	+0445
2¾M NE Zeebrugge Mole	-0140	+0440

Tidal differences and ranges (based on HW Dover)

Place	*HW (time)*	*Springs/Neaps (range in metres)*
Gravelines	+0048	5.4/3.5
Dunkerque	+0058	5.2/3.3
Oostende	+0120	4.8/3.1
Zeebrugge	+0137	4.4/2.7

Major lights

Name of light	*Characteristics*	*Position*	*Structure*
Offshore			
Dunkerque Lanby	Fl.3s10m25M	51°03′.1N 1°51′.9E	R cylinder on circular buoy, Racon
West Hinder LtV	Fl(4)30s14m17M Horn Mo(U)30s	51°23′.0N 2°26′.3E	R hull, 2 masts, R can on mainmast, RC
Onshore			
Gravelines	Oc(4)WG.12s29m27/22M	51°00′.3N 2°06′.6E	Wh round tower, B diagonal stripes 186°-G intens-193°-W intens-200°
Dunkerque, Port Ouest Ldg Lts 120°, Front by day Rear by day	Dir.F.G.16m19M Dir.F.W Dir.F.G.30m22M Dir.F.W	51°01′.7N 2°12′.0E	Wh column, G top, 119°-intens-121° Wh column, G top, 119°-intens-121°
Dunkerque	Fl(2)10s59m28M	51°03′.0N 2°21′.9E	Wh tower, B top
Dunkerque, Jetée Ouest	Oc(2+1)WG.12s35m17/13M Dia(2+1)60s	51°03′.7N 2°21′.2E	Wh tower, brown top 252°-G-310°-W-252°
Dunkerque, Ecluse Charles de Gaulle, auxiliary	Fl.5s9m22M	51°03′.4N 2°20′.9E	Wh tower, G top, 200°-vis-080°
Nieuwpoort, E root	Fl(2)R.14s26m21M	51°09′.3N 2°43′.8E	R round concrete tower, Wh bands
Oostende	Fl(3)10s63m27M	51°14′.2N 2°55′.9E	Grey concrete tower, RC 069.5°-obscd-071°
Blankenberge	Oc(2)8s30m20M	51°18′.8N 3°06′.9E	Wh concrete tower, B top 065°-vis-245°
Zeebrugge, Mole Hd	Oc.WR.15s22mW20/18M Horn(3+1)90s	51°20′.9N 3°12′.3E	Grey round tower, RC 068°-W-145°-R-212°-W-296°

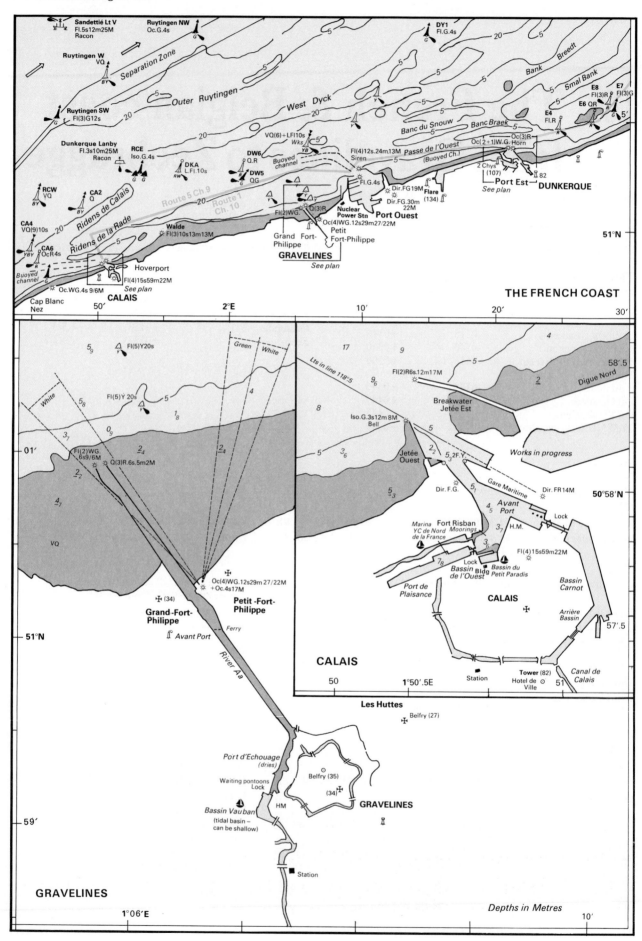

Sandettié Lt V
Fl.5s12m25M
Racon

Ruytingen NW
Oc.G.4s

DY1
Fl.G.4s

Ruytingen W
VQ

Separation Zone

Bank

Breedt

Smal Bank

Outer Ruytingen

West Dyck

Ruytingen SW
Fl(3)G12s

E8
Fl(3)R

E7
Fl(3)G

E6 QR

E4
Fl.R

Banc du Snouw

Banc Braek

Oc(3)R

Dunkerque Lanby
Fl.3s10m25M
Racon

RCE
Iso.G.4s

VQ(6)+LFl10s
Wks

DW6
Q.R

DKA
L.Fl.10s

DW5
QG

Fl(4)12s.24m13M
Siren

Passe de l'Ouest
(Buoyed Ch.)

Oc(2+1)W.G. Horn

Buoyed
channel

2 Chys
(107)

82

RCW
VQ

CA2
Q

Fl.G.4s

Dir.FG19M

Port Est

DUNKERQUE

Ridens de Calais

Buoyed channel

Fl(2)W.G.

Q(3)R

Dir.FG.30m
22M

Flare
(134)

Port Ouest

See plan

Ridens de la Rade

Walde
Fl(3)10s13m13M

Oc(4)WG.12s29m27/22M

Nuclear
Power Stn

Petit
Fort-Philippe

51°N

CA4
VQ(9)10s

Grand Fort-
Philippe

GRAVELINES

CA6
OcR4s

Buoyed
channel

Hoverport

Oc.WG.4s 9/6M

Fl(4)15s59m22M

See plan

See plan

CALAIS

Cap Blanc
Nez

THE FRENCH COAST

Route 5 Ch 9

Route 1
Ch. 10

50'

2°E

10'

20'

30'

Fl(5)Y20s

Green

White

4

Digue Nord

Fl(2)R6s.12m17M

58'.5

White

Fl(5)Y 20s

5

Breakwater
Jetée Est

Works in progress

Fl(2)WG.
6s9/6M

Q(3)R.6s.5m2M

Iso.G.3s12m 8M
Bell

Jetée
Ouest

Lts in line 118°·5

2F.Y

Dir. FR14M

50°58'N

VQ

Dir. F.G.

Avant
Port

Lock

Marina
YC de Nord
de la France

Fort Risban

Moorings

H.M.

Fl(4)15s59m22M

Oc(4)WG.12s29m 27/22M
+ Oc.4s17M

Petit -Fort-
Philippe

Lock

Bassin
de l'Ouest

Bldg

Bassin du
Petit Paradis

Bassin
Carnot

(34)

Port de
Plaisance

CALAIS

Grand-Fort-
Philippe

Avant Port

Ferry

Arrière
Bassin

57'.5

51°N

River Aa

CALAIS

Tower (82)

Canal de
Calais

Station

Hotel de
Ville

1°50'.5E

50

51

Les Huttes

Belfry (27)

Port d'Echouage
(dries)

Belfry (35)

Waiting pontoons
Lock

(34)

GRAVELINES

59'

Bassin Vauban
(tidal basin –
can be shallow)

HM

Station

GRAVELINES

1°06'E

Depths in Metres

10'

Radiobeacons

Name	Freq. (kHz)	Ident.	Range (miles)	Seq.	Position
Marine radiobeacons					
Falls Group	305.7				
Falls LtF		FS	50	1	51°18′.1N 1°48′.5E
W Hinder LtV		WH	20	3	51°23′.1N 2°26′.4E
Oostende		OE	30	4	51°14′.2N 2°55′.9E
Calais		CS	20	5	50°57′.7N 1°51′.3E
S Foreland Lt		SD	30	6	51°08′.4N 1°22′.4E
N Foreland Lt	301.1	NF	50	Cont	51°22′.5N 1°26′.8E
Zeebrugge Mole Group	296.5				
Zeebrugge Mole Lt		ZB	5	1,2	51°20′.9N 3°12′.3E
Nieuwpoort West Pier Lt		NP	5	4,5	51°09′.4N 2°43′.1E
Aero beacon					
Calais/Dunkerque	275	MK	15	Cont	50°59′.9N 2°03′.3E

Coast radio stations

Station	Transmits (kHz)[2] (VHF Ch)	Receives (kHz) (VHF Ch)	Freq[7] VHF Ch[7]	Traffic lists (times)	Storm warnings (times)	Weather messages (times)	Navigational warnings (times)
Dunkerque[1]	16,**24**,61	24[3],61[3]	61		On receipt	0633,1133	
Oostende[1] (OST)(OSU)	**1817**,1820[4], 2182,2484, **2761**[4],2817[5], **3632**[5],3684[4]	2182[3],2191[6] 2484[34], 3178[34]	2761	even H+20	On receipt	0820,1720	0233,0633,1033 1433,1833,2233
Vicinity of							
Oostende	16,27,**85**,88	16[3],27[3],**85**,88	27	H+20	On receipt	0820,1720	0233,0633,1033 1433,1833,2233
La Panne	16,**23**,78	16[3],**23**,78					
Zeebrugge	16,27,**87**	16[3],27[3],**87**					
Antwerpen[1] (OSA)	1649.5, 1652.5 **1901**,2182	1649.5[34], 1652.5[34] 2182[3]	2182	H+05			
Located at							
Antwerp	7,27,**28**,81	7,27,**28**,81					
Antwerp & Ghent	16,24,26	16[3],**24**[3],26[3]	24	H+05	H+03,H+48		H+03,H+48
Doornzele	16,83	16,83					

Notes
[1] 24hr service
[2] Other non-working transmitting frequencies that do not receive are not listed.
[3] 24hr watch
[4] For Belgian vessels
[5] For foreign vessels
[6] When 2182 is occupied by distress traffic
[7] Frequency (kHz) and VHF Channel on which traffic lists, storm warnings, weather messages and navigational warnings are given.

The nuclear power station near Dunkerque Port Ouest which is a major conspicuous object on this coast

Major fixed daylight marks (W to E)

Calais to Gravelines

Walde Lt structure 1M offshore (18m B pylon on hut)
Oye-Plage pointed belfry (39m) 3M SW of the ends of
  Gravelines piers
Grand-Fort-Philippe square belfry (34m)
Gravelines entrance piers
Gravelines town: square church tower (34m) and close by a
  domed belfry (35m)
Petit-Fort-Philippe church spire (43m conspic)

Dunkerque Ouest to Dunkerque Est

Nuclear power station 1M NE of Petit-Fort-Philippe close to
  Dunkerque Ouest (Jetée des Huttes)
Dunkerque Ouest: leading lights (16m and 30m Wh columns
  with G tops), and the entrance columns of Jetée du Dyck
  (16m Wh with G top, lit) and Jetée Clipon (24m Wh with
  R top, lit), 2 chimneys (134m and 107m) 3M E of entrance
Dunkerque Est: from 3M W of the entrance up to the entrance are silos, blast furnaces, cooling towers, a gasometer,
  2 chimneys (107m conspic) and numerous oil refinery
  tanks
Light tower on Jetée Ouest (35m Wh tower brown top)
1¾M SE of the entrance is Hotel de Ville tower (76m with
  pointed roof) and a second tower (82m) close NE

Dunkerque Est to Oostende

3¼M E of Dunkerque Est entrance: a water tower (conspic)
  and steelworks chimneys close by
Bray-dunes-Plage belfry (32m)
De Panne square church and water tower
Hotel Royal Plage at Koksijde Bad with church tower 1M
  inland
Buildings of Oostduinkerke Bad
Long, tall blocks of flats close W of Nieuwpoort entrance
Nieuwpoort entrance piers
Nieuwpoort light (26m R round tower, Wh bands) ½M E of
  entrance
Tower block (conspic) close W of Oostende entrance
Oostende: entrance piers, St Jozef church steeple, twin spires
  of the cathedral, main light (63m grey tower)

Oostende to Netherlands frontier (considerably built-up)

Den Haan water tower
Water tower and Wenduine church close ENE
Blankenberge: entrance piers, main light tower (30m Wh
  tower B top) and large building blocks E of entrance
Zeebrugge: immense breakwater of concrete rubble blocks,
  radar on the inner mole, container gantries on Westhoofd,
  church with water tower ½M to the SW
Duinbergen water tower
Westkappelle church 2M inland
Row of tall buildings at Knokke
Isolated hotel (conspic) at Wielingen close E of the frontier

Approach routes and tidal timing

The offshore area from the Walde light E to Zeebrugge and the Westerschelde approaches
is a maze of longitudinal banks and intervening zigzag channels, particularly difficult between Dunkerque Port Ouest and Nieuwpoort – the graveyard of many of the Spanish
Armada – where many patches dry out or are very shallow. The banks Hills, Trapegeer,
Broers, Banc Smal and Banc Breedte near Dunkerque Est are the worst of these, with many
banks of less than 5m fanning out E and NE from Dunkerque Ouest to the West Hinder TSS
and Zeebrugge; e.g. Inner Ruytingen, Oost Dyck, Buiten Ratel, Kwinte Bank, and Middelkerke Bank offshore, and Nieuwpoort Bank, Stroom Bank and Wenduine Bank closer
inshore. In the intervening inshore channels, the 'rades' or 'redes', typical depths are 5 to
15m whilst in the offshore channels over 20m is common.

Despite the wider, deeper offshore channels, the TSS gives the area a wide berth to the N,
a plus for yacht cruising, but by the same token there is only one adequately buoyed/lit channel mainly inshore; this also benefits navigationally from the many charted industrial buildings, chimneys, towers and spires rising from the flat dunes. Further offshore there are a few
isolated approach buoys requiring very careful dead-reckoning navigation where Decca is
extremely useful. Crossing any of the banks in bad weather is dangerous with breakers and
swells which can roll boats over, and in the shallower areas sand movements cast doubts on
their charted accuracy, with frequent areas having 'less water reported'.

The S part of the area is well provided with long-range shore lights, and Oostende is one
of the transmitting stations for NAVAREA I feeding weather and other hazard information into
the NAVTEX system. In fog conditions anchorage away from traffic can usually be found on the
edge of one of the many nearby banks.

The offshore approach from the N can be made over a wide area from the West Hinder LtV
eastwards, although in bad weather the scattered deeper banks such as the Akkaert and
Goote with 10m patches are worth avoiding. From the Kwintebank (N card) buoy, a useful
marker at the end of the TSS, there are three main routes needing careful compass courses,
and utilising the several buoys near to the routes for position checks. In good weather short
cuts can be made across the banks to reduce distance.

Given the complications of the TSS between the Sandettié and W Hinder LtVs, together
with the disposition of shallow-water areas, there are four main yacht routes to be recommended into and through the area.

ROUTE 1 COASTING FROM CALAIS TO ZEEBRUGGE
(see also Chapter 9, Route 5)

Distance 60M

Commentary

The dredged, lit channel from Dunkerque Lanby to the newly constructed Port Ouest has over 20m depth but should be avoided by yachts; approach or departure from the W can be anywhere between the channel and the wide area of drying banks with their many wrecks between Calais and Dunkerque Port Ouest. The Gravelines breakwaters extend well out across the drying area and there are some Y Lt buoys approaching Port Ouest entrance.

Between Port Ouest and Oostende the inshore route gives access to all the harbours but is a devious one, still taking care to keep to the starboard edges of a narrower fairway. Having carefully crossed Port Ouest entrance (banned to yachts except in emergency) at right angles to the dredged channel and giving a wide berth to the N arm, follow the buoyed inshore channel, Passe de l'Ouest and Rade de Dunkerque, which crosses Dunkerque Est entrance and peters out near Zuydcote village with its nearby water tower, chimney and sanatorium building. Then NE through the Passe de Zuydcote, a 'saddle' in the Hills and Trapegeer banks with Lt buoys at each end and 2.7m least sounding. Eastwards down the Nieuwpoort Rede with widely spaced Lt buoys on each side and down into Nieuwpoort (there is a wreck with a N cardinal Lt buoy S of the Nieuwpoort approach), or continuing on to Oostende along the close inshore (now a steeply shelving shore) Kleine Rede, S of Stroom Bank (2.2m on its S edge).

Opposite Oostende there is a gap through Stroom Bank, Direct Pass, marked by 2 Lt buoys, into Grote Rede, and the route then turns E either past 2 obstruction-marking Lt buoys and across Wenduine Bank (in an area with least depths of 4m) or in bad weather giving the bank a wider berth to the N of its buoyed edge (the buoys at the W end are lit), and finally heading for the Lt buoys off the entrance to Zeebrugge. Blankenberge, W of Zeebrugge, is actually inside the Wenduine Bank and in onshore winds can be a dangerous entrance.

Tidal timing

Tidal streams tend to follow the coast, turning eastwards 1 to 2hrs before HW Dover, westwards 4 to 5hrs after HW Dover. They are not as strong as on the English east coast; 2 to 2½ knots spring rates (faster nearer to the Westerschelde). Calais is 73M from Vlissingen requiring at least 2 tides to complete in a single passage, whilst Oostende is approximately 44M from Calais again requiring the pushing of the second half of an opposing tidal stream at the beginning of the passage to achieve a final 6hrs of fair tide. Intermediate passages are of more convenient one-tide distances; Calais–Dunkerque 20M, Dunkerque–Oostende 24M, Oostende–Vlissingen 29M.

Calais and Dunkerque have inner harbours available to yachts, entered through locks, and Blankenberge has a marina which is shallow at LW so only here are tidal arrival heights critical. Moving E along the coast with the E-going stream and with much of the falling tide in the second half of the passage it is not usually possible to arrive at these harbours at the best time of tide and a choice has to be made of pushing the tide to arrive at the right time or waiting on arrival in the outer harbours. Moving W with the stream and with a rising tide in the second half of the passage it is easier to time a convenient arrival.

ROUTE 2 KWINTEBANK BUOY TO NIEUWPOORT ENTRANCE

Distance 16½M

Commentary

The route is from the Kwintebank (N card) Lt buoy through the Negenvaam channel between the Kwinte and Middelkerke Banks to close W or E of D1 (E card) Lt buoy, then close W of Nieuwpoortbank (W card) Lt buoy at the W end of the bank itself (2.8m least depth but can be crossed in light weather) and then SE to the harbour entrance avoiding the wreck (N card) Lt buoy to the S.

Tidal timing

Inwards from Kwintebank buoy about 3hrs of tide are needed within the favourable SW-going stream from 5½hrs after to 1hr before HW Dover, and on North Sea Passages 10 and 14 (Chapter 4) this is not difficult to achieve. Outwards similarly 3hrs are needed preferably within the favourable NE-going stream from 1hr before HW Dover to 5½hrs after, and on Passages 10 and 14 2hrs of favourable to slack and 1hr of weak adverse tide follow the suggested start time of HW Dover +4hrs.

ROUTE 3   KWINTEBANK BUOY TO OOSTENDE ENTRANCE

Distance   11M

Commentary

From Kwintebank (N card) Lt buoy to close S of Middelkerkebank N (N card) Lt buoy, then, crossing the 8m end of Oostende Bank, to close S of Oostendebank E (R) Lt buoy and finally straight for Oostende via Direct Pass.

Tidal timing

This route is cross-tide so departure and arrival timing is not critical.

ROUTE 4   KWINTEBANK BUOY TO ZEEBRUGGE ENTRANCE

Distance   18M

Commentary

A very easy route eastwards with no banks, and from close N of Middelkerkebank (N card) Lt buoy heading direct for the entrance but keeping to the S of all the approach Lt buoys, particularly the Oostendebank N wreck buoys.

Tidal timing

The inward-bound requirement is for 3 to 4hrs of NE-going tide within the period from 1hr before to 5½hrs after HW Dover, whilst outward bound the remaining 6hr period is ideal. Unfortunately in order to obtain the best streams on the English coast, North Sea Passages 11 and 15 (Chapter 4) result in: inwards about 2hrs of adverse stream and outwards about 1hr of adverse tide.

The conspicuous tower and buildings of Oostende

The harbours and coast

The Flanders coast is flat sand-dune country with towns and industrial buildings rising up like rows of candles and children's building blocks, and with dangerous sandbanks offshore. It was the graveyard of many of the Spanish Armada fleet in the sixteenth century, home of the Dunkerque pirates in the seventeenth and eighteenth centuries, base of a flourishing fishing industry in the nineteenth century, and devastated by two world wars and industrial development in the twentieth, particularly on the French side of the frontier. There are many museums associated with the two wars. All the towns have sandy beaches nearby and in many cases are, compared with Britain, sophisticated seaside resorts.

On this smooth coast the harbours are artificial and many are canalised river entrances, with primary protection from the prevailing southwesterlies. Winds from W through N to NE of over Force 4 can create dangerous lee-shore entrance conditions at some harbours. Many of the harbours have entrances to the Belgian and French canal systems.

GRAVELINES

Port radio and entry signals None

Customs

Quai Vauban ☎ 28 23 08 52

Entrance

If the wind is onshore and much over Force 5, then forget it, do not enter!

The entrance piers with port and starboard beacons at the ends are 1M SW of the unmistakable long building with 6 reactor domes of Dunkerque Nuclear Power Station. Approached directly parallel to the piers across a drying area, the whole channel up to the Port d'Echouage dries, and is available from about half-tide either side of HW.

The bridge and lock into the marina in the Bassin Vauban open between approximately 1½hrs before and after HW at springs and for somewhat shorter periods with smaller tidal rises.

Mooring and facilities

There are private pontoon moorings which dry out on the Petit-Fort-Philippe side of the River Aa near the town. The Grand-Fort-Philippe side is used by fishing boats. There are drying pontoon moorings on the E side of the Port d'Echouage just outside the lock and bridge where visitors may wait for opening.

The marina (office: Quai Vauban ☎ 28 23 06 12) run by the Union Sportive Gravelinoise and supported by the local authority was in course of development in early 1986 to increase the number of berths and provide club and toilet facilities. The Bassin Vauban is shallow but 1½m-draught vessels can usually float although with the sluicing of floodwater through the locks boats sometimes take the mud. There are repair and lifting facilities and the Yacht Club Gravelinois. Gravelines is an interesting historic fortified town.

Gravelines entrance

Moorings on the River Aa *Photo* Lester McCarthy, Motor Boat & Yachting

Entrance to the French/Belgian canal system can be made through the lock at the inland end of Bassin Vauban, but you will need to leave your mast.

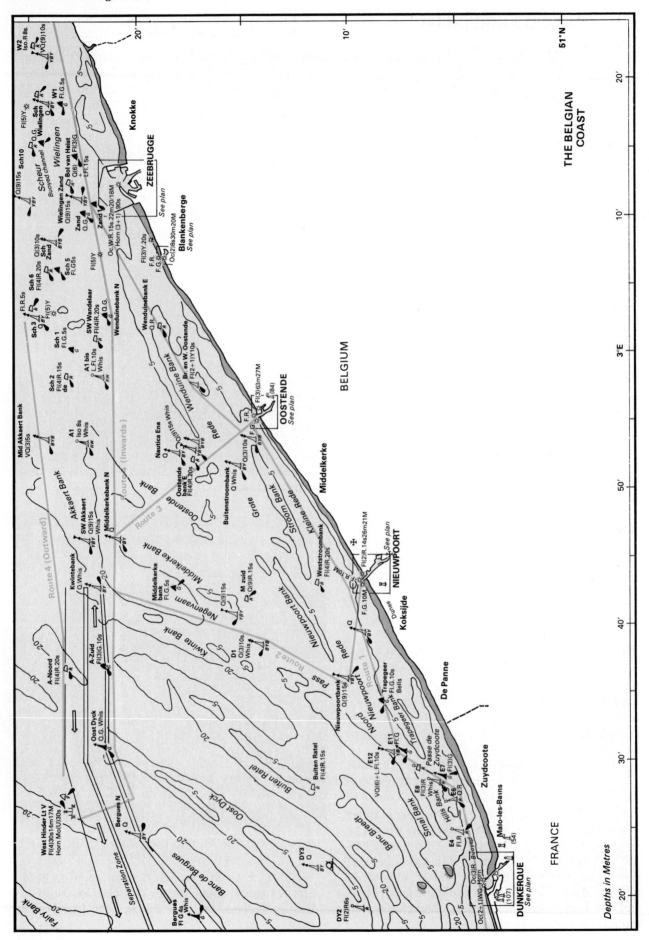

THE BELGIAN
COAST

BELGIUM

FRANCE

Depths in Metres

PORT OUEST, DUNKERQUE

This is a useful harbour; but only for crew to use for the Sally Line crossing from Ramsgate, and then taking a bus, taxi or train to meet their skippers in either Gravelines or Dunkerque Est, since, as a deep-sea tanker terminal this port is out of bounds for yachts except in emergency. It is useful to have Admiralty chart *1350* in the case of such emergency so that port control instructions can be followed after calling on Channel 16.

DUNKERQUE

Port radio (VHF)

Dunkerque	VHF Ch	Hrs listening
☎ 28 65 99 22	12, 16, **73**	24

Entry signals

Contact harbour control by VHF.

Main LtHo on W jetty shows full international port traffic signals.

Ecluse Watier (lock opposite harbour entrance) lock signals, upper pair denoting Ecluse Watier, lower pair for Trystram lock: F.G and Fl.G (hor) Lts – enter lock, moor near Fl Lt; 2 R (hor) Lts – no entry. 3 Lts in triangle (1 W, 2 G) – enter, all gates open.

4 blasts – request Bridge Mole No 2 open; 2 long 1 short blasts – request Bridge Darse No 1 open.

Customs

Rue l'Hermite ☎ 28 66 87 14

Entrance

This is France's back yard. The entrance is at the W end of the largest coastal industrial complex I have ever seen – blast furnaces, chimneys, cooling towers, and numerous oil refinery tanks – usually topped with a cloud of orange-coloured smoke.

Entrance is just beyond the two Rade de Dunkerque channel Lt buoys, DW30 (R) and DW29 (G), and in the entrance the shoal at the end of Jetée Ouest should be given a wide berth. Then follow E jetty (there are leading lights to help at night) and either peel off to starboard to go through Trystram lock and two opening bridges (synchronised with lock-opening) to the Bassin de Commerce, or continue ahead to the Port d'Echouage and the tidal YC de la Mer du Nord Marina. Trystram lock opens at any time between 0800 and 1930 daily for boats coming in or out. You should turn in circles before the lock gates to be seen from the control tower, or blow your horn, or call them on Ch 73 or 12. Then continue slightly to port through Mole No 2 swing bridge (video camera control from the tower), then a dogleg southwards and the Pertuis Amont swing bridge will open and you should tie up at No 4 pontoon at the YC de Dunkerque marina.

Mooring and facilities

YC de la Mer du Nord Marina (☎ 28 66 79 90) is the tidal pontoon marina in the Port d'Echouage with toilets, showers, scrubbing grid and crane, repair and chandlery facilities nearby. 10-minute walk to town. The two smaller marinas further down the Port d'Echouage have no facilities for visitors.

YC de Dunkerque (☎ 28 66 11 06) runs the municipal marina with 120 berths in the Bassin du Commerce, entered through Trystram lock and the two swing bridges. There are no permanent staff and the secretary comes in three times a week. The fees are advertised on the club office in English (first night stop free) and visitors should put the appropriate fees with the boat's name in an envelope and post it in the office letter box. Shower and toilet facilities were under development in 1986 and a new clubhouse was to be opened in an old lightship in 1987. A welcoming and non-profit club very convenient for the town.

The town has been largely rebuilt since the war, but still has some historic churches and buildings,

Dunkerque approach

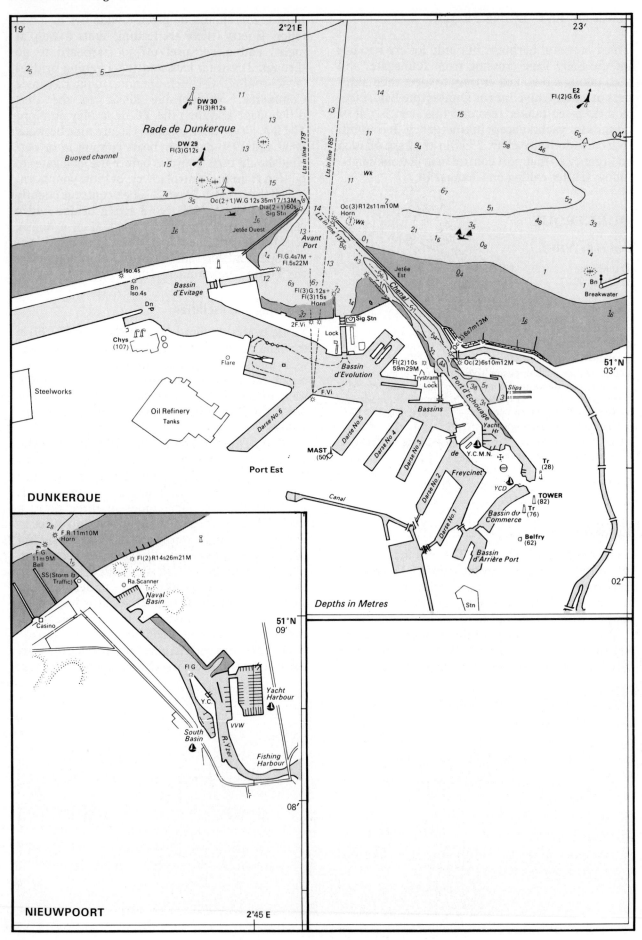

19'   2°21 E   23'

2_5

5

11

1_4

DW 30
Fl(3)R12s

E2
Fl.(2)G.6s

Rade de Dunkerque

13   13 1_5

DW 29
Fl(3)G12s

13   13

15

9_4 5_8 4_6 6_5

Buoyed channel

15

Wk

6_7 5_1 5_2 3_3

7_4 3_5 1_8
Oc(2+1)W.G 12s 35m17/13M
Dia(2+1)60s
Sig Stn

11

7
Oc(3)R12s11m10M
Horn

2_1 1_6 0_8 4_8 1_4

1_6
Jetée Ouest

14   Wk

3_5

0_1

1   Bn
Breakwater

Iso.4s

1_4 *Avant Port*
Fl.G.4s7M +
Fl.5s22M   13 3_6

4_3 Jetée
Est

1_6

Bn
Iso.4s

12   6_3 13
Fl(3)G.12s+
Fl(3)15s
Horn 7_2

1_4 5_6 5_1
Chenal 1_6 1_6

Dn

Bassin d'Evitage

3_2
2F.Vi

Sig Stn 5_4 Oc(3)6s7m12M

Chys
(107)

Flare   Lock 4_4 5_1
Oc(2)6s10m12M

Fl(2)10s
59m29M 3_8 Slips
3_1 3

Bassin d'Evolution   Trystram
Lock

Port d'Echouage

Steelworks   F.Vi   *Bassins*   Yacht
Hr

Oil Refinery
Tanks   Darse No 6   Darse No 5   de   Y.C.M.N.   Tr
(28)

MAST
(50)   Darse No 4   Darse No 3   *Freycinet*   YCD   TOWER
(82)

DUNKERQUE   Canal   Darse No 2   Bassin du
Commerce   Tr
(76)

Belfry
(62)

Darse No 1   *Bassin d'Arrière Port*

Depths in Metres

2_8 F.R 11m10M
Horn

Stn

F.G.
11m 9M
Bell   Fl(2)R14s26m21M

1_5 51°N
09'

SS(Storm &
Traffic)   Ra Scanner

Naval Basin

Casino

Fl G

Yacht
Harbour

Y.C.

VVW

South Basin   R. Yzer

Fishing
Harbour

08'

NIEUWPOORT   2°45 E

51°N
03'

02'

51°N
09'

Dunkerque approach. Note the industrial haze from the steelworks

Dunkerque harbour entrance

Dunkerque Bassin de Commerce

bistros, restaurants, a good shopping area and market, a hypermarket, a museum of the Second World War British evacuation of Dunkerque and Malo-les-Bains seaside resort next door.

Entrance to the French/Belgian inland water-way system is obtained through locks at Darses Nos 1 and 2, but advance contact with and information from the harbour authority is essential and again masts need to be lowered, and possibly left behind in Dunkerque.

NIEUWPOORT

Port radio (VHF)

Nieuwpoort	VHF Ch	Hrs listening
	09, 16	24

Entry signals

Contact port control by VHF.

Signal station root of W pier showing R flag or R Lt – no entry, G flag or G Lt – no exit; R over G flags or Lts – no entry or exit.

Tide signals, B cone point up or G over W Lt – rising tide; B cone point down or W over G Lt – falling tide. Sphere or W Lt – 5m; cylinder or R Lt – 1m. Cone point down or G Lt – 0.2m above datum; cone point up or Vi Lt – 0.2m below datum.

2 cones points together or Fl.Bu Lt – craft of 5m or less not to leave harbour (onshore wind Force 3 or over, offshore Force 4 or over).

Customs

If you have anything to declare, ☎ (058) 23 34 51.

Entrance

No entrance problems except in onshore winds of over Force 5. Minimum depth in the entrance channel 1.5m, so entrance around LW should be avoided, and scrutinise the tide signals on the signal station (see above). Keep clear of wreck (swept depth 2.6m) ½M N of entrance and marked by N cardinal Lt buoy. LtHo to E of entrance is unmistakable. Enter directly parallel with piers between the two white entrance towers (port and starboard Lts).

There are black and white dolphins to keep you off the sides of the channel and the junction. Keep clear of the usually fairly empty Naval Basin to port, and at the junction turn to starboard for the South Basin or to port for VVW and the Yacht Harbour. The latter is entered round another black and white dolphin with a wide turn to port.

Nieuwpoort entrance. Note position of signal station

Mooring and facilities

The large new marina to the E off the River Yzer
has pontoon moorings in both an inner and outer
basin with all the usual facilities. It is run by VVW
(Vlaamse Vereniging voor Watersport), ☎ (058)
23 52 32, with extensive repair and yard facilities
nearby.

In the South Basin another excellent marina is
run by Koninklijk Jachtclub Nieuwpoort, ☎
(058) 23 44 13, and close S of this Watersportkring
van de Luchtmacht, ☎ (058) 23 36 41.

All of these facilities are within half a mile to a
mile's walk from the shopping area. Not many historic buildings remain but the town has been
rebuilt very well. In 1914 the Belgians halted the
German advance by flooding the Lower Yzer region.

Yachts are not allowed to moor in the harbour
'canal' proper nor further down near the town in
the fishing harbour. Continuing on, after lowering
the mast because of the fixed bridges, and possibly
leaving it stored, entry to the canal system is possible through either a lock to Oostende or another to
Dunkerque.

OOSTENDE

Port radio (VHF)

Oostende	VHF Ch	Hrs listening
	09, 16	24
Mercator Marina	14	24
☎ (059) 70 57 62		

Entry signals

Contact in advance with harbour control on VHF is advisable.

Most significant signal is that for Cross Channel Ferries
entering or leaving, all other traffic to keep clear: Q.Or Lt
from E pier – keep clear of harbour entrance and channel,
to De Mey lock; Q.Or from Pilot Station or
Montgomerydok – keep clear of channel from
Montgomerydok and from tidal harbour.

E pier and railway station traffic signals: sphere/cone/sphere
or RWR Lts – no entry; cone down/cone up/cone down or
GWG Lts – no departure; cone down/cone up/sphere/
sphere or GWR Lts – no entry or departure.

If installations damaged: R flag or R Lt – no entry; G flag
or G Lt – no departure, R over G flag or Lts – no entry or
departure.

Montgomerydok Pilot Station: 2 B cones points together or
Fl.Bu Lt – craft of 5m or less not to leave harbour (onshore
wind Force 3 or over, offshore Force 4 or over).

Customs

If you have anything to declare, ☎ (059) 70 20 09.

Oostende entrance. Note position of signal station

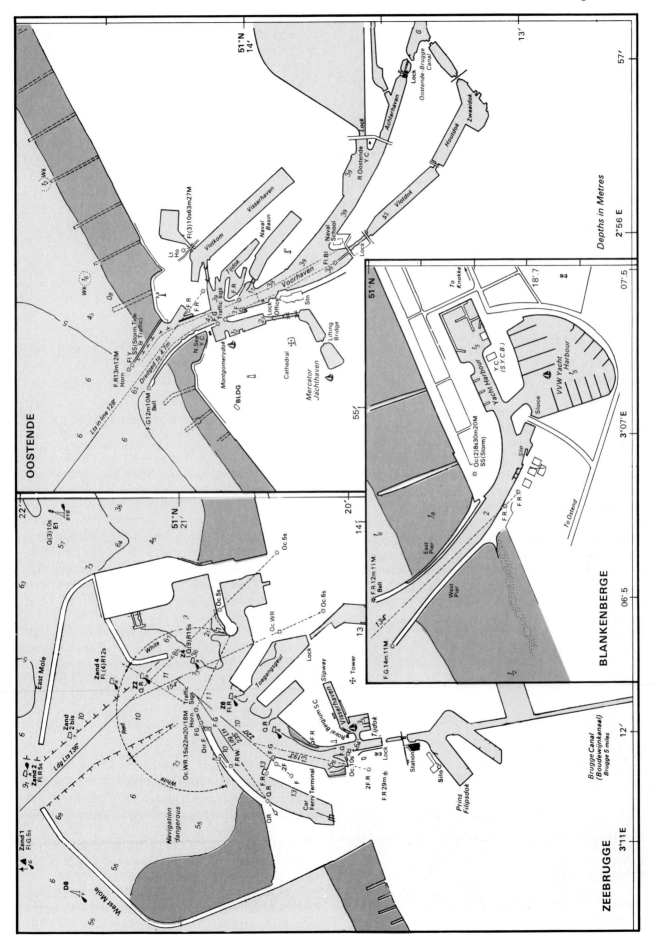

OOSTENDE

51°N
14'

13'

57'

Lock
Oostende-Brugge Canal

Zwaaidok

Houtdok

Lock

R. Oostende Y.C.

Vlotdok

39
39
4½
39

Achterhaven

Fl Bl

Naval School

39 Fl Bl
Voorhaven

Fl(3)10s63m27M

Visserhaven

Vlotkom
Naval Basin

Tijdok

Lt Ho

Traffic Sigs
F.R
F.G
Lock Office
Stn

Depths in Metres

2°56 E

SS(Storm, Tide & Traffic)
Fl Y
F.R13m12M
Horn

N Sea

Dredged to 4.7m
6²

Montgomerydok

Lifting Bridge

Lts in line 128°

F.G12m10M
Bell

BLDG

Cathedral

Mercator Jachthaven

55'

6

6

6

BLANKENBERGE

51°N

07'.5

To Knokke

18·7

Yacht Harbour

Y.C (S.Y.C.B.)

VVW Yacht Harbour

Oc(2)8s30m20M
SS(Storm)

Sluice

3°07' E

Slip

To Ostend

F.R

East Pier

West Pier

F.R.12m11M
Bell

F.R

F.G.14m11M

13⁴F

06'·5

55'

ZEEBRUGGE

22'

Q(3)10s E1
BYB

36

6⁴

45

51°N
21'

Oc.5s

13

6³

Oc.5s

Oc.6s.

East Mole

Oc. WR

White

Fl(4)R12s
Zand 4

Z4
Q(9)R15s

Toegangsgeul

Lock

13

Zand 2 bis
Fl.R5s

22
Q.R

Traffic Sigs

Tower

Ldg Lts 136°

26
Fl.R

Zand 2
Fl.R5s

Oc.WR 15s22m20/18M
Horn

Dir F

F.G

Royal Belgium SC

Slipway
Visserhaven

F.R

Oc.10s

Lock

Tijdok

F.R.W

F.R.G

Car Ferry Terminal

2F.R

Q.R

F.R.G
Q.R

F

Navigation dangerous

2F

F.R 29m

Station

Silo

Prins Filipsdok

Brugge Canal (Boudewijnkanaal)
Brugge 5 miles

Zand1
Fl.G.5s

West Mole

D8

3°11E

20'

14'

12'

155

Entrance

An uncomfortable entrance with onshore winds over Force 5 even though dredged to 4.7m.

A very tall tower block just over 3 cables SSW of the W pier is unmistakable. From close to Binnenstroombank (E card) Lt buoy straight into the entrance between the 2 white pier-end towers (port and starboard lit), the E pier-end with a large signal mast. There are leading lights to help at night.

Turn to starboard for Montgomerydok 4 cables inside, and then to port for the lock and two opening bridges into Handels Dokken and Mercator Jachthaven; contact on VHF to arrange the most convenient bridge and lock openings.

Continuing on past Montgomerydok entrance and taking the port fork of the Voorhaven, the Royal Yacht Club of Ostend is on the peninsula at the E end near the bridge into the Achterhaven.

Mooring and facilities

North Sea Yacht Club, ☎ (059) 70 27 54, runs a pontoon marina in Montgomerydok, which is usually very crowded and visiting yachts raft to the main central pontoon. There are toilets, showers, restaurant and drying/scrubbing facilities.

Mercator Jachthaven, ☎ (059) 70 57 62, in the Handels Dokken is a large pontoon marina in the busy part of town next to the station, with toilet and shower facilities.

Royal Yacht Club of Ostend, ☎ (059) 70 14 52, is extremely friendly, is a pontoon marina, and has good club facilities including drying out, crane and slipway, and can certainly direct vessels to good repair facilities if not available on the spot. The town is well over a mile away and the bus (or a folding bicycle) helps.

The Royal Yacht Club of Ostend

Entrance to the Belgian inland waterways necessitates dismasting, continuing on under the bridge into the Achterhaven and through the locks into the Oostende-Brugge canal.

Oostende is an extremely lively seaside resort with casino, restaurants, museums and some attractive parks. It is also extremely handy for delivering crew via hoverfoil or ferry to and from Dover. Brugge, the 'Venice' of Flanders, and one of Europe's best preserved medieval cities, is only a short train (or canal) journey away, and the station is on the doorstep.

BLANKENBERGE

Port radio (VHF)

None

Entry signals

Storm and traffic signals are shown from a semaphore mast near the LtHo: 2 B cones points together or Fl.Vi Lt – no vessel under 5m may leave harbour.

Customs

If you have anything to declare, ☎ (050) 54 42 23.

Entrance

Very like Nieuwpoort, the charts indicate varying minimum depths from drying to 1m, but at time of going to press dredging operations were reported in progress to increase the entrance depth to 2.5m so if you have a deep draught it is worth checking depths with the yacht club, ☎ (050) 41 14 20. However, to be on the safe side entrance is not advisable in onshore winds of over Force 5, nor about 1½hrs either side of LW. Entrance is preferable on the flood.

There are large square building blocks to E of entrance slightly higher than the LtHo. Enter parallel to piers between white round towers, port and starboard lit. At night leading lights help.

Fork to port for the town's old N harbour, and starboard for VVW Yacht Harbour.

Mooring and facilities

Both harbours have pontoon moorings, and the minimum depth in both is claimed to be 1.5 to 2m, but the new and commodious VVW Marina tends to be a little shallow at LW and deeper draught vessels may touch.

There are all facilities; slipping, lifting, repair and chandlery, particularly near the S side of the entrance E of the leading lights.

I could find no freely available toilet or shower facilities, but there are three clubs VVW near the marina, Scarphout YC Blankenberge, ☎ (050) 41

Blankenberge. VVW Marina

Blankenberge. S side of entrance showing chandleries and slipway

14 20, on the island between the two harbours which is usually open for long periods, and VNZ (Vrije Noordzeezeilers – a grand title!) on the N side of the harbour.

A picturesque town with a market, interesting shopping streets, and a superb beach and promenade.

ZEEBRUGGE

Port radio (VHF)

Zeebrugge	VHF Ch	Hrs listening
Hbr ent. pilotage	09	24
Port Control	13, 71	24

☎ (050) 54 32 33

Entry signals

It is compulsory to contact Port Control on VHF.
Traffic signals on W Mole head.
Pilot Station W entrance near lock: 2 B cones points together or Fl.Bu Lt – craft of 5m or less not to leave harbour (onshore wind Force 3 or over, offshore Force 4 or over).

Customs

If you have anything to declare, ☎ (050) 54 42 23.

Entrance

No problems in approach due to depth or weather, except with strong wind blowing directly into outer entrance.

Recently this harbour has been very considerably enlarged with an additional encircling outer mole of concrete rubble blocks protruding over 1½M offshore, with a central northern entrance. When I visited early in 1986 this entrance had yet to be completed and lacked pierhead lights, the opening being simply buoyed with Zand 1 (G) and

Zeebrugge. Royal Belgian Sailing Club moorings

Zand 2 (R) Lt buoys, so make sure your chart is up to date when you visit as it is almost sure to have changed, and make sure your radio works!

Entrance is from between Zand 1 and 2 buoys, but closer to Zand 2 (about 1 cable from it), then directly across to the second, inner entrance between the older W Mole head and Z4 (R) Lt buoy (which is S of a new inner E wall, LNG-DAM), taking care not to stray out of the charted 2 to 3-cable-wide channel into the delimited area 'Navigation Dangerous'. Then hard to starboard following and keeping NW of the R channel Lt buoys to yet a third, inner entrance between Westhoofd and Z8 (R) Lt buoy. Then down the inner harbour channel, giving a wide berth to the drying area on the E side and S of Z8 buoy. Then hard to port and port again into the Visserhaven to the pontoons of the Royal Belgian Sailing Club on the curved wall at the W end of the Visserhaven.

There are no less than 4 sets of leading lights for following the above approach at night. Do not stray out of the initial approach channel across the outer harbour, watch out for buoys being moved, and for Y Lt buoys used for current measurement.

Mooring and facilities

The S and E end of the Visserhaven, is what it says – a fish dock, so keep to the pontoons of the Royal Belgian Sailing Club, ☎ (050) 54 49 03; a very friendly club with good facilities, some room for visitors, and all boatyard, scrubbing, and chandlery facilities nearby, as well as a number of restaurants. There is, however, very little else worth seeing in Zeebrugge, but it is another convenient spot for taking a train into Belgium proper, or making a short pilgrimage to Brugge (see under Oostende), or for delivering crew by North Sea ferry (to Felixstowe and Dover).

Finally there is a lock entrance, S of the Visserhaven entrance in the inner harbour, into the Boudewijnkanaal to Brugge – yet another dismasting job but worthwhile if you have time.

11. The Schelde Delta

Charts   Admiralty *110, 120, 122, 132, 133, 139, 192, 325, 1406, 1872, 2322, 3371*
Imray *C30*
Dutch Hydrographic *1801, 1803, 1805, 1807, 1809*

Tidal atlases   Admiralty *North Sea – southern portion*
Dutch *Stroomatlassen a, c, e, f* and *n*

Tidal streams

Westerschelde approaches
(based on HW Vlissingen and HW Dover)

Position	VLISSINGEN		DOVER	
	East	*West*	*East*	*West*
4M NE Kwintebank buoy	−0230	+0330	−0030	+0530
Neths/Belg. frontier★	−0400	+0215	−0200	+0415
Westerschelde ent. (Nieuwe Sluis)	−0500	+0115	−0300	+0315
Oostgat (N end)	+0540	−0130	−0445	+0030

Note
★ NE and SW-going

Westerschelde River
(based on HW Vlissingen and HW Antwerp)

Position	VLISSINGEN		ANTWERP	
	In	*Out*	*In*	*Out*
Vlissingen Road	−0515	+0100		
Terneuzen	−0430	+0130		
Ellewoutsdijk	−0500	+0130		
Hansweert	−0400	+0200	−0610	−0010
Bath	−0310	+0230	−0540	+0020
Lillo	−0315	+0230	−0525	+0020
Antwerp	−0230	+0300	−0440	+0050

Oosterschelde
(based on HW Vlissingen and HW Dover)

Position	VLISSINGEN		DOVER	
	In	*Out*	*In*	*Out*
Westgat approach to Oosterschelde★	−0400	+0200	−0200	+0400
Roompot	−0500	+0115		
Zierikzee	−0430	+0125		
Wemeldinge	−0450	+0135		
Zijpe	−0410	+0205		

Note
★ Strong SW-going (in) and NE-going (out) offshore

North of Oosterschelde
(based on HW Hoek van Holland and HW Dover)

Position	H. VAN HOLLAND		DOVER	
	NE	*SW*	*NE*	*SW*
Offshore N of Oosterschelde to close S of Europoort	−0305	+0310	HW	−0530

	In	*Out*	*In*	*Out*
Mid Brouwershavensche Gat	−0305	+0255	HW	+0600
SW point of Goeree★	−0550	+0115	−0245	+0320
In Slijkgat	−0330	+0330	−0025	−0550
Off Maasmond entrance	−0200	+0430	+0105	−0450

Note
★ Flood SSE, ebb NNW

Nieuwe Waterweg, Hoek van Holland to Rotterdam
(based on HW Hoek van Holland and HW Dover)

Position	H. VAN HOLLAND		DOVER	
	In	*Out*	*In*	*Out*
In Maasmond entrance	−0230	+0215	+0035	+0520
Maassluis	−0215	+0230	+0050	+0535
Rotterdam	−0145	+0300		

Tidal differences and ranges   (based on HW Dover)

Place	HW (time)	Springs/Neaps (range in metres)
Vlissingen	+0212	4.4/3.0
Terneuzen	+0236	4.7/3.2
Hansweert	+0314	
Antwerp	+0342	5.4/4.0
Zierikzee	+0322	3.3/2.4
Wemeldinge	+0347	3.9/2.7
Hoek van Holland	+0251	1.9/1.5
Rotterdam	+0414	1.8/1.6

Major lights

Name of light	Characteristics	Position	Structure
Offshore			
Noord Hinder LtV	Fl(2)10s16m27M	52°00'.1N 2°51'.2E	R hull, Wh upperworks, RC, Racon
	Horn(2)30s		
Goeree	Fl(4)20s31m28M	51°55'.5N 3°40'.2E	RW chequered tower on platform on piles,
	Horn(4)30s		RC, Racon, helicopter platform
Onshore			
Hoek van Holland			
Maasvlakte	Fl(5)20s66m28M	51°58'.2N 4°00'.9E	B 8-sided concrete tower, orange bands
Maasmond Ldg Lts 112°			
Front	Iso.4s29m21M	51°58'.9N 4°04'.9E	Wh concrete tower, B bands, 101°-vis-123°
	By day 11M		
Rear	Iso.4s46m21M	51°58'.5N 4°06'.0E	Wh concrete tower, B bands, 101°-vis-123°
	By day 11M		
Maasmond Ldg Lts 107°			
Front	Iso.R.6s29m18M	51°58'.6N 4°07'.6E	R tower, Wh bands, 099.5°-vis-114.5°
	By day 7M		
Rear	Iso.R.6s43m18M	51°58'.5N 4°08'.0E	R tower, Wh bands, 099.5°-vis-114.5°
	By day 7M		
Europoort Calandkanaal			
Ldg Lts 116° Front	Oc.G.6s29m16M	51°57'.6N 4°08'.8E	Wh concrete tower, R bands, 108.5°-vis-123.5°
	By day 7M		
Rear	Oc.G.6s43m16M		Wh concrete tower, R bands, 108.5°-vis-123.5°
	By day 7M		
Westhoofd	Fl(3)15s55m30M	51°48'.8N 3°51'.9E	R sq stone tower
West Schouwen	Fl(2+1)15s57m30M	51°42'.6N 3°41'.6E	Grey round stone tower, R diagonal stripes on upper part
Westkapelle	Fl.3s48m28M	51°31'.8N 3°26'.9E	Sq stone tower, R metal superstructure
Nieuwe Sluis	Oc.WRG.10s27m14-10M	51°24'.5N 3°31'.3E	B 8-sided metal tower, Wh bands 055°-R-084°-W-091°-G-132°-W-238°-G-244°, 244°-W-258°-G-264°-R-292°-W-055°
	Horn(3)30s	51°24'.4N 3°30'.4E	Wh metal framework tower, B bands

Radiobeacons

Name	Freq. (kHz)	Ident.	Range (miles)	Seq.	Position
Marine radiobeacons					
Smith's Knoll Group	287.3				
Smith's Knoll LtV		SK	50	1	52°43'.5N 2°18'.0E
Goeree Lt		GR	50	2	51°55'.5N 3°40'.2E
Dudgeon LtV		LV	50	3	53°15'.5N 1°13'.5E
Outer Gabbard LtV		GA	50	4	51°59'.4N 2°04'.6E
Cromer Lt		CM	50	5	52°55'.5N 1°19'.1E
Noord Hinder LtV		NR	50	6	52°00'.2N 2°51'.2E
IJmuiden Group	294.2				
IJmuiden		YM	20	1,4	52°27'.8N 4°34'.6E
Hoek van Holland		HH	20	2,5	51°58'.9N 4°06'.8E
Eierland Lt (fog only)		ER	20	3,6	53°11'.0N 4°51'.4E
Aero beacon					
Valkenburg/Scheveningen	364	GV	25	Cont	52°05'.6N 4°15'.2E

Coast radio stations

Station	Transmits (kHz)[2] (VHF Ch)	Receives (kHz) (VHF Ch)	Freq[7] VHF Ch[7]	Traffic lists (times)	Storm warnings (times)	Weather messages (times)	Navigational warnings (times)
Antwerpen[1] (OSA)	1649.5, 1652.5 **1901**,2182	1649.5[34], 1652.5[34] 2182[3]	2182	H+05			
Located at Antwerp	7,27,**28**,81	7,27,**28**,81					
Antwerp & Ghent	16,**24**,26	16[3],**24**[3],26[3]	24	H+05	H+03,H+48		H+03,H+48
Doornzele	16,83	16,83					
Scheveningen[1] (PCG)(PCH)	**1764**,2824 2182 & Nes	2049[3] 2182[3]& Nes 2520[3]	1862[8]	odd H+05	On receipt	0340,0940 1540,2140	0333,0733,1133 1533,1933,2333
			1890	odd H+05	On receipt	0340,0940 1540,2140	0333,0733,1133 1533,1933,2333
			2824	0105,0305, 0505,2305		0340	0333,2333

Notes
[1] 24hr service. Scheveningen frequencies without 24hr service are not listed.
[2] Other non-working transmitting frequencies that do not receive are not listed.
[3] 24hr watch. Scheveningen 2049 when 2182 is distress working. 2520 is calling frequency, alternative is 2182.
[4] For Belgian vessels
[7] Frequency (kHz) and VHF Channel on which traffic lists, storm warnings, weather messages and navigational warnings are given.
[8] Located at Nes 53°24′N 6°04′E.

The following stations are remotely controlled by Scheveningen; call is *Scheveningen Radio* in all cases.

Station	VHF Channel	Position
Goes	Ch 16,**23**[9],**25**,**78**,**84**	51°31′N 3°54′E
Rotterdam	Ch 16,**24**,**27**,**28**,**87**[9]	51°56′N 4°28′E
Scheveningen	Ch 16,**26**,**83**[9]	52°06′N 4°16′E
Haarlem	Ch 16,**23**,**25**[9],**85**	52°23′N 4°38′E
Wieringerwerf	Ch 16,**27**[9],**87**	52°55′N 5°04′E
Location L7	Ch 16,**28**[9],**84**	53°32′N 4°13′E
Terschelling	Ch 16,**25**[9],**78**	53°22′N 5°13′E
Nes	Ch 16,**23**[9]	53°24′N 6°04′E
Appingedam	Ch 16,**24**,**27**[9]	53°18′N 6°52′E

Notes
Ch 16 and all working channels (bold) have 24hr watch.
Traffic lists on all working channels except Ch 16 at every H+05.
[9] VHF Channel for storm warnings at every H+05; weather messages at 0605, 1205, 1805, 2305 (in Dutch). Broadcasts given 1hr earlier when DST is in force.

Major fixed daylight marks

Belgian frontier to Breskens
Isolated hotel (conspic) at Wielingen close E of the frontier
Nieuwe Sluis Lt (22m, B octagonal tower, Wh bands)
Breskens grain silo between Westhaven and Oosthaven

N bank of Westerschelde – Vlissingen to Noorderhoofd
Vlissingen: 2 chimneys (125m) 1M inland on Kanaal door Walcheren, a windmill, St James Church spire, and Vlissingen main Lt (10m, brown metal framework tower at entrance to Koopmanshaven)
Kaapduinen: radio mast and two Ldg Lt towers (14 & 13m, yellow square stone, R bands)
Westkapelle Lt (52m, square stone tower, R metal superstructure)

Walcheren coast – Noorderhoofd to Roompot
Noorderhoofd Lt (16m, R round metal tower, Wh band)
Domburg church spire and nearby water tower
Oosterhoofd, chimney
Veere church tower (square with dome) and town hall spire

Schouwen coast (S to N)
Concrete lookout tower
West Schouwen Lt (50m, grey round stone tower, R diagonal stripes on upper part)
Westhoofd Lt (52m, R square stone tower)
Goedereede church tower
Kwade Hoek Lt (4m, B mast, Wh bands)

Voorne coast (S to N)
Radio mast (75m, R Lts) NE of Zwarte Hoek
Brielle church tower
Oostvoorne church tower

Maasvlakte and Hoek van Holland
2 chimneys (175m, R Lts)
Maasvlakte Lt (62m, B octagonal tower, orange bands)
Europoort entrance Lts
Nieuwe Zuiderdam & Nieuwe Noorderdam entrance towers to Maasmond (31m, orange towers, B bands, helicopter landing platforms on top)
Church towers of 's-Gravenzande, Monster and Ter Heijde

Approach routes and tidal timing

With the closing of the Oosterschelde at the Roompot the Delta Scheme is now virtually complete. The Delta now has two freely open entrances at its N and S extremities leading to two of Europe's largest port complexes: Europoort and Nieuwe (Rotterdamsche) Waterweg; and Westerschelde and Antwerp. In between are two dammed entrances with lock access: the Haringvliet, a dam with an occasional water-sluicing function; and the Roompot which closes the entrance and acts as an occasional tidal flood barrier with water normally flowing freely through, and having the Roompotsluis. Between these entrances the Grevelingenmeer has a completely dammed entrance as does the Veerse Meer close W of the Roompot.

Areas of less than 5m soundings and drying shallows stretch W and S from the point of each of the main islands: Voorne (near Europoort, the Bollen), Goeree (the Ooster), Schouwen (the Banjaard), and Walcheren (the Rassen and Raan).

Offshore the Delta is screened by three broken longitudinal banks generally with 7 to 8m depths and of danger to yachts only in Force 4 winds and upwards: the outer and northernmost is the Schouwenbank; SE of that is a string of banks, Middelbank, Rabsbank and Thorntonbank, stretching from Schouwen to the Westerschelde approaches; and finally the Steenbanken off the Roompot which has some patches of less than 4m definitely to be avoided.

Navigationally the area is the world's best-provided. Offshore from the Delta to guide traffic to and past Europoort are the two major precautionary areas at Noord Hinder Junction and round Maas Center, the latter with a radar surveillance area stretching for a 6 to 7M radius from the Hoek. It is easy to navigate outside these areas since the whole offshore and inshore region is profusely buoyed and lit, with offshore lights and long-distance onshore lights on the point of each island. There is also a screen of Fl.Y.5s beacons at varying, but frequently 3 to 5M, intervals strung from island point to island point across the closed Delta estuaries, and usually not far from the 5m contour. Finally Scheveningen is one of NAVAREA I's special transmitting stations feeding frequent weather, hazard and other navigational information into the NAVTEX network.

The nearly completed Oosterschelde barrage. The photo was taken whilst work was still in progress. In the foreground is the partially complete barrage across the Roompot and the service bridge in the left-hand corner is to be removed now that work is complete.   *Photo:* Hofmeester

Tidal streams N of the Westerschelde are straightforward offshore but complicated inshore, even without the distorting effects of the Delta Scheme, and in the latter respect some of the Dutch *Stroomatlassen* are somewhat dated and a little sparse of information.

Offshore in the 2hr period between 3½ and 1½hrs before HW Hoek van Holland there is a large area of slack water moving N after which the NE stream sets in for about 4hrs until the period from 2½ to 4hrs after HW Hoek during which the stream first turns and runs outwards into the North Sea from the Delta channels and then turns and runs SW for another 5hrs.

Inshore in the channels N of the Oosterschelde the NE and SW offshore streams tend to dam up the streams inside the Delta inlets for 4 to 5hr periods, delaying the times of the turns of tide and giving considerable periods of slack water in these channels.

At the Hoek van Holland this results in a prolonged double LW and a 4½hr rise of tide, with nearly 7½hrs ebb stream from Rotterdam. At the Maasmond entrance of the Nieuwe Waterweg near the Hoek the ingoing tide does not start until 5hrs after the first LW and 2½hrs before HW (Hoek) when the tide is slack off the Delta and about to turn NE. Similarly the outgoing tide at Maasmond is delayed until 2½hrs after HW Hoek at the beginning of the second period when the offshore stream turns outwards into the North Sea before running SW.

The timings of streams in the Brouwershavensche Gat off the Grevelingen Dam and the Slijkgat from the Haringvlietdam are similar to Maasmond, ingoing starting 3 to 3½hrs before HW Hoek and outgoing 2 to 3hrs after HW Hoek.

The net effect of all these tidal timings is that it is easier to travel from N to S across the Delta entrances than in the opposite direction, as Routes 3, 4, 5 and 6 below illustrate.

Three offshore routes into the area, and four coastal routes are described below.

ROUTE 1   OFFSHORE – KWINTEBANK BUOY TO WESTERSCHELDE ENTRANCE

Distance   32M to Breskens

Commentary

The very simple deep-water route from the Kwintebank buoy S of the proliferously buoyed Scheur and Wielingen channels has been described in Chapter 10, Route 4, as far as Zeebrugge. Crossing the outer entrance of the new Zeebrugge harbour at right angles it enters the Westerschelde close to the Nieuwe Sluis Lt, continuing along the well buoyed river, or into Breskens, or into Vlissingen, taking care to cross the river at right angles. The outward route can be the same or alternatively N of the Scheur channel.

Tidal timing

Nieuwe Sluis at the Westerschelde entrance is 30M from Kwinte Bank, a good single tide's sail, but streams off the Sluis turn some 2½hrs earlier than near Kwinte Bank. The outward passage from the Westerschelde therefore is easier than the inward passage; starting 1hr after HW Vlissingen gives 9hrs of favourable stream to get well beyond the Kwintebank buoy and start the crossing to Harwich or the Essex rivers. Unfortunately in the latter case it means towards the end of the passage starting to fight a full adverse tide somewhere down King's Channel into the Crouch or Blackwater.

The inward passage to the Westerschelde cannot avoid some adverse tide, since E-going tide starts near Kwinte Bank at −2½hrs and at Nieuwe Sluis at +1¼hrs on HW Vlissingen giving only 3 to 4hrs of fair tide, but always plan to have some flood into the entrance since the ebb is very strong.

Approaching the Roompotsluis

ROUTE 2   OFFSHORE – SCHOUWENBANK TO THE ROOMPOT

Distance   16M Middelbank buoy to Roompotsluis

Commentary

Approach is from directly across the North Sea and the Noord Hinder south TSS, then pick-ing up one or more of the group of buoys marking the outer banks, then crossing or wending through the banks depending on weather, and finally picking up the channel-marking buoys of the Roompot or Oude Roompot to the Roompotsluis.

The best approach route is to aim for the Middelbank buoy, the middle of a NW–SE string of 5 Lt buoys stretching just over 7M from the S end of Schouwenbank to the Steenbanken and straddling Middelbank. It is very difficult to miss all of these buoys, from N to S: Schouwenbank (RW, Mo(A)8s), SBZ (S card), Middelbank (RW, Iso.8s), Magne (N card, 9.6m swept wreck marker) and MSB (W card). Westkapelle Lt (Fl.3s28M) and Noor-derhoofd Lt (Oc.WRG.10s13-10M, and Wh in this sector) on the end of Walcheren are lead-ing lights (149.5°) which line up with all 5 of these buoys, with some slight deviations.

From Middelbank buoy head along the leading line (taking care to miss the buoys!) for the southernmost of the 5 buoys, MSB, near which the Steenbanken can be crossed (6m at this point so the weather must not be much more than Force 4 to 5, otherwise a detour may be necessary depending on wind direction). E of this there is a choice of two routes (in daytime, only one at night) around each side of the Hompels bank direct to the Roompot lock entrance (which opens on request), and each with closely spaced G, and R channel-marking buoys and occasional cardinal buoys and Y beacons:

1.   The Westgat and Oude Roompot channel in which many of the marks are lit.
2.   The Roompot proper in which the marks are not lit and at the end of which there is a short buoyed dogleg N to the lock.

Tidal timing

From 16M offshore after a North Sea passage about 3hrs are needed to get to Roompotsluis. The ingoing tide in the Roompot starts at HW Zierikzee −0600 (Dover −0300) so ideal arri-val timing at Middelbank is between HW Dover −0600 and −0300, giving from a full to a half flood start at the Roompot, depending on how far into the Oosterschelde you wish to sail and whether or not you wish to push some tide initially in the Roompot approaches. During the whole of this period the offshore tide is running SW, continuing to do so until HW Dover, so the last part of the passage from the TSS is port-bowing into the tide.

Leaving the Roompotsluis on the return journey across the North Sea, the outgoing tide starts just after HW Zierikzee +0100 (HW Dover +0300), and depending again on whether or not you wish to push some tide from Zierikzee or Colijnsplaat best timing to reach the lock is between HW and +0300 on HW Zierikzee, reaching the Middelbank between +0600 and −0330 on HW Dover when the tide is running SW giving a good quartering tidal stream across to the Noord Hinder S TSS.

ROUTE 3 COASTING – WESTERSCHELDE TO HOEK VAN HOLLAND

Distance 50M

Commentary

The coastal route from the Westerschelde to the Hoek can be made in either a single 50M passage from Vlissingen to Maasmond entrance around the outermost buoys and beacons of the 3 banks stretching from the 3 northern islands after negotiating the Oostgat off Walcheren, or in a series of 3 shorter passages (Routes 4, 5 and 6 below). The Oostgat, which 2 of the 4 routes use, on the W side of Walcheren is very well lit and buoyed, and is the only easily navigable channel across any of the sandbank extensions of the 4 main islands.

After the Oostgat, between OG3 (G) and OG (RW) Lt buoys, the outer route continues to follow a string of Lt buoys: WG1 (G) Westgat (watch out for unlit WG2 ½M N), MD3 (G) Middeldiep (watch out for unlit OSB 1.7M S), Ooster (W card) with a Fl.Y beacon 1.8M SW, and then an 11M gap across the Slijkgat approach and crossing a 7m patch to Adriana (W card) leaving this buoy about 1M to the N and then following the recommended pleasure craft route (see Chapter 2) across or into the Europoort entrance.

Tidal timing

The 50M takes 11 to 12hrs for a modest yacht. Northwards starting from Vlissingen 4 to 5hrs after HW gives about an hour, not quite enough, of fair tide up the Oostgat, then 4 to 5hrs pushing the tide offshore followed by 5 to 6hrs fair tide to Maasmond and probably taking over 12hrs in all. In the opposite direction southwards leaving Maasmond at 3hrs after HW Hoek gives a 30 to 35M passage with the tide for 6½hrs to the outer end of the Westgat (leading into the Roompot), 1 to 2hrs crossing the tide to the Oostgat and 2 to 3hrs of tide against into Vlissingen, possibly completing the passage in less than 11hrs.

ROUTE 4 COASTING – WESTERSCHELDE TO THE ROOMPOT

Distance 20M

Commentary

The short passages are more closely buoyed with the addition of the Fl.Y beacons. Between the Oostgat OG (RW) Lt buoy and the first of the Roompot channel buoys (see Route 2 above, these buoys are unlit) there is a string of unlit G buoys and two usefully spaced Fl.Y beacons, OS15 Fl.Y.5s, and closer inshore Mo Fl(4)Y.10s, followed by an E cardinal Lt buoy on the dogleg up to the Roompot lock.

Tidal timing

On the northward route a fair tide throughout is impossible for a normal yacht. From about ½hr before to 2½hrs after HW Vlissingen the tide sweeps northwards round the Westkapelle end of Walcheren, but prior to this period there is a foul tide down the Oostgat and after this period a foul tide out of the Oosterschelde. Leaving Vlissingen 1hr before HW gives 2 to 3hrs of fair tide and 1 to 2hrs of foul at the end. In the opposite direction southwards from the Roompot a fair tide all the way round Walcheren can be obtained by starting about 3hrs after HW Vlissingen (this also follows an hour of fair tide out of the Oosterschelde travelling from, say, Colijnsplaat or Zierikzee.

The entrance to Zierikzee on the Roompot

ROUTE 5   COASTING – ROOMPOT TO GOEREESESLUIS, HARINGVLIETDAM

Distance   30M

Commentary

From the lock the route is along the well lit Oude Roompot channel and then the Geul van de Banjaard. The latter is G-buoyed but unlit, easy by day but difficult by night, although there is one Fl.Y.5s beacon (OS13) in the next channel to the E which is helpful but still difficult, whilst the Banjaard shoals are a maze of patches often drying or less than 1m, so a daylight passage is called for. From NBJ (W card, but the shoals are to the S!) unlit buoy follow the offshore route (Route 3 above) outside Ooster (W card) Lt buoy, then yet another Fl.Y beacon on the way, picking up one of the group of Slijkgat outer channel buoys and another Fl.Y beacon. The route then follows the R, G, and Y channel buoys, many of which are lit, finally turning S into the Buitenhaven at Stellendam and the Goereesesluis through the dam.

Tidal timing

The northward passage is again difficult and is likely to take 7hrs or more; to enter the Haringvliet on an ingoing tide requires pushing the tide out of the Roompot. Leaving the Roompotsluis 2 to 3hrs before HW Vlissingen and pushing and crossing the tide for about 2hrs through the Banjaard gives a fair tide for the rest of the route. The southward route is easiest, and starting 3½hrs after HW Hoek van Holland gives a fair tide throughout with a continuing ingoing tide in the Oosterschelde.

Haringvliet lock and bridge – inland side

ROUTE 6   COASTING – GOEREESESLUIS TO HOEK VAN HOLLAND

Distance   15M

Commentary

Forget the short cut which used to exist along the Rak van Scheelhoek and the shallow Gat van de Hawk; a land extension (no problem to the Dutch) of the Maasvlakte side of Europoort has now cut off this channel for evermore. The only route to the Hoek is to leave the Slijkgat at about SG6 buoy (unless you work the tide and cheat to cross the shallow S end of the Ribben) heading for Hinder (W card) Lt buoy, keeping to the W of Ha1 Fl.Y beacon, and then for MV (W card) Lt buoy and joining the recommended route (see Chapter 2) across or into the Europoort entrance.

Tidal timing

The short distance gives some timing flexibility but the usual problem northwards, and starting 3hrs after HW Hoek gives a fair tide out of the Slijkgat for an hour, an hour of slackish turning tide offshore and a hard push to and into the Maasmond entrance for the last hour or so. Southwards a start from the Hoek at between 6 and 3hrs before HW gives a fair tide throughout.

ROUTE 7 COASTING – SCHEVENINGEN TO HOEK VAN HOLLAND

Distances 10½M to Maasmond entrance
14M to Berghaven

Commentary

This route from Scheveningen avoids the Maas N TSS and the precautionary area and is direct into or out of the Nieuwe Waterweg taking care to cross to the correct side after contacting Maasmond Radar (Ch 2). Alternatively, also requiring VHF contact on Ch 2, there is a route past Europoort to the rest of the Delta directly across the precautionary area as described in Chapter 2, at a sharp angle to the Europoort entrance and inside the recommended 1½M band marked on the Dutch Hydrographic charts.

Tidal timing

Best departure at Maasmond entrance is HW Hoek −0200 (−0300 at Berghaven) at the end of a falling tide down the Nieuwe Waterweg, and at the beginning of the N-going tide to Scheveningen. Starting from Scheveningen at HW Hoek −0330 gives 1½hrs of the end of the S-going tide to reach the entrance in time for the start of the ingoing stream along the Waterweg.

Bad weather routes inland

In bad weather for crossing the Delta by the outer routes it makes sense for fixed-mast vessels to take inland diversions, which are by no means 'short cuts' and usually require short waits at bridges and locks. Although outside the more detailed scope of this book, following is a brief description of each, coinciding with the departures and destinations of Routes 4, 5, and 6 above.

ROUTE 4 INLAND – VLISSINGEN TO ROOMPOTSLUIS
Distance 28M
Charts Dutch Hydrographic *1803, 1805*

The first part is non-tidal; lock through at Vlissingen, follow the Kanaal door Walcheren (several opening bridges) to Veere, lock into the Veerse Meer and turn E, lock out of E end of Veerse Meer into Zandkreek. The second part is tidal (use *Stroomatlas c*, Oosterschelde); from Zandkreek into the Oosterschelde, then turn W through or under the Zeelandbrug and then to the Roompotsluis.

ROUTE 5 INLAND – ROOMPOTSLUIS TO GOEREESESLUIS
Distance 46M
Charts Dutch Hydrographic *1805, 1807*

The first part is tidal so needs careful timing (use *Stroomatlas c*, Oosterschelde); lock through the Roompotsluis, through or under the Zeelandbrug, follow the Keeten, Mastgat, Zijpe, Krammer, Volkerak, and lock out at the Jachtensluis at the end of the Volkerak. The second part behind the Haringvlietsluizen is effectively non-tidal, the seaward source being ultimately at Maasmond but complicated by the Rhine discharge and sluicing operations at the dam; W through or under the Haringvlietbrug, then Haringvliet, to Goereesesluis in the Haringvlietsluizen.

ROUTE 6 INLAND – GOEREESESLUIS TO MAASMOND OR ROTTERDAM
Distance 36M
Charts Dutch Hydrographic *1807, 1809*

This route from the Spui is tidal so needs careful timing (use *Stroomatlas e*, Benedenrivieren). Lock through Goereesesluis, then along the Haringvliet to the Spui entrance, N along the Spui, W at the T-junction and along the Oude Maas, with 2 opening bridges at Spijkenisse (the first one is high enough to be negotiated by smaller fixed-mast yachts without opening), and finally W at the T-junction and along the Nieuwe Waterweg to Maasmond entrance. Alternatively you can turn E at the T-junction and go to Rotterdam. There is very heavy sea-going shipping and barge traffic along this route.

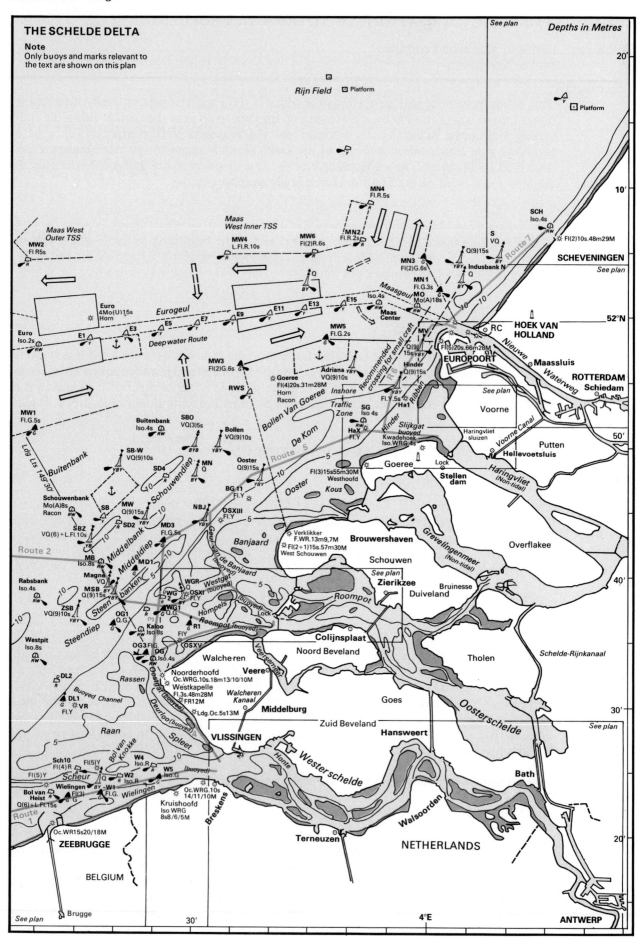

THE SCHELDE DELTA

Note
Only buoys and marks relevant to
the text are shown on this plan

Depths in Metres

Rijn Field ☐ Platform

☐ Platform

MN4
Fl.R.5s

*Maas West
Outer TSS*

*Maas
West Inner TSS*

MW4
L.Fl.R.10s

MW6
Fl(2)R.6s

MN2
Fl.R.2s

MN3
Fl(2)G.6s

Indusbank N
Q

Route 7

SCH
Iso.4s

Fl(2)10s.48m29M

SCHEVENINGEN

MW2
Fl R5s

S
VQ

MN1
Fl.G.3s

MO
Mo(A)18s

Q(9)15s

See plan

52°N

Euro
4Mo(U)15s
Horn

Eurogeul

E11

E13

E15

Maas
Center

Maasgeul

Iso.4s

MV

**HOEK VAN
HOLLAND**

Euro
Iso.2s

E1

E3

E5

E7

E9

Deep water Route

MW5
Fl.G.2s

Q(9)
15s

Fl(5)20s.66m28M

EUROPOORT

RC

Nieuwe

Maassluis

ROTTERDAM

Schiedam

MW3
Fl(2)G.6s

Adriana
VQ(9)10s

Hinder
Q(9)15s

Waterweg

RWS

Inshore

Fl.Y.5s

Ha1

See plan

Voorne

Goeree
Fl(4)20s.31m28M
Horn
Racon

*Traffic
Zone*

SG
Iso 4s

HaX
Fl.Y

Hinder

Voorne Canal

MW1
Fl.G.5s

Buitenbank
Iso.4s

SBO
VQ(3)5s

Bollen
VQ(9)10s

Ooster
Q(9)15s

De Kom

Route 5

Slijkgat
buoyed
Kwadehoek
Iso.WRG.4s

Goeree

Lock

*Haringvliet
sluizen*

Hellevoetsluis

Putten

*Haringvliet
(Non-tidal)*

**Stellen
dam**

Buitenbank

SB-W
VQ(9)10s

SD4

MN
Q

Fl(3)15s55m30M
Westhoofd

Ooster

Kous

**Stellen
dam**

Overflakee

Schouwenbank
Mo(A)8s
Racon

SB

MW
Q(9)15s

Schouwendiep

BG11
Fl.Y

Verklikker
F.WR.13m9,7M
Fl(2+1)15s.57m30M
West Schouwen

Brouwershaven

Schouwen

*Grevelingenmeer
(Non-tidal)*

Ldg Lts 149°30′

SBZ
VQ(6)+L.Fl.10s

SD2

MD3
Fl.G.5s

NBJ

OSXIII
Fl.Y

Banjaard

Duiveland

Bruinisse

Route 2

MB
Iso.8s

Magne
VQ

MD1

van de Banjaard
(buoyed)

WGR-
OSXI
Q

Westgat
(buoyed)

Zierikzee

See plan

Roompot

Rabsbank
Iso.4s

MSB
Q(9)15s

WG
2
Fl.Y

WG1

Hompels

Lock

Noord Beveland

Colijnsplaat

Tholen

Schelde-Rijnkanaal

ZSB
VQ(9)10s

OG1
Q.G.

Kaloo
Iso.8s

R1

Fl.Y

OSXV

Roompot (buoyed)

Westpit
Iso.8s

OG3 Fl.G

OG
Iso.4s

Walcheren

Veersemeer

Goes

Zuid Beveland

DL2

Noorderhoofd
Oc.WRG.10s.18m13/10/10M
Westkapelle
Fl.3s.48m28M
FR12M

*Walcheren
Kanaal*

Veere

DL1
Fl.Y

VR

Rassen

Ldg.Oc.5s13M

Middelburg

Hansweert

Bath

Raan

Spleet

Deurloo (buoyed)

VLISSINGEN

Honte

Wester schelde

Oosterschelde

Sch10
Fl(4)R

Fl(5)Y

W4
Iso.R

W5
Iso.G

(buoyed)

Fl(5)Y

Scheur

Wielingen
Fl(3)

W2
Iso.R

W1

Wielingen

Oc.WRG.10s
14/11/10M
Kruishoofd
Iso WRG
8s8/6/5M

Breskens

Walsoorden

NETHERLANDS

Bol van
Heist
Q(6)+L.Fl.15s

Route 1

Oc.WR15s20/18M

ZEEBRUGGE

Terneuzen

BELGIUM

See plan

Brugge

30′

4°E

ANTWERP

20′

10′

52°N

50′

40′

30′

20′

The harbours and rivers

The area covered here is quite a mixed one. The islands of Zeeland constitute the major part of it including an outlying area on the S side of the Westerschelde with Breskens and Terneuzen. Antwerp is in Belgium's Limbourg province, whilst the industrial part of Zuid Holland spills over into the N of the Delta. Antwerp and Rotterdam are large, interesting cities full of entertainment and tourist interest, balanced by the quieter landscapes of the islands; gabled villages, polders and dykes, lakes, woods and tree-lined roads.

This guidebook only brushes the outer edges of the cruising ground, missing places such as Middelburg, Goes, Veere, Willemstad, Dordrecht, and many a dozen others. Above all there is always a well appointed marina nearby, or if you prefer a sheltered, often tideless anchorage. After a while this may become boring to some, but following an uncomfortable thrash across the North Sea it is usually more than welcome to the normal cruising family.

THE WESTERSCHELDE TO ANTWERP

Port radio (VHF)

Schelde Information Service Shore stations	VHF Ch	Hrs listening
Vlissingen		
Schelde entrance		
Call *Vlissingen Radio*	14	24
Vlissingen area		
Call *Post Vlissingen*	21	24
Terneuzen		
Call *Post Terneuzen*	**03**, 14	24
Hansweert		
Call *Hansweert Radio*	71	24
Zandvliet		
Call *Zandvliet Radio*	12, 14	24
Waarde Radar	19	24
Saaftinge Radar	21	24
Zandvliet Radar	04	24
Kruisschans Radar	03	24

Broadcasts (Schelde Shipping Reports in Dutch and English) by: Vlissingen on Ch 14 H+35; Terneuzen on Ch 03 H+55; Zandvliet on Ch 12 H+35.

Vlissingen port	09
Vlissingen locks	22

☎ (01184) 1 23 72

Terneuzen		
Call *Havendienst Terneuzen*	11	24

☎ (01150) 1 21 61

Note Also Terneuzen–Ghent canal, listening watch on Ch 11 compulsory. Information broadcasts, Ch 11 H+00, for vessels in outer basins and on the canal.

Hansweert locks	22

Antwerpen		
Call *Antwerpen Havendienst*	18	24
Kattendijksluis and Royerssluis locks	22	

Entry signals

VHF contact is desirable for entry of any of the above harbours available to yachts, with the exception of Imalso Yacht Harbour at Antwerp.

Vlissingen, signal station on S pier, R Flag/R Lt – port closed; R and G Flag/RG Lt – port closed to vessels over 6m draught.

Breskens, no entry signals or VHF channel

Terneuzen, no entry signals to yacht harbour. 2F.R Lt from E side of main entrance – basins closed to traffic. Fl.Or Lt shown from pilot station when vessel 3000 GRT or over is about to enter river from Zeevaarthaven or Binnenvaarthaven.

Hansweert, signal station on W pier, Iso.Y.2s over F.W – ingoing navigation at entrance; F.W over Iso.Y.2s – outgoing navigation at entrance; Iso.Y.2s over F.W over Iso.Y.2s – two-way traffic at entrance; Iso.Y.2s only – vessel leaving the lock.

Imalso Yacht Harbour signal station, R flag/R Lt – exit only; G flag/G Lt – entry only; Bu flag/Bu Lt – closed.

Customs

Vlissingen: Westerhavenweg near the lock
☎ (01184) 6 00 00

Breskens: Deltahoek 7, ☎ (01172) 18 37

Terneuzen: Ambtenarenwacht at the lock, or enquire at Jachthaven Terneuzen, or ☎ (01150) 1 23 77

Hansweert: ☎ (01130) 17 23

Imalso Yacht Harbour: enquire from HrMr Belgian customs procedures (see Chapter 1)

Entrance and river passage

Distance 44M Breskens to Antwerp (Imalso Yacht Harbour)
33M Terneuzen to Antwerp (Imalso Yacht Harbour)

Because of the constricted neck of the estuary between Vlissingen and Breskens tidal streams at springs can average 4 knots, i.e. they can run even faster, and further along on many of the bends between the sandbanks rates are almost as high.

Given the stream rates and the above distances to Antwerp tidal timing on the full passage is critical, and it is advisable to break the journey at Terneuzen, the best of the intermediate harbours. At Nieuwe Sluis the ingoing stream starts at HW Vlissingen −0500 and at Antwerp 45M up the river the outgoing stream starts at HW Vlissingen +0300, giving 8hrs for the upriver passage. From

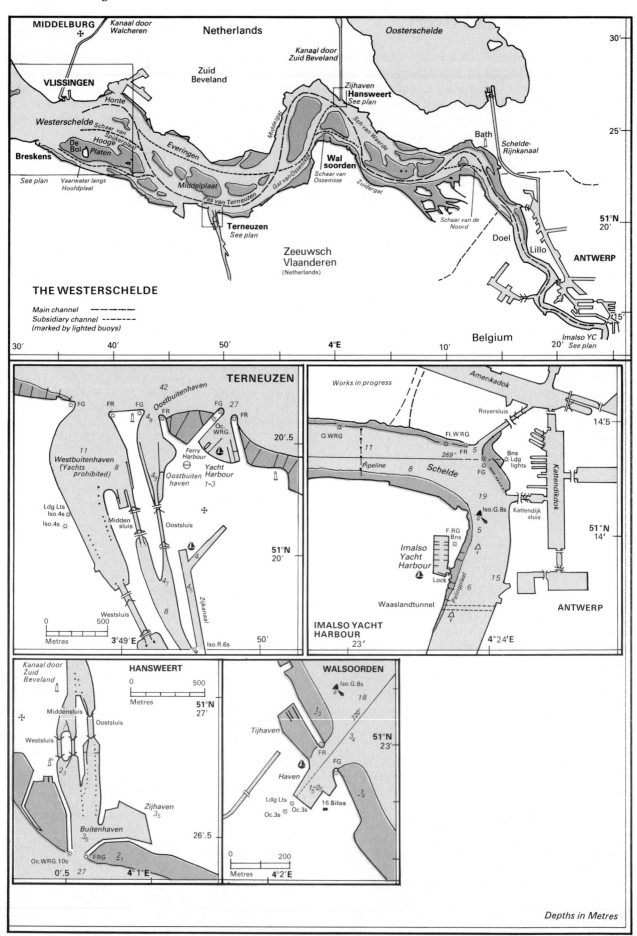

MIDDELBURG
Kanaal door Walcheren
Netherlands
Oosterschelde
30'

VLISSINGEN

Zuid Beveland

Kanaal door Zuid Beveland

Zijhaven
Hansweert
See plan

Honte

Westerschelde
Schaar van Spijkerplaat
Middelgat
Sch van Waarde
Bath
Schelde-Rijnkanaal
25'

Breskens
De Bol
Hooge Platen
Everingen

See plan
Vaarwater langs Hoofdplaat
Middelplaat
Pas van Terneuzen
Gat van Ossenisse
Wal soorden
Schaar van Ossenisse
Zuidergat

51°N 20'

Schaar van de Noord
Doel
Lillo
ANTWERP

Terneuzen
See plan

Zeeuwsch Vlaanderen
(Netherlands)

Belgium

Imalso YC
See plan

THE WESTERSCHELDE

Main channel — — —
Subsidiary channel - - - - -
(marked by lighted buoys)

30'   40'   50'   **4°E**   10'   20'   15'

TERNEUZEN

42
Oostbuitenhaven
27
FG FR FG
FR
FG FR
Oc. WRG.
20'.5

11
Westbuitenhaven (Yachts prohibited)
8
4₉
4₉
Ferry Harbour
Oostbuiten haven
Yacht Harbour 1–3

Ldg Lts
Iso.4s
Iso.4s
Midden sluis
Oostsluis

51°N 20'

4₇
4₅
8
Zijkanaal
4

0   500
Metres
Westsluis
3°49'E
Iso.R.6s   50'

Works in progress
Amerikadok
Royersluis

Q.WRG
Fl.WRG
14'.5

11
269° FR 5
Bns
Ldg lights
Kattendijkdok

Pipeline
8
Schelde
FG

19
Iso.G.8s
Kattendijk sluis

51°N 14'

F.RG
Bns
5
Imalso Yacht Harbour
Lock
6
Palingplaat
15

Waaslandtunnel
ANTWERP

IMALSO YACHT HARBOUR
23'
4°24'E

HANSWEERT

Kanaal door Zuid Beveland

0   500
Metres
51°N 27'

Middensluis
Oostsluis
Westsluis

2₃
Zijhaven 3₅

26'.5

Buitenhaven 3₅
Oc.WRG.10s   FRG   *2₁*
0'.5   27   **4°1'E**

WALSOORDEN

Iso.G.8s
18
1₃
220°
Tijhaven
3₄

51°N 23'

FR
Haven
FG
1₅–2₅
-1₄
Ldg Lts
Oc.3s Oc.3s
16 Silos

0   200
Metres   **4°2'E**

Depths in Metres

Typical Westerschelde scenery – the dyked S bank

Terneuzen the ingoing stream starts a little later than Vlissingen at HW −0430 but still gives 7½hrs to do 33M. On the return journey there is 5hrs of stream to Terneuzen, but only 4½ to Vlissingen so a break of journey is even more essential and engine power is usually required in any case. The tidal lock at Imalso Yacht Harbour, Antwerp, is open from 1hr before to 1hr after HW Antwerp so it is advisable to depart as soon as it opens.

The river entrance is deep (10–20m soundings), and a buoyed (lit) shipping entrance channel leading to a precautionary area traffic scheme off Vlissingen delimits the Wielingen N and S anchorage areas on each side of the channel. There is heavy seagoing shipping and barge traffic, and on one occasion I counted some 40 ships anchored off the entrance. Yachts approaching Vlissingen from seaward normally enter through the S anchorage area past Nieuwe Sluis Lt, turn northwards when opposite the entrance, cross the traffic lanes at right angles and then enter the harbour. Two chimneys inland, St James Church spire and a windmill (see *Major fixed daylight marks* above) are unmistakable landmarks. The Breskens–Vlissingen ferry also crosses the river at this point.

The river is well buoyed and lit to Antwerp and the buoyed channels through the wide meanders and drying patches are generally at least 5m and as much as 20m deep. There are leading lights along all the main straights except Middelgat, e.g. Pas van Terneuzen, Terneuzen to Eendragtpolder, Overloop van Hansweert, Zuidergat, Nauw van Bath (bend, 2 sets), a series round the Plaat van Doel and a set on each straight from then on to Antwerp.

There is a choice of two channels N and S of the banks between Breskens and Hansweert. From Breskens the S one via the Vaarwater langs Hoofdplaat and the Springergeul into the Pas van Terneuzen is usually more convenient. From Vlissingen the N channel, Schaar van Spijkerplaat, also leading into the Pas van Terneuzen, is quicker. On the rising tide a short cut can be taken across the drying patch S of Hansweert through the buoyed Schaar van Ossenisse (0.4m dry), but on the return journey with a falling tide it can be a little risky. Another buoyed channel, Schaar van de Noord (0.6m least depth), is available across the bend S of Bath, and there are a few more tempting but unbuoyed drying bends, but for the newcomer it is best to keep to the edges of the channel. The final hazard is the Plaat van Doel on the W bank with a drying, dammed outer edge, an 82m pylon at the N end spectacularly rising from apparently the middle of the river, and a sector Lt halfway along the dam. Needless to say the channel is well buoyed (lit) – the Belgians do not do things by halves! There are a few more bends with modest drying banks on their convex sides before a sharp bend around a windmill to Imalso opposite Antwerp's waterfront.

Finally, it cannot be stressed enough that a good lookout should be kept for ships from both directions. If you are sailing – and the banks are not heavily built-up except for occasional factory installations – then be ready with the engine.

Mooring and facilities

VLISSINGEN Pass between the port and starboard entrance Lts (2¼ cables apart) into the main E harbour (Buitenhaven). Watch out for the Breskens and the larger Olau Line (from Sheerness) ferries which tie up N of the locks. The locks are serviced throughout the day, have traffic signals and a megaphone, and whilst waiting it is possible to tie up to pilings on the S wall, although it is preferable to motor around or tie alongside any suitably sized commercial vessel which is waiting at the pilings. Inside carry on past the two Binnenhavens, turn to starboard into the Kanaal door Walcheren and to starboard again before the bridge into the Jachthaven (VVW Schelde Vlissingen ☎ (01184) 6 59 12). Mooring is head and stern to pontoons and posts, up to 3m depth. It is a small marina and busy in season but usually

Vlissingen Jachthaven entrance to the right, Kanaal door Walcheren to the left

places are found and there are all facilities (some of which are rather basic!) – showers, toilets, slipway, diesel, nearby repair facilities, a chandlery across the road, and a laundrette 15 minutes away (Hobeinstraat 15). The town centre itself is a good half hour's walk away. The harbour is very convenient for meeting crew from the ferry, and for continuing on along the Kanaal door Walcheren to the Veerse Meer, through a usually open lock and several opening bridges (times can be found out in the yacht harbour). Vlissingen succeeds in being a pleasant seaside and fishing town with a 14th century church and 16th century gate, as well as being a large shipbuilding centre.

BRESKENS has a large building with a silo tower as the main landmark in the centre of the harbour, which can be seen well offshore. Past the Veerhaven (ferry harbour) entrance on the S bank of the estuary there is a simple starboard turn into the harbour, and then to port into the municipal Jachthaven (☎ (01172) 19 02), 1–3m minimum depth, taking care to keep clear of fishing vessels using the entrance (no signals or VHF). This is a very sophisticated and fully equipped marina with all facilities, including the yachting wife's paradise, a laundrette. The nearby town has a good shopping street but is rather suburban.

TERNEUZEN is undoubtedly the yachting jewel of the Westerschelde. Continuing down the Pas van Terneuzen, past the white tanks of the huge Dow Chemical works, then past the two entrances to the commercial harbour with the locks into the Ghent–Terneuzen canal, the Veerhaven and Jachthaven entrance is a simple turn to starboard between two sea walls with port and

starboard Lts on top and the B and Wh metal framework tower of the main harbour Lt on the W mole. The pontoons (with fingers) are in a triangular area in the S corner. Jachthaven Terneuzen (for information contact the secretariat, Grand Hotel Rotterdam, ☎ (01150) 1 20 41) has a small clubhouse on the marina pontoon with showers, toilets and washing machines. This is the marina of two clubs, the Royal Belgian Sailing Club and WV Honte. To the E in the harbour is another Jachthaven, this time of the WV Neusen, with toilet and washing facilities only, which also has a second marina at the top of Zijkanaal A inside the town, but which has to be entered via the locks in the Oostbuitenhaven and Terneuzen–Ghent canal. Please note the Westbuitenhaven, which also has a lock, is banned to yachts.

Terneuzen

Hansweert. The approach in rough weather

HANSWEERT's Zijhaven is an uncomfortable berth for yachts, and the harbour's main function is as protection for locking through into the Kanaal door Zuid-Beveland, which is navigable with mast up, and is a 5M short cut to Wemeldinge and the Oosterschelde. This is an alternative to Route 4 Inland above. There are pilings to tie up to on the river and on the inland sides of the three locks. Entrance is between an occulting sector Lt on the W mole on a R metal framework tower with a Wh band, and a Lt (G to the river, R to the harbour) on a grey column on the E mole.

WALSOORDEN is a tiny ferry harbour 3½M S of Hansweert, with a least depth of 3.4m and a northern drying arm. Entrance is between port and starboard-hand Lts on columns on each side wall. The HrMr will allocate a berth on arrival, but the swell makes this an uncomfortable stopover.

LILLO on the E bank, 7M before Antwerp, is a picturesque little harbour and town with of course a windmill, but the harbour dries out and is only available 3hrs either side of HW. The HrMr, ☎ (031) 68 64 56, will allocate a berth, and Yachtclub Scaldis have showers and toilet facilities. Entrance is through a gap in the bank with a black beacon, R square topmark and the R lantern of an occulting sector Lt.

ANTWERP Imalso Yacht Harbour's address is Imalso, Technisch Bureau, Thonetlaan 102, 2050 Antwerpen, ☎ (03) 219 08 95. Having rounded the final bend unveiling the magnificent vista of Antwerp's churches and waterfront, the lock, open from 1hr before to 1hr after HW, is to starboard. If it is necessary to wait it may be possible to pick up a mooring buoy just off the entrance, or to anchor. In 1986 the ownership of the marina was in process of transfer to the municipality. Mooring was head and stern between posts and pontoons. There were no social or washing facilities at that time, but four yacht clubs were a possible source of help: Liberty YC ☎ (03) 219 11 47, Royal YC of Belgium ☎ (03) 219 26 82, VVW, and Société Royale Nautique Anversoise ☎ (03) 219 14 25, and there are considerable repair and lifting facilities. About ¾M S of the marina is a foot and cycle tunnel under the river to Antwerp with a 550m walk through into the main part of the town, so unless you get a taxi, a folding bicycle is useful. Antwerp is an historic city and port, a seedbed of Flemish culture, with many churches, museums and places of interest, including Reuben's house, a National Maritime Museum, and the Sunday morning *Voglmarkt* dating from the 16th century. There is an opera house, several theatres and many good restaurants.

Entrance to the Belgian canal system is through either Royerssluis or Kattendijksluis on the other side of the river. Useful telephone numbers to contact regarding entry to the canal system are: Antwerp, Inningskantoor-Straatsburgdok ☎ (03) 41 06 58, 41 06 59 and 41 09 84.

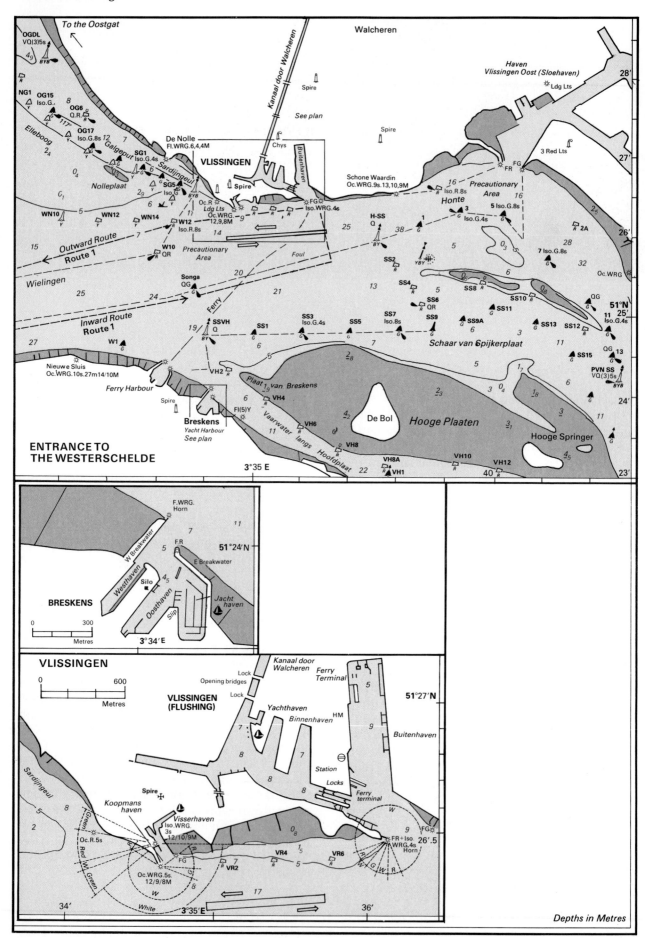

To the Oostgat

Walcheren

OGDL
VQ(3)5s

Haven
Vlissingen Oost (Sloehaven)

28'

Ldg Lts

Spire

Kanaal door Walcheren

See plan

BYB

NG1   OG15
Iso.G.

OG6
Q.R.

Spire

3 Red Lts

27'

OG17
Iso.G.8s

De Nolle
Fl.WRG.6,4,4M

117°

Elleboog

Galgeput

VLISSINGEN

Chys

Spire

Schone Waardin
Oc.WRG.9s.13,10,9M

Honte

FR   FG

Sardijngeul

SG1
Iso.G.4s

Iso.R.8s

16

Precautionary
Area

Nolleplaat

SG5
Iso.G

Oc.R
Ldg Lts
Oc.WRG.
12,9,8M

BYB

FG
Iso.WRG.4s

5 Iso.G.8s

16

2A

26'

WN10

WN12

WN14

W12
Iso.R.8s

H-SS
Q

38

1

3
Iso.G.4s

7 Iso.G.8s

28

32

Oc.WRG

Outward Route
Route 1

W10
QR

Precautionary
Area

Foul

25

BY

SS2

0₃

Oc.WRG

51°N

Wielingen

Songa
QG

20

21

13

SS4

SS8

5

SS6
QR

SS10

QG

SS11

11

Iso.G.4s
25'

Inward Route
Route 1

SSVH
Q

SS3
Iso.G.4s

SS5

SS7
Iso.8s

SS9

SS9A

SS13

SS12

QG   13

27

W1

BY

6

7

Schaar van Spijkerplaat

11

SS15

PVN SS
VQ(3)5s
BYB

24'

Nieuwe Sluis
Oc.WRG.10s.27m14/10M

VH2

Plaat   van Breskens

De Bol

Hooge Plaaten

Hooge Springer

Ferry Harbour

Spire

VH4

VH6

Vaarwater langs Hoofdplaat

Breskens
Yacht Harbour
See plan

Fl(5)Y

VH8

**ENTRANCE TO
THE WESTERSCHELDE**

3°35 E

VH8A

VH10

VH12

VH1

22

40

23'

F.WRG.
Horn

11

W Breakwater

F.R

51°24'N

E Breakwater

Westhaven

Silo

BRESKENS

Oosthaven

Slip

Jacht
haven

0   300
Metres

3°34'E

VLISSINGEN

Kanaal door
Walcheren

Ferry
Terminal

Lock

Opening bridges

5

51°27'N

0   600
Metres

Lock

**VLISSINGEN
(FLUSHING)**

Yachthaven

Binnenhaven

HM

Buitenhaven

9

Station

Locks

Spire

Koopmans
haven

Green

Visserhaven
Iso.WRG.
3s
12/10/9M

Ferry
terminal

FR+Iso
WRG.4s
Horn

FG

26'.5

Oc.R.5s

Red W. Green

VR4

VR6

FG

VR2

Oc.WRG.5s.
12/9/8M

White

3°35'E

17

36'

34'

Imalso Yacht Harbour, Antwerp. Photo on the left is the view across the entrance northwards and on right, the entrance lock

THE OOSTERSCHELDE ENTRANCE

Port radio (VHF)

Oosterschelde	*VHF Ch*
Roompotsluis lock	18
Call *Roompotsluis*	
Zeelandbrug	18
☎ (011) 10 32 37	

Entry signals

Contact with the locks on VHF is desirable. No entry signals other than traffic-light signals into the locks.

Customs

Clearance can be obtained at the Roompotsluis lock in the tower.

Tides in the Oosterschelde

Streams at the Oosterschelde entrance are similar to the Westerschelde starting in the Roompot at the same times as at Nieuwe Sluis (Westerschelde). In the main river which is much wider than the Westerschelde the streams tend to turn almost simultaneously within about half an hour right up to the E end near Bergen op Zoom. Ingoing starts around 6hrs before and outgoing just after HW Zierikzee.

The stream rates at the entrance similarly are almost as fast as those at the Westerschelde entrance, reaching 3½ to 4 knots average (note they can be stronger) at springs. Again therefore in certain conditions a normal auxiliary yacht trying to enter against the stream is well advised to anchor off by the outer banks, the Banjaard, the Hompels or near Oostkapelle well off the entrance and wait for a few hours. Once through the lock, timing is not so critical since good harbours at Colijnsplaat and Zierikzee are only 6 to 8M away.

The final completion of the Oosterschelde storm-surge barrier (autumn 1986) and of the Oesterdam at the eastern end and its northern offlier cutting off the Volkerak, the Philipsdam, will reduce the range of tide in the Oosterschelde, but effects of this upon the tidal stream rates in the Oosterschelde estuary and its entrance will also be distorted by the narrowing of the water outlet through the storm-surge barrier at the W end of the estuary by the many pillars and artificial islands. At some point it seems likely that a new tidal atlas will have to be published. In the meantime your navigation in the area including the Keeten and Mastgat offshoot should be cautious, particularly near the barrier itself.

The Delta Scheme

Inaugurated by a Dutch parliamentary act in 1958 the Delta Scheme will be complete by about autumn 1987, nearly thirty years later, after a series of technological trials, errors and successes which places the Netherlands in the forefront of world hydraulic engineering. The Roompot storm-surge barrier completed in 1986 is a line of guillotine gates with a motorway across the top and a shipping lock on an island in the middle. The gates remain lifted except during danger of storm tidal surges like that of 1953 which caused considerable havoc and loss of life in the Delta. Even more spectacular are the last three projects associated with the 'compartmentalisation' principle of separating fresh water and tidal salt water into specific compartments. The three projects due for completion in 1987 are the Oesterdam at the E end of the estuary and its northern offlier cutting off the Volkerak, the Philipsdam, both aligned along the W side of the Schelde-Rijnkanaal, and finally the discharge canal at the S end to Bath to control the level and quality of the fresh water behind the dams. The dams will have shipping locks with saltwater reservoirs which use a system of saltwater and freshwater separation based on the fact that salt water is heavier than fresh flushing the lock with salt or fresh water at each locking through.

As described above the progressive completion of the Scheme in 1986 and 1987 will change the range of tide and may affect the rates of stream in the Oosterschelde estuary, so in visiting the area it is essential to have up-to-date Dutch Hydrografische charts (and read all the footnotes!), the latest ANWB *Almanak voor watertoerisme deel 2*, check whether any new tidal atlas has been published, and contact the address below. The final completion of the project is scheduled for autumn 1987, but there is always the possibility of delay, and in summer to autumn 1986 and spring 1987 periods are scheduled for temporary closures of the storm-surge barrier to allow construction activity on both barriers. For up-to-date tide tables of the estuary contact: Rijkswaterstaat Deltadienst, Afd. Voorlichting Oosterscheldewerken, Van Veenlaan 1, 4301 NN Zierikzee. ☎ (01110) 8000; for daily information ☎ (01110) 7058.

Entrance

The estuary approach is 14M wide so it is essential to pick up the buoyed channel, since the only significant marks on each shore to help in good visibility in daylight are the West Schouwen Lt (a R diagonally striped tower like a helter-skelter) to the N, and the Westkapelle Lt (a square stone tower with a R metal superstructure) to the S, but both are 50–52m high. At night in reasonable visibility with their 28–30M ranges they are even more useful. In addition the Noorderhoofd Lt on Walcheren (13-10M range) provides a leading line with Westkapelle Lt to pick up the outer approach buoys (see Route 2), but the tower is only 16m high (R with a Wh band). There are also a number of prominent churches on Walcheren.

Entrance to the Buitenhaven of the Roompotsluis is assisted by a number of Lt buoys and beacons at the end of the Oude Roompot channel, as well as by leading Lts, and having passed between the port and starboard-hand Buitenhaven entrance Lts on ends of the high curving walls, there is a turn to starboard and S to round the Q.R Lt into the lock, which has traffic lights. There are pilings for waiting on each side of the locks, but entrance is available throughout the day. Approaching the dam from either side, however,

Roompotsluis lock and road bridge

it cannot be stressed enough to keep in the buoyed channels and keep well out of the charted danger areas, over ½M wide, on each side of the storm-surge barriers.

It is essential for vessels with high air draught to note that there is a fixed bridge over the Roompotsluis lock not shown on Admiralty charts, but shown on the Dutch charts as having 18.2m minimum clearance and in *deel 2* of the *Almanak* with a clearance of NAP (Normal Amsterdam Standard) +20m, so if you have a marginal air draught above 18.2m it is essential to work out the NAP height of tide from Dutch tide tables (see also under Zeelandbrug).

Once through the lock, then past the Binnenhaven with its Q.G Lt and between the inland entrance Lts into the estuary. The route is then S of the buoys marking the shoals behind the central dam and either N or S around the Vuilbaard shoal along the buoyed channels to Zierikzee or Colijnsplaat.

Mooring and facilities

There is an easy buoyed channel to Zierikzee N of the Vuilbaard shoal. The town, which is inland, is distinguishable from well offshore in daylight by the huge square building of Sint Lievens Monstertoren (60m), a belfry tower which would have been larger had the money not run out! There can be a fast stream running across this entrance. Entrance is from the SE between port and starboard beacons on the breakwaters, and on the W jetty there is also a distinctive sector Lt (6-4M range) on a 6m R round pedestal with a Wh band, then along a 1M canal through a lock, usually left open. To port is a pontoon marina (least depth

Zierikzee. Visitors pontoon in the distance beyond the fishing vessels

2.3m) and then about ½M further on below the 15th century harbour gateway is a single long pontoon for visitors, the Stadshaven or Nieuwe Haven, which is shallower, from just drying to 1.8m depending on position, so check your depth sounder. There are toilets and showers and all boat repair and chandlery facilities. There is a selection of HrMrs' telephone numbers you can try: ☎ (01110) 3174, 4716, 4700.

ZIERIKZEE is one of the best-preserved old fishing harbour towns in the Netherlands, with attractive narrow streets, gabled houses, churches, pleasant walks around the moated fortifications, and modest shopping and restaurant/café facilities near the square.

COLIJNSPLAAT is even easier to reach from the Roompotsluis, following the deep S bank a few cables offshore. A windmill W of the town is the most prominent landmark. There can be fast streams across the entrance which is parallel to the coast, and between stone rubble walls with port and starboard Lts. The E end of the harbour is for fishing boats only. The W end is a floating pontoon marina of WV Noord-Beveland, ☎ (01199) 7 62, with 2 to 2.5m minimum depth. There are shower, toilet and laundrette facilities with lifting and repair facilities nearby. The village is of a simple square pattern with wide tree-lined streets and limited shopping facilities.

Colijnsplaat entrance

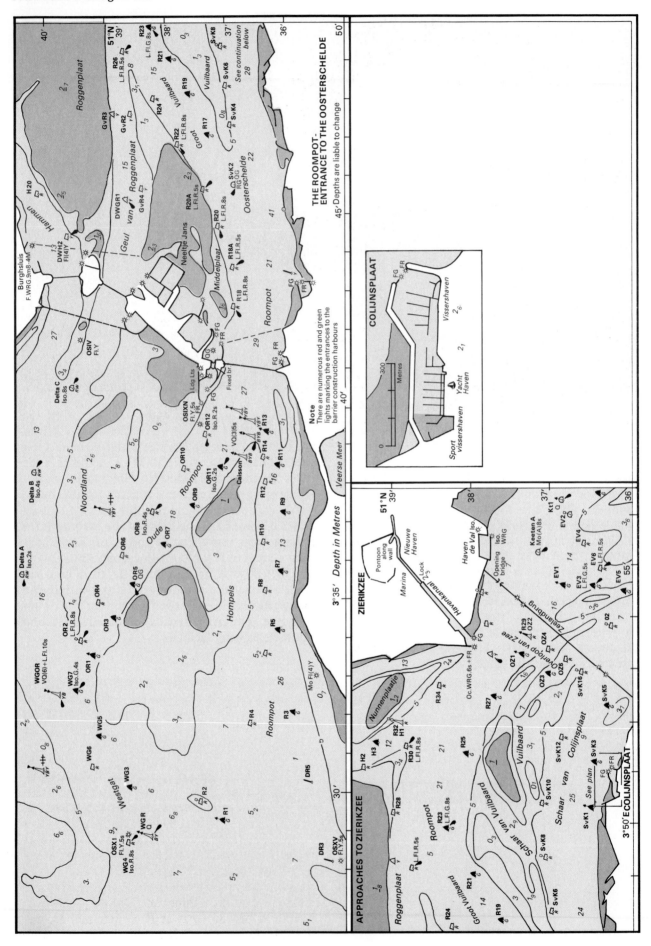

THE ROOMPOT -
ENTRANCE TO THE OOSTERSCHELDE
45′ Depths are liable to change

Note
There are numerous red and green
lights marking the entrances to the
barrier construction harbours

COLIJNSPLAAT

APPROACHES TO ZIERIKZEE   3°35′   *Depth in Metres*

ZIERIKZEE

Passing under Zeelandbrug it looks easy until you are close to the span which is only cleared by what seems like inches

ZEELANDBRUG It is essential to know your 'air draught' above the waterline if you wish to continue on into the estuary. The ANWB *Almanak deel 2* gives full details, but in summary between pillars 3 and 52 each span of the bridge has a clearance above NAP (Normal Amsterdam Standard) of 15.1m in the middle and only 11.5m in the corners. HWS is NAP +1.8m, and LWS NAP −1.8m, so the average range of clearance in the middle is between 16.9m and 13.3m. However there are spring tides which exceed these mean levels, so it always pays to check on the day of your passage with NAP-based tide tables of the estuary obtained from the *Getijtafels voor Nederlands* or from the address listed earlier in this chapter under *The Delta Scheme*, extract the day's predicted NAP +/− levels for HW and LW and apply them to the above NAP-based clearances. If in doubt ring up the bridge and arrange to go through the opening bridge at the N end.

THE HARINGVLIET ENTRANCE

Port radio (VHF)

	VHF Ch
Stellendam	
Call *Goereesesluis*	13
☎ HrMr (01879) 10 00	
Haringvlietbrug	13
☎ (01862) 18 65	

Entry signals

Contact with the locks on VHF is desirable. No entry signals other than traffic-light signals into the locks.

Customs

Clearance cannot be obtained at the Goereesesluis into the Haringvliet, nor at Hellevoetsluis. The nearest ports of clearance are at the Roompotsluis, Vlissingen and Berghaven (Hoek van Holland).

Entrance

Haringvlietdam was completed in 1971 relatively early in the history of the Delta Scheme and cuts off the freshwater tideless Haringvliet from the sea. Unlike the Oosterschelde storm-surge barrier the Haringvlietdam has rotating sluice gates which are normally left closed, except for a few which are occasionally opened on a falling tide to let out fresh water when there is an excessive discharge from the Rhine and Maas. The lock, Goereesesluis and that at Roompotsluis are the only two locks in the Scheme leading directly into the open sea. The Brouwersdam, which cuts off the stagnant salt or freshwater Grevelingenmeer between the Haringvliet and Oosterschelde entrances, was completed in 1972 and has a sluice (completed in 1978). The stagnant saltwater Veerse Meer south of the Oosterschelde has a simple but spectacular blocking dam, completed in 1961. With hindsight now that the effects on the natural environment are known it seems unlikely that the latter two projects would have been completed in the same form in the 1980s.

Admiralty charts do not cover the Slijkgat so the Dutch chart *1801* is needed. There is no tidal atlas coverage of this particular area but streams start running into and out of the Slijkgat at −0330 and +0330 on HW Hoek van Holland, but can be affected by sluicing operations through the dam which occur particularly when there is a heavy discharge from the Rhine and Maas. Red Lts are shown on the Haringvlietdam when sluicing operations are in progress and these are more likely near to low water when you are unlikely to be leaving or arriving at the dam. But on all occasions it is essential to keep to the buoyed channels on either side of the dam and keep well away from the sluices, at least 300m away on the W side and 500m on the E.

Since customs clearance is not available it is unlikely that entrance to the Slijkgat is made after a North Sea crossing; it is more likely to be after a coasting passage. The channel is 7½M long from the offing Lt buoy, SG (RW) close to a Fl.Y beacon, and is simple and well buoyed with closely spaced R and G buoys, some lit, as well as several Y conical Lt buoys on the S side of the channel, and a Fl.Y beacon at the southwards turn towards the final buoys leading up to the Buitenhaven entrance. Entrance is parallel to the high E wall past the starboard Lt at its end, and then a sharp swing to port round the port Lt on the S breakwater and through the Buitenhaven to the lock (traffic lights). On the other side follow the N wall and a sharp swing to port through the entrance following the buoyed channel out and then E along the S side of the R buoys (some lit) across to Hellevoetsluis. On each side of the locks there are waiting pilings, and across each end of the lock an opening bridge. The seaward main road bridge has NAP +14.3m clearance (see above under Zeelandbrug) so is, if at all possible, left closed for yachts. The inland bridge has only NAP +6.14m so has to be opened.

STELLENDAM   On the seaward side in the Buitenhaven (4.7m least depth) at the Goereesesluis is 'Aqua Pesch' marina, so if you want to pay a visit to the Delta Exhibition near the lock, together with a film (in several languages) and a guided tour inside the 'Spuisluizen' then this is your chance. There are toilet, shower and limited repair facilities, but Stellendam town is several miles away. At weekends the only space is to lie in the Binnenhaven fishing harbour (5m least depth) on the inland side of the lock probably alongside a fishing boat.

Hellevoetsluis entrance

HELLEVOETSLUIS   This is now a vast yachting complex, a far cry from the days when the dam did not exist and the harbour was the Dutch end of the North Sea packet service to Harwich. The tiny white LtHo on the W side of the Industriehaven is the outstanding landmark. There are three simple entrances into the harbours.

Port and starboard-hand Lts lead directly into Heliushaven (2.5m to 5m least depth), which has a complex of marinas on the W side of the harbour: WV Helius ☎ (01883) 1 65 63 or (01881) 17 03, WV Haringvliet ☎ (01883) 1 40 39, WV Hellevoetsluis ☎ (01883) 1 58 68. There are all facilities. The harbour however is a long walk from the town and it is exposed at the outer pontoons to swell in southerly weather.

Only the LtHo marks the port side of the entrance to Industriehaven (1.8m to 4.3m least depth), although there are angled wooden pilings on each side. There are, again somewhat southerly exposed, moorings along the quayside of the entrance canal, and it is better to tie up temporarily and go on through the bridge and lock which opens at frequent intervals. Inside there are two pontoon marinas, again WV Hellevoetsluis ☎ (01883) 1 46 40, and Jachthaven Arie de Boom ☎ (01883) 1 21 66. There are showers and toilets. The pleasant village centre is nearby with restaurants and limited shopping facilities, somewhat overpowered by yachts, but there is a very well provided precinct shopping centre about half an hour's walk away on the E side of the Voornsche Kanaal; you can walk along the W bank and there is a bridge across.

The Koopvaardijhaven (5m minimum depth) leading to the Voornsche Kanaal (3.5m to 4.5m least depth) is entered between port and starboard Lts so take care not to confuse them with Heliushaven at night. On the W side there are moorings with shower and toilet facilities (again the warning about exposure to S winds), or you can lock through (approximately hourly intervals) into the canal where there are two marinas: near the lock on the E side is WV Waterman, and 500m along on the W side is WV Haringvliet ☎ (01883) 1 29 24.

Finally for vessels continuing on down the Haringvliet the next major obstacle is the Haringvlietbrug some 10M from Hellevoetsluis harbour entrance. There are 10 piers numbered from S to N, and clearance above NAP (Normal Amsterdam Standard) between numbers 5 and 6 in the middle is +14m, at piers 1 and 10 at each end +10.64m, 2 and 9 +12.02m, 3 and 8 +13.01m, and 4 and 7 +13.67m. It is advisable however to make a reduction of about 1m on each of these figures for safety's sake, since the water level in the Haringvliet varies normally between +0.2m and +0.5m above NAP depending on the twice daily sluicing programme which occurs on the ebb or at LW on the seaward side of the Haringvlietdam. In exceptional flood conditions the tidal height can reach NAP +1.55m, reducing the above clearance based on NAP proportionally. If in any doubt contact the bridge and go through the opening section at the N end.

Hellevoetsluis.

HOEK VAN HOLLAND AND THE NIEUWE (ROTTERDAMSCHE) WATERWEG

Traffic control and port radio (VHF)

Call*	VHF Ch	Location
Maasmond Radar†	2	
Maas Approach	20	51°58′.9N 4°06′.8E
Maas Entrance	2	51°58′.9N 4°06′.8E
Hoek Radar Rivier‡	4	51°58′.9N 4°06′.8E
Europoort Radar	3	51°58′.9N 4°06′.8E
Rozenburg Radar	5	51°57′.0N 4°10′.0E
Maassluis Radar West	7	51°55′.0N 4°14′.8E
Maassluis Radar Oost	21	51°55′.0N 4°14′.8E
Tankhoofd Radar	22	51°53′.6N 4°20′.3E
Pernis Radar	19	51°53′.7N 4°23′.4E
Lekhaven Radar	2	51°54′.3N 4°25′.8E
Charlois Radar	4	51°53′.9N 4°28′.0E
Centrale Post (pilots)	1	51°54′.8N 4°25′.9E
Rijkshavendienst	14	51°58′.9N 4°06′.8E &
(State Hbr Service)		51°54′.8N 4°25′.9E
Gemeentehavendienst	11	51°58′.9N 4°06′.8E &
(Municipal Hbr Service)		51°54′.8N 4°25′.9E

* All operate 24hrs. All vessels on the Nieuwe Waterweg should maintain a continuous listening watch on the appropriate radar station.
† Report to and monitor when crossing the Maasmond entrance.
‡ Report to and monitor when approaching Maasmond entrance from inland.

Oude Maas	VHF Ch
Call *RHD Post Hartel*	10, 13
☎ (010) 438 12 12	
Botlekbrug	13
Spijkenisserbrug	13

Entry signals

It is essential to make contact on VHF when arriving, departing or crossing the entrance (Ch 2 at entrance, Ch 4 inland).

N mole tidal signals, G over Wh Lt – rising; Wh over G Lt – falling.

Traffic signals for Europoort, Nieuwe Waterweg and Oude Maas are shown from the radar station at Noorderhoofd, and at least give the yachtsman an idea of what is happening around him, whilst he must dutifully obey the orders of harbour control on VHF.

From seaward

R R R
W no entry Europoort nor Waterweg
R R R

R R
W no entry Europoort
R R

R R
W no entry Waterweg
R R

R R
W no entry Oude Maas
R R

W
W ship leaving Europoort for sea
W

From upstream (Rotterdam)

R R
W navigation to sea prohibited
R R

W
W ship leaving Europoort for sea
W

Customs

Clearance can be obtained alongside at either Berghaven (the Hoek van Holland), or Maassluis (Buitenhaven), or Vlaardingen (Buitenhaven), or Spuihaven (Schiedam), or Rotterdam Veerhaven.

Entrance

Distance Maasmond entrance to Rotterdam 18M

Entrance after permission from the harbour control can be confusing by day or night, since there are three sets of daylight intensity leading lights (as well as the many other harbour lights and ship lights at night), so make sure you are thoroughly familiar with them. They are in fact very simple; the middle set are white marking the direct line of the Splitsingsdam (112°) separating the Nieuwe Waterweg and Europoort, the port set are (naturally) red (107°) leading into the Waterweg, which is what you want, and the starboard set green (116°) lead into Europoort, which you definitely do not want.

This is a straightforward deep-water route with typical 10 to 15m depths in the main channel with generally very steep edges and a few narrow shoal banks between the entrance and the Oude Maas junction. However, it has the world's heaviest sea-going and barge traffic. A reliable engine is essential, and it is not advisable to put up sail given the fluky wind conditions created by the traffic and the varied shore installations and built-up area near Rotterdam. Care should be taken to keep to the letter of the International Collision Regulations; large numbers of tugs, barges and launches leave and depart from the many side channels often at a high speed and from behind blind corners, whilst there is constant overtaking by often quite large ships.

Approaching from offshore, and if Berghaven which is uncomfortable for yachts is bypassed, this route is better attempted from Scheveningen and the N. Arriving at Maasmond entrance at slack water, 2hrs before HW Hoek, follows a long period of SW-going tide offshore, and precedes the full 4 to 5hrs of ingoing tide to Rotterdam. Approaching Maasmond from the S requires either pushing a complete foul tide to obtain the full ingoing stream at the above time at the entr-

Maassluis from the air – looking southeastwards with Binnenhaven at the top of the photograph   *Photo* VVV Maassluis

ance or taking a fair tide offshore and accepting a hard push up the Nieuwe Waterweg. The tide is outgoing along the Waterweg for about 7½hrs, starting at Rotterdam at +0300 on HW Hoek. Leaving Rotterdam about 4½hrs later at around −0500 HW Hoek gives time to reach the entrance before the tide turns ingoing and just catches the start of the N-going tide offshore ready for Route 7 above. Starting at HW +0100 and pushing the tide for 1½hrs for part of the way means reaching the entrance on an outgoing tide with the offshore tide just turning S, nicely placed for a 6½hr push round the outer Route 3 above.

Mooring and facilities

BERGHAVEN, Hoek van Holland is entered to port between port and starboard Lts just before the front leading Lt tower of the Waterweg (R with Wh bands). This is a working harbour in frequent use by patrol boats, so apart from clearing customs, picking up and dropping crew from the ferries, or in emergency it is not recommended for yachts, so if you do enter make sure the HrMr approves the spot. For the HrMr ☎ (01747) 37 51.

MAASSLUIS Buitenhaven is entered between port and starboard Lts just after the radar tower. Customs clearance, ☎ (010) 76 16 66, is again one of the few reasons for tying up to the quayside, unless you are continuing on through the lock and opening bridges (bridgekeeper ☎ (01899) 1 28 52) into the Binnenhaven where there are moorings. This is an industrial area.

VLAARDINGEN, entered between port and starboard Lts on the N bank about 1M beyond the junction with Oude Maas, is a very similar harbour to Maassluis, with a Buitenhaven and a Binnenhaven (the Oude haven) entered through a lock with two opening bridges. Both have yacht berths, but the inner one is infinitely more comfortable. ☎, all preceded by (010): customs 35 73 33/76 16 66; HrMr 34 00 44; Prinses Julianabrug 34 23 70; lock-keeper 34 38 50; WV Vlaardingen on the E quay of the Oude haven 34 67 86/34 95 35 or 74 23 32.

SPUIHAVEN, Schiedam is 2M beyond Vlaardingen on the N bank, entered past a single F.G Lt to port (believe it or not!), with the Jachtclub Schiedam standing proudly on the small promontory to starboard. The pontoon marina is on the E side of the harbour. Jachtclub Schiedam, WV Nieuwe Waterweg and WV De Schie provide

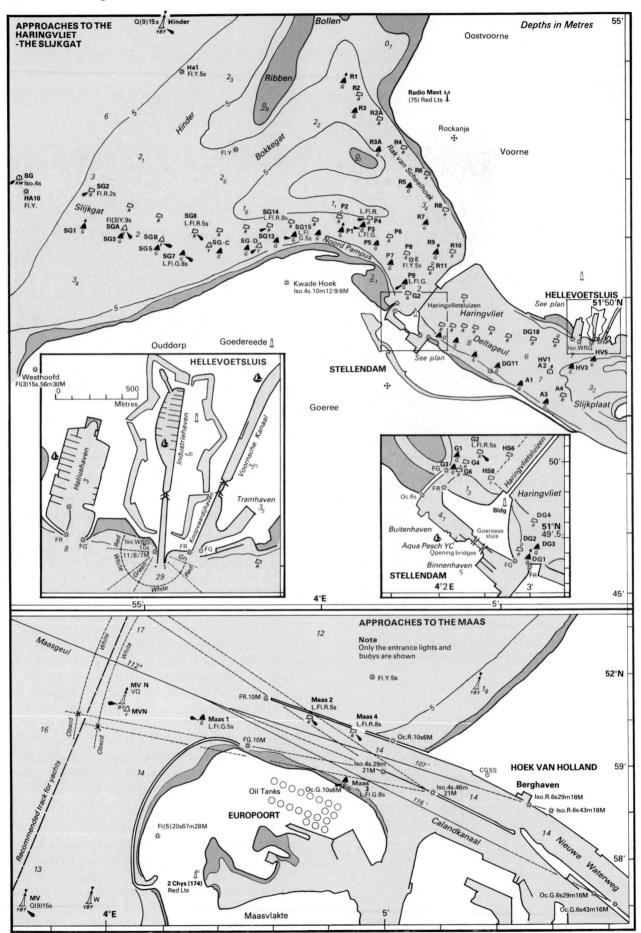

APPROACHES TO THE HARINGVLIET -THE SLIJKGAT

Depths in Metres 55'

Q(9)15s Hinder
YBY

Bollen
Oostvoorne

Ribben

Ha1 Fl.Y.5s

R1 G
R2 R
R3 G
R2A R

Radio Mast (75) Red Lts

Rockanje

Hinder

Bokkegat

Fl.Y

R3A G
R4 R
Rak van Scheelhoek
R6 R
R5 G
R7 R
R8 R

Voorne

SG RW Iso.4s
HA10 Fl.Y.

SG2 R Fl.R.2s

Slijkgat

Fl(3)Y.9s
SG8 L.Fl.R.5s
SGA
SG14 L.Fl.R.8s
SG15
SG1 G
SG3 G SGB
SG-C
SG-D
SG13
SG5 G SG7 L.Fl.G.8s
P2
P1 L.Fl.
G.5s
P3 L.Fl.G.
P4
L.Fl.R.
P5
P6
Noord Pampus
P7 G
P8 G
E Fl.Y.5s
R9 G
R10
R11 R
P9 L.Fl.G.

Kwade Hoek Iso.4s.10m12/9/8M

G2
Haringvlietsluizen
Haringvliet
See plan

HELLEVOETSLUIS 51°50'N
See plan

DG18
Deltageul
DG11
A2
HV1
A1
A3
A4
HV3
HV5
Slijkplaat
Iso.WRG

STELLENDAM

HELLEVOETSLUIS

Ouddorp
Goedereede

Westhoofd Fl(3)15s.56m30M

0   500
Metres

Heliushaven
Industriehaven
Voornsche Kanaal
Koopvaardijhaven
Tramhaven

FR
FG
Iso.WRG 10s 11/8/7M
FR FG

G2 L.Fl.R.5s
G1
G3 FG
G4
G6
HS6
HS8
Haringvlietsluizen
Oc.6s
FR
Haringvliet
Bldg
DG4 51°N 49'.5
Buitenhaven
Aqua Pesch YC
Opening bridges
Goereese sluis
DG2 DG3
DG1
FG
FR
Binnenhaven

STELLENDAM

APPROACHES TO THE MAAS

Maasgeul
White
112°

Note
Only the entrance lights and buoys are shown

Fl.Y.5s

52°N

MV N VQ
MVN
Obscd Obscd
FR.10M
Maas 2 L.Fl.R.5s
Maas 1 L.Fl.G.5s
Maas 4 L.Fl.R.8s
Oc.R.10s6M
FG.10M
Iso.4s.29m 21M
107°
Maas 3 L.Fl.G.8s
Iso.4s.46m 21M
116°
14

HOEK VAN HOLLAND 59'
Berghaven
Iso.R.6s29m18M
Iso.R.6s 43m18M
CGSS

Recommended track for yachts

Oil Tanks
Oc.G.10s6M

EUROPOORT

Fl(5)20s67m28M

Calandkanaal
Nieuwe Waterweg

Oc.G.6s29m16M 58'

2 Chys (174) Red Lts

MV Q(9)15s YBY
W YBY

4°E
Maasvlakte 5'
Oc.G.6s43m16M

Rotterdam. Royal Rowing and Sailing Club De Maas

showers and toilet facilities, there is a boatyard and most repair facilities, and, of course, a laundrette near the harbour. ☎ both preceded by (010): customs 76 16 66; HrMr 62 22 22.

VEERHAVEN, Rotterdam is 3M beyond Spuihaven just after the famous Euromast with its parkland to the E. It is a somewhat concealed turning off to port usually hidden by barges, ships or tugs moored along the Rotterdam quaysides – Parkkade and Westerkade are the first two quay walls and Willemskade is beyond the entrance. There are no marks on either side of the entrance other than black and white posts. On the starboard promontory is the Royal Rowing and Sailing Club De Maas, to give it its full title. The pontoon moorings are round three sides of the harbour, but with rather more pontoon and stern post moorings on the E side where there are shower and toilet facilities in a somewhat basic ancient wooden boathouse. The clubhouse itself is welcoming and very palatial. There are no repair facilities on the spot and the nearest yard is 5km away at Overschie, but there are chandlery and engine repair facilities in Rotterdam. There is a magnificent park next door, across which you can walk to the Euromast (183m) for a restaurant meal which can be occasionally in the clouds but hopefully with clear visibility to get a bird's-eye view of the busy Nieuwe Waterweg and the city. Largely rebuilt since the devastation of the Second World War, there are magnificent shopping precincts, a wide variety of eating places, museums, and, dare I say, you can even take a river trip. It is also fascinating to walk along the Cool haven and admire the beautifully kept barges of all nationalities which gather there. An interesting excursion is to the picturesque 16th century Delfshaven. ☎ all preceded by (010): customs 76 51 44/76 16 66; Chief HrMr 89 69 11; De Maas 13 76 81 (moorings), 13 85 14 (club).

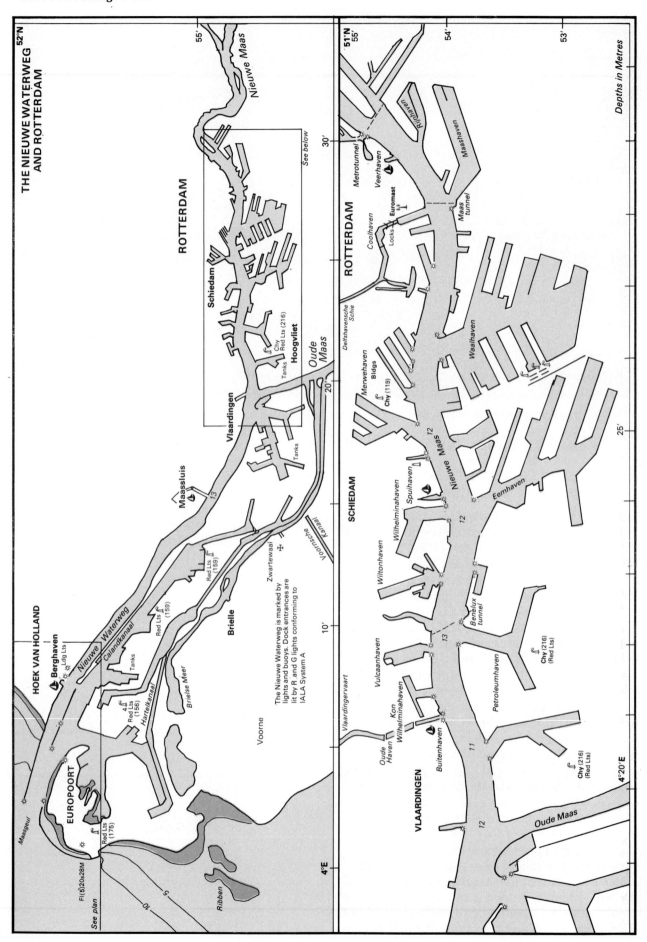

THE NIEUWE WATERWEG AND ROTTERDAM

ROTTERDAM

HOEK VAN HOLLAND

Berghaven

EUROPOORT

Maasgeul

See plan

Fl(5)20s28M

Ribben

Red Lts (175)

Red Lts (156)

Hartelkanaal

Tanks

Brielse Meer

Voorne

Red Lts (159)

Red Lts (159)

Brielle

Zwartewaal

Voornsche Kanaal

The Nieuwe Waterweg is marked by lights and buoys. Dock entrances are lit by R and G lights conforming to IALA System A.

Nieuwe Waterweg

Calandkanaal

Ldg Lts

Maassluis

Tanks

Vlaardingen

Schiedam

Chy Red Lts (216)

Tanks

Hoogvliet

Oude Maas

Nieuwe Maas

ROTTERDAM

See below

Depths in Metres

Lower panel

ROTTERDAM

SCHIEDAM

VLAARDINGEN

Metrotunnel

Veerhaven

Euromast

Coolhaven

Locks

Maas tunnel

Rijnhaven

Maashaven

Waalhaven

Delfshavensche Schie

Merwehaven

Bldgs

Chy (119)

Eemhaven

Nieuwe Maas

Wiltonhaven

Wilhelminahaven

Spuihaven

Benelux tunnel

Petroleumhaven

Chy (216) (Red Lts)

Vulcaanhaven

Kon Wilhelminahaven

Buitenhaven

Oude Haven

Vlaardingervaart

Chy (216) (Red Lts)

Oude Maas

4°E

4°20'E

10'

20'

30'

55'

52°N

51°N

55'

54'

53'

25'

11

12

12

13

12

13

12. Noord & Zuid Holland

Charts Admiralty *122, 124, 191, 1408, 2322*
Imray *Y5*
Dutch Hydrographic *1801, 1810, 1811*

Tidal streams (based on HW Hoek van Holland, HW Helgoland and HW Dover)

Position	Start times			
	H.VAN HOLLAND		DOVER	
	North	*South*	*North*	*South*
Off Maasmond ent.	−0200	+0430	+0105	−0450
3 M W of IJmuiden	−0120	+0430	+0145	−0420
Off IJmuiden ent.	−0210	+0350	+0055	−0530
	*In**	*Out*	*In**	*Out*
In IJmuiden ent.	−0110	+0250	+0155	+0555

* Runs clockwise round Buitenhaven

	HELGOLAND		DOVER	
Schulpengat	+0325	−0330	+0305	−0350

Tidal atlases Admiralty *North Sea – southern portion*
Dutch *Stroomatlassen j* and *n*

Tidal differences and ranges (based on HW Dover)

Place	HW (time)	Springs/Neaps (range in metres)
Scheveningen	+0321	1.9/1.5
IJmuiden	+0401	1.8/1.6
Den Helder	−0438	1.5/1.1

Major lights

Name of light	Characteristics	Position	Structure
Texel LtV	Fl(3+1)20s16m26M Horn(3)30s	52°47'.1N 4°06.'6E	R hull, Wh band, RC, Racon. RW buoy, Oc.10s, Racon, during maintenance

Onshore

Name of light	Characteristics	Position	Structure
Scheveningen Ldg Lts			
156° Front	Iso.4s17m14M (6M day)	52°05'.8N 4°15'.7E	Grey metal mast
Rear	Iso.4s21m14M (6M day)		Grey metal mast
Scheveningen	Fl(2)10s48m29M	52°06'.3N 4°16'.2E	Brown metal tower, 014°-vis-244°
Noordwijk aan Zee	Oc(3)20s32m18M	52°14'.9N 4°26'.1E	Wh square stone tower
IJmuiden Ldg Lts			
100.5° Front	F.WR.30m16/13M	52°27'.8N 4°34'.5E	Dark R round metal tower, RC 050°-W-122°-R-145°-W-160°
by day	F.4M		090.5°-vis-110.5°
Rear	Fl.5s52m29M		Dark R round metal tower, 019°-vis-199°
by day	F.4M		090.5°-vis-110.5°
Egmond aan Zee	Iso.WR.10s36m18/14M	52°37'.3N 4°37'.6E	Wh round stone tower, 010°-W-175°-R-188°
Huisduinen	F.WR.27m14/11M	52°57'.2N 4°43'.3E	Square stone tower 070°-W-113°-R-158°-W-208°
Kijkduin	Fl(4)20s56m30M	52°57'.4N 4°43'.6E	Brown metal tower, rear Ldg Lt 253.5° with Den Helder Marinehaven Willemsoord, Harssens Island, W breakwater head
Schulpengat Ldg Lts			
026.5° Front	Iso.4s18M (9M day)	53°00'.9N 4°44'.5E	Metal pedestal, 024.5°-vis-028.5°
Rear, Den Hoorn	Oc.8s18M (9M day)		Church spire
Schilbolsnol	F.WRG.27m15-11M	53°00'.6N 4°45'.8E	Green tower, 338°-W-002°-G-035° 035°-W-038° Ldg for Schulpengat 038°-R-051°-W-068.5°
Den Helder Wierhoofdhaven, Ferry Ldg Lts			
207° Front	Iso.2s10m14M (5M day)	52°57'.8N 4°46'.8E	Grey mast, 199°-vis-215°
Rear	Iso.2s14m14M (5M day)		Grey mast
Den Helder Ldg Lts			
191° Front	Oc.G.5s16m14M (6M day)	52°57'.4N 4°47'.2E	B triangle pt up on building, 161°-vis-221°
Rear	Oc.G.5s25m14M (6M day)		B triangle pt down on B framework tower

Radiobeacons

Name	Freq. (kHz)	Ident.	Range (miles)	Seq.	Position
Marine radiobeacons					
IJmuiden Group	294.2				
IJmuiden		YM	20	1,4	52°27'.8N 4°34'.6E
Hoek van Holland		HH	20	2,5	51°58'.9N 4°06'.8E
Eierland Lt (fog only)		ER	20	3,6	53°11'.0N 4°51'.4E
Texel LtV	308	HK	50	5	52°47'.1N 4°06'.6E
Aero beacons					
Valkenburg/Scheveningen	364	GV	25	H24	52°05'.6N 4°15'.2E
Amsterdam/Spijkerboor	381	SPY	75	H24	52°32'.5N 4°50'.5E

Coast radio stations

Station	Transmits (kHz) (VHF Ch)	Receives (kHz) (VHF Ch)	Freq[7]	Traffic lists (times)	Storm warnings (times)	Weather messages (times)	Navigational warnings (times)
Scheveningen[1] (PCG)(PCH)	**1764**,**2824** **2182** & Nes	2049[3] 2182[3] & Nes 2520[3]	1862[8]	odd H+05	On receipt	0340,0940 1540,2140	0333,0733,1133 1533,1933,2333
			1890	odd H+05	On receipt	0340,0940 1540,2140	0333,0733,1133 1533,1933,2333
			2824	0105,0305, 0505,2305		0340	0333,2333

Notes
[1] Only frequencies with 24hr service are listed.
[3] 24hr watch; 2049 when 2182 is distress working. 2520 is calling frequency, alternative is 2182.
[7] Frequency (kHz) on which traffic lists, storm warnings, weather messages and navigational warnings are given (in English and Dutch).
[8] Located at Nes 53°24'N 6°04'E.

The following stations are remotely controlled by Scheveningen; call is *Scheveningen Radio* in all cases.

Station	VHF Channel	Position
Goes	Ch 16,**23**[9],**25**,**78**,**84**	51°31'N 3°54'E
Rotterdam	Ch 16,**24**,**27**,**28**,**87**[9]	51°56'N 4°28'E
Scheveningen	Ch 16,**26**,**83**[9]	52°06'N 4°16'E
Haarlem	Ch 16,**23**,**25**[9],**85**	52°23'N 4°38'E
Wieringerwerf	Ch 16,**27**[9],**87**	52°55'N 5°04'E
Location L7	Ch 16,**28**[9],**84**	53°32'N 4°13'E
Terschelling	Ch 16,**25**[9],**78**	53°22'N 5°13'E
Nes	Ch 16,**23**[9]	53°24'N 6°04'E
Appingedam	Ch 16,**24**,**27**[9]	53°18'N 6°52'E

Notes
Ch 16 and all working channels (bold) have 24hr watch.
Traffic lists on all working channels except Ch 16 at every H+05.
[9] VHF Channel for storm warnings at every H+05; weather messages at 0605, 1205, 1805, 2305 (in Dutch). Broadcasts given 1hr earlier when DST is in force.

Major fixed daylight marks

Church towers of 's-Gravenzande, Monster and Ter Heijde S of Scheveningen
Kijkduin tower blocks
Radio masts (73m) close SW of Scheveningen
Scheveningen Lt (30m brown metal tower)
The Vredespaleis tower behind Scheveningen
Scheveningen pier N of the town and a water tower to its E
Katwijk aan Zee: 2 church towers, offshore platform 6M NW of Katwijk
Noordwijk aan Zee: Main Lt (25m Wh square tower)
Zandvoort: water tower, 2 churches and tall buildings
IJmuiden: Main Lt (43m dark R tower), 2 steelworks chimneys N of entrance (138m and 166m, R Lts), 3 chimneys further inland (157m, 155m, 115m, R Lts), radio masts further N (76m)

Wijk aan Zee: 2 churches
Egmond aan Zee: Main Lt (28m Wh round tower), church spire
Bergen aan Zee: houses on dunes
Petten: nuclear power station with 2 chimneys (45m, R Lts)
Zanddijk: Grote Kaap Lt (17m, Lt 31m, brown round tower)
Huisduinen Lt (18m R square tower) and Kijkduin Lt (55m brown tower)
Texel: Schulpengat Ldg Lts can be seen by day, the rear being a church spire, Schilbolsnol Lt (21m G tower)
Den Helder: Town Hall midway between Kaap Hoofd and harbour entrance, a water tower inland behind it, and a church further E.

Approach routes and tidal timing

The 65 miles of smooth sandy coast from Maasmond breakwater to Kijkduin is virtually unbroken. Throughout most of its length it falls steeply to soundings of 10m at a distance of only 1½M offshore, but then slopes very gently to 20m 25M offshore in the S and slopes more steeply to reach 20m 5–10M offshore in the N, near Petten nuclear power station. The North Sea beyond is a flat basin with wide areas of 20 to 40m soundings and a steeper rise on the opposite English coast.

The prevalent southwesterlies rake this coast at an acute angle and are deflected along it, and although there are no offshore banks or boulders to create tide-rips the above underwater configuration in winds of over Force 4 creates rolling, often breaking seas which can be dangerous especially in wind-over-tide conditions with the S-going stream, and particularly nearing the 2M offshore strip and northwards from IJmuiden to the Schulpengat with its funnelling effect. Strong northerly winds can be almost as bad when they are channelled along the coast in opposition to the N-going tidal stream.

There are only 3 harbours of refuge. Scheveningen and IJmuiden have N-facing entrances and provide shelter from winds from the SW quadrant but in northerlies the swell running into the entrances can be dangerous. Den Helder is well protected from most quarters but approach through the Schulpengat can be extremely difficult in southerly winds funnelled into the channel. If conditions are extremely bad a yacht should not hesitate to stand off well beyond the 2M coastal range leaving plenty of lee and wait for an improvement before running in, even though this means prolonging the discomfort.

Onshore marks, as on the rest of the southern North Sea coast are unmistakable and well charted by the Dutch Hydrographic Office; towns, factories, water towers and church towers and spires rising from the flat shore – the steelworks N of IJmuiden is typical, also Petten nuclear power station, and the various high light towers. There are also a number of offshore platforms; some near Scheveningen, the Rijn Field near IJmuiden, and some outliers of the Helder Field to the SW of the Texel LtV. Many of these features are well lit and since the long-distance onshore lights are at 10–15M intervals there is usually no necessity to navigate blind at night even some distance offshore.

In fog anchoring is not usually feasible on most of this coast without running dangerously close inshore, other than inside the Schulpengat near Noorderhaaks island and the Westgat. When caught by fog, standing well off is often advisable, making sure to keep well away from the approach channels to the harbours, and particularly from the IJmuiden-Geul.

But the TSSs are not obtrusive, and in the right weather yacht access and departure is easy in all directions; direct to or from Norfolk, Suffolk and Essex, picking up the marks and lights, coastal cruising, visiting Amsterdam and the IJsselmeer via the Noordzeekanaal, entering the protected Waddenzee channels via the Schulpengat, and heading S from Scheveningen into the Delta by the inshore traffic zone.

The distance from Maasmond entrance to Den Helder is 67M, a 15hr passage taking almost three tidal streams. Intervening distances are as usual more convenient for single-tide passages; Hoek to Scheveningen 14M, Scheveningen to IJmuiden 25M, and IJmuiden to Den Helder 34M.

On the 67M passage there is no particular advantage in either direction. To achieve a full stream into the Schulpengat means starting from Maasmond entrance 3hrs after HW Hoek van Holland with 2hrs of fair to slack tide, 6hrs of foul and a further 7hrs of fair to slack tide into Den Helder, arriving just after HW Den Helder. In the opposite direction starting with a full S-going tide at HW Den Helder (6hrs before HW Hoek) or possibly an hour earlier gives a slant of 6hrs of fair tide to begin, 6hrs against and a final 3hrs of fair tide.

Of the three shorter passages Hoek to Scheveningen (14M) is dealt with in Chapter 11, Route 7.

From Scheveningen to IJmuiden (25M) a start at Maasmond entrance −0200 HW Hoek holds 6 to 6½hrs of N-going tide, and in the reverse direction a start +0300 on local HW (+0430 HW Hoek) means a similar period of S-going tide.

Finally from IJmuiden to Den Helder (34M) requires starting at −0300 HW IJmuiden (−0200 HW Hoek and −0100 LW Den Helder) which means half an hour of pushing the tide and then a full flood up to Den Helder. In reverse it is best to start −0100 HW Den Helder with the same tidal pattern as from IJmuiden; pushing a little then a full flood, arriving −0230 HW IJmuiden (−0130 HW Hoek).

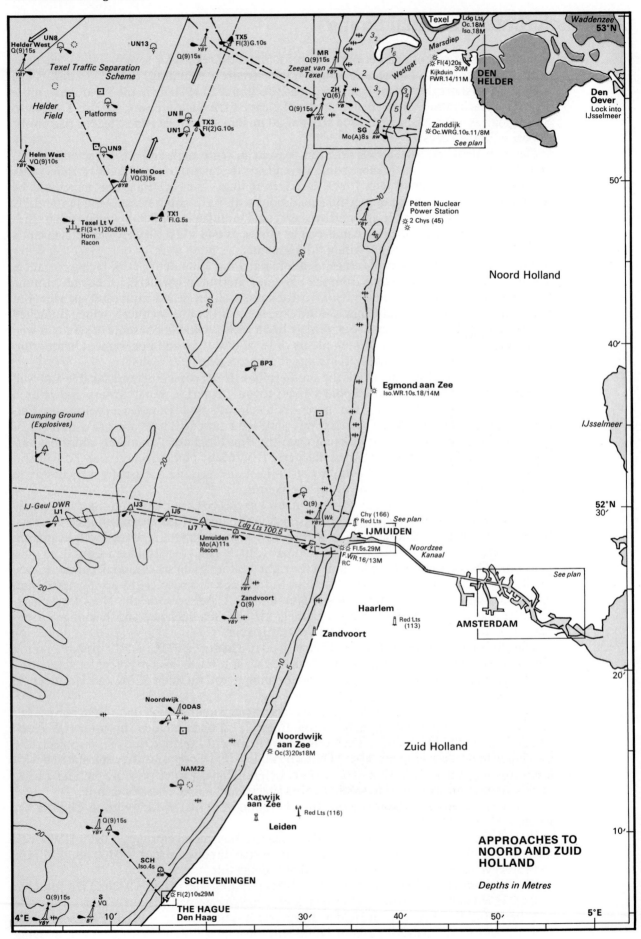

UN8
Helder West
Q(9)15s
YBY

Texel Traffic Separation
Scheme

UN13
Q(9)15s

TX5
Fl(3)G.10s

Texel

Ldg Lts
Oc.18M
Iso.18M

Waddenzee
53°N

Marsdiep

MR
Q(9)15s
YBY

Zeegat van
Texel

Westgat

3₂

1₆

2

3₇

Kijkduin
Fl(4)20s
30M
Oc.18M
FWR.14/11M

DEN
HELDER

Den
Oever
Lock into
IJsselmeer

Helder
Field

Platforms

UN II
Y

UN1
Y

TX3
G Fl(2)G.10s

ZH
VQ(6)
YB

Q(9)15s
YBY

3₄

5

4

SG
Mo(A)8s
RW

Zanddijk
Oc.WRG.10s.11/8M
See plan

Helm West
VQ(9)10s
YBY

UN9

Helm Oost
VQ(3)5s
BYB

50′

TX1
G Fl.G.5s

Texel Lt V
Fl(3+1)20s26M
Horn
Racon

Petten Nuclear
Power Station
2 Chys (45)

YBY

10

4₈

Noord Holland

20

20

BP3

Egmond aan Zee
Iso.WR.10s.18/14M

40′

Dumping Ground
(Explosives)
Y

20

IJ-Geul DWR
IJ1
Y

IJ3
Y

IJ5
Y

Ldg Lts 100.5°

IJ7
RW

52°N
30′

Q(9)
YBY

Wk

Chy (166)
Red Lts
See plan

IJMUIDEN

Noordzee
Kanaal

See plan

20

IJmuiden
Mo(A)11s
Racon

Y

Fl.5s.29M
F.WR.16/13M
RC

AMSTERDAM

YBY

Zandvoort
Q(9)
YBY

Haarlem

Red Lts
(113)

Zandvoort

20′

20

10

5

Noordwijk
ODAS
Y

Y

Noordwijk
aan Zee
Oc(3)20s18M

Zuid Holland

NAM22
Y

Katwijk
aan Zee
Red Lts (116)

Leiden

APPROACHES TO
NOORD AND ZUID
HOLLAND

10′

Q(9)15s
YBY
Y

SCH
Iso.4s
RW

SCHEVENINGEN

Depths in Metres

Q(9)15s
YBY

S
VQ
BY

Fl(2)10s29M

THE HAGUE
Den Haag

4°E

10′

30′

40′

50′

5°E

Bad weather routes inland

The inland detours required to reach some of the above departure and arrival points are particularly long but very rewarding. There are many possibilities, two of which are summarised below.

Rotterdam to Amsterdam via Gouda and Ringvaart (distance 40M); Dutch ANWB *waterkaarts J, H,* and *I* and the *Almanak, deel 2* are essential for this route. Continue E from Rotterdam's Veerhaven, turn N along the Hollandse IJssel to Gouda then N along the Gouwe to Alphen a.d. Rijn, then via the Braassemermeer lake and the Ringvaart van de Haarlemmermeerpolder (E side), eventually emerging in the Noordzeekanaal via the Wester Kanaal close N of Amsterdam. There are many opening bridges en route but some delightful wide open sailing lakes in Dutch windmill countryside.

Amsterdam to Den Helder via the IJsselmeer (distance 55M); Dutch charts *1810* and *1811* are needed. The route is tideless as far as Den Oever and locks out of the Noordzeekanaal at the Oranjesluizen and through the (lifting) Schellingwouderbrug, follows the Buiten IJ channel out into the IJsselmeer, the Pampus buoyed channel, then a direct course to Enkhuizen locking through, heading N to the Kreil channel then NW to pick up the channel-marking buoys into the Binnenhaven at Den Oever, locking through into the Waddenzee, and finally following the Wierbalg and Malzwin buoyed (lit in places) channels to the entrance of Den Helder. Crossing the IJsselmeer requires caution particularly for deeper draught yachts since there are many soundings of less than 3m. In strong winds a nasty choppy sea can build up, and it is possible to become embayed on one of the many potential lee shores. However, there are many interesting harbours of refuge. From Den Oever to Den Helder it is again necessary to consult the *Stroomatlas* (*k*, Waddenzee, Westelijk).

Sailing on the IJsselmeer – a traditional Dutch sailing barge
Photo Amsterdam Tourist Office

The harbours and coastal region

The British have erroneously named the whole of the Netherlands after these two provinces of Noord and Zuid Holland, and to the majority of foreign visitors this is probably all they will have time to see, since it includes the seat of government and the royal family in The Hague (Den Haag), the historic water-city of Amsterdam, and, above all, the bulb fields. But as a yachtsman you will have to work hard to see all of this from the limited coastal bases without cycling or using the comprehensive road and rail system.

Of the three harbours, only Scheveningen is worth a long visit, since as well as having its own seaside resort it is extremely convenient for Den Haag. IJmuiden is industrial, and Den Helder a somewhat stark naval harbour. There are, however, a number of other pleasant seaside resorts with first-class beaches and facilities which do not have harbours but can be reached by bus, tram or train – Katwijk, Noordwijk, and Zandvoort in the flower-growing district between Scheveningen and IJmuiden, and Egmond and Bergen N of IJmuiden. Lisse and the nearby world-famous Keukenhof Gardens are about halfway between Scheveningen and Amsterdam not far from Noordwijk and are best seen early in the yachting season between end-March and mid-May.

Finally a major attraction of this area is the ease of getting right into the centre of Amsterdam via the Noordzeekanaal.

SCHEVENINGEN

Port radio (VHF)

Scheveningen	VHF Ch	Hrs listening
Call *Scheveningen Haven*	14	24

☎ (070) 51 40 31

Entry signals

Contact on VHF to obtain instructions for entering harbour.

Shown from signal station: R over Wh Lt – entry prohibited; Wh over R Lt – departure prohibited; Fl.Y Lt – large vessel leaving or entering, shown seaward if vessel leaving and landward if vessel entering.

Shown from W corner of fish market: Q.R Lt – vessel inward bound in the outer harbour.

Customs

Kranenburgerweg 202
☎ (070) 51 44 81

Approach and entrance

Approach is made easy by the 29M long-range light, as well as leading lights (daylight intensity 6M, night 14M) into the outer entrance and a second set into the Voorhaven (4M day, 11M night). There are a number of outstanding landmarks (see *Major fixed daylight marks* above) and there are no offshore hazards.

The yacht harbour at Scheveningen

The entrance faces N and is uncomfortable of course in northerlies. Streams run extremely fast across the entrance at certain times so it pays to consult *Stroomatlas j.*

The entrance is a triple one, passing between 3 sets of port and starboard Lts on the encircling walls and into the Voorhaven, then hard to port into the fish dock (1st Haven, least depth 4.8m) and hard to starboard and along the short 'canal' into the 2nd Haven (least depth 2.8m), turning to starboard again for the large pontoon marina at the S end.

Mooring and facilities

WV Marina Scheveningen, ☎ (070) 55 02 75, is very well provided with showers, toilets, washing machines, and alongside is the comfortable Jachtclub Scheveningen, ☎ (070) 52 03 08, with its own restaurant, as well as several others along the waterfront. There are all repair, lifting and chandlery facilities nearby. This is a very good spot for tourism; a lively seaside resort with a Victorian casino, recreation centre, bathing beaches, and pier, as well as an interesting shopping street and historic fishing harbour. Den Haag, the Dutch seat of government and of the International Court of Justice is only a short tram journey away along the delightful, tree-fringed Scheveningen-weg, with a plethora of museums, art galleries, royal palaces and historic buildings. The Hoek ferries are only a short taxi or train journey away from Den Haag for meeting crew.

Scheveningen. The casino

IJMUIDEN

Port radio (VHF)

Port Operations Centre	VHF Ch	Hrs listening
Call *Traffic Centre*		
IJmuiden (pilots)	12	24

Noordzeekanaal

IJmuiden

Call *IJmuiden Port Control*	09	24
IJmuiden Locks for		
locks direct	11	24
☎ (02550) 1 57 03		
Amsterdam (port basins)	14	24
☎ (020) 22 15 15		
Oranjesluizen (locks)	18	
☎ (020) 36 07 44		

Broadcasts
H+30, Ch 14 H+00. Maintain continuous listening watch on the appropriate channel when in the Noordzeekanaal; mandatory during poor visibility.

Entry signals

It is important to make contact by VHF before entry, as the canal and lock entry signals are extremely complicated and there are four locks. If you have no VHF check the signal frame inside the harbour (see below), and if you are clear to do so go to the Kleinesluis and motor around or tie to the pilings and listen for the megaphone. Yachts will usually be directed to the two southern smallest locks (of the four), the Zuidersluis and the Kleinesluis which are serviced only in

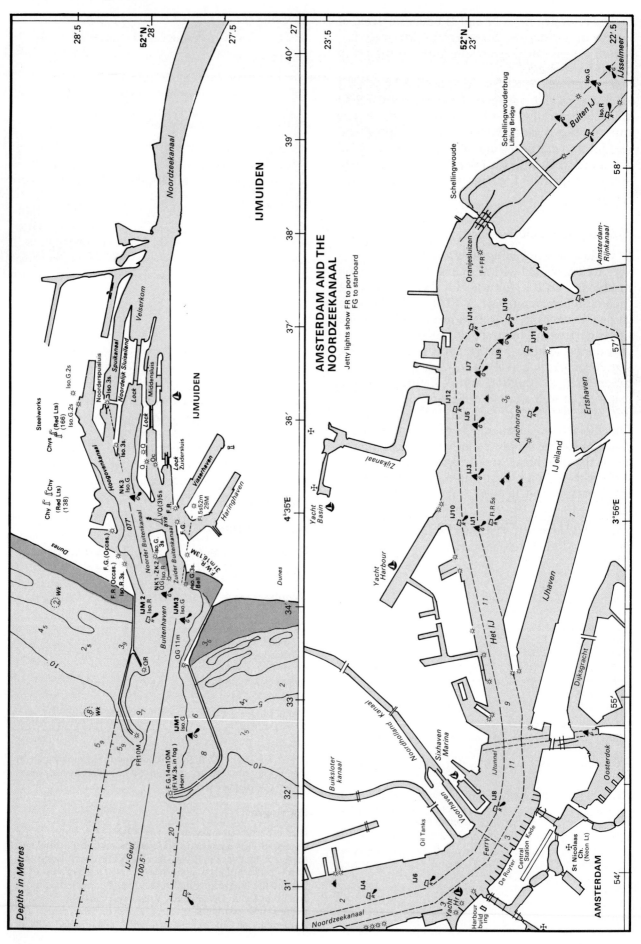

IJmuiden locks   *Photo* VVV Haarlem

daytime from 0600 to 2100. At night you may be directed through one of the other locks with a large ship, unless you decide to wait at the pilings near the smaller locks.

Tidal signals from the signal station on the S wall of the Zuider Buitenkanaal; vertical R and G Lts relate to depths of water over 11m, so you can ignore them. The most useful ones for yachts are, G over Wh Lt – rising tide; Wh over G Lt – falling tide.

Close to the signal station is a frame with a 3 by 3 square of 9 sets of lights for ingoing traffic. 6 of the sets relate to signals you will hopefully never need to use unless you arrive at night and are requested to go through the larger locks, in which case I am afraid you will have to frantically leaf through *Reed's Nautical Almanac*! The 3 right-hand sets apply to, from top to bottom, the Zuidersluis (there are none for the Kleinesluis), the Zuider Buitenkanaal, and fishing vessels and coastal craft.

Zuidersluis, Fl.G – lock being prepared; F.G – lock ready; Fl.R – outgoing traffic; F.R – lock out of use.

Zuider Buitenkanaal, Fl.R – outgoing traffic; F.R – traffic prohibited.

Fishing and coastal craft, F.R – entry prohibited except with permission.

For outgoing traffic the top left-hand set of 4 sets of signals applies to the Zuider Buitenkanaal, Fl.R – incoming traffic from sea; F.R – Zuider Buitenkanaal closed.

The Zuidersluis has a swing bridge. Lock entrances have traffic signals, G on each side – enter; R on each side – entry prohibited; RG horizontal on each side – wait, lock being prepared; and a G over G – make fast on this side of the lock.

Customs

On the N side of the entrance to the Zuidersluis it is possible to tie up alongside the customs building for clearance. After clearance it is then necessary to clear Immigration (they are the ones with guns!) whose office is on the road above the S bank close E of the Zuidersluis.

Approach and entrance

Approach to IJmuiden is assisted by its 29M long-range light and in daylight by the complex of chimneys as well as smoke rising from the steel-works N of the entrance visible many miles out to sea. The IJmuiden-Geul approach area and channel has Y conical Lt buoys on its S side as well as a RW offing Lt buoy 5M from the entrance, so it is easy for yachts to cross the channel at right angles or when approaching the harbour to keep out of the channel until the entrance. There are also Ldg Lts with 4M daylight and 29/16M night range.

The entrance faces N so can be uncomfortable in strong winds from this direction. Tides also run fast across it, and it pays to consult *Stroomatlas j*. Like Scheveningen, the entrance is a triple one, between an outer set, an inner set, and the Zuider Buitenkanaal entrance set of port and starboard Lts. Between the inner set and the canal there is a channel with a buoy on the S side and one on the N side off the end of the wall projecting from Forteiland. Then straight along the channel, past the fishing harbour entrance, to the locks.

Mooring and facilities

I have a confession to make. On the many occasions I have tied up in IJmuiden I have never walked into the town. The problem is that the available moorings on the S bank of the canal at the approach to the locks can only be regarded as temporary because of danger and discomfort of wash. The tidal Haringhaven and Vissershaven are extremely busy – this is the Netherlands' largest fishing port – but it is sometimes possible to tie alongside a fishing vessel or to the pontoons in the Haringhaven. If you succeed in achieving an acceptable mooring, places of interest are the huge fish market, the beaches and the Velserbeek Park.

THE NOORDZEEKANAAL & AMSTERDAM

Mooring and facilities

The distance from the locks to Amsterdam's Sixhaven is 13½M. Sheltered deep water (8m throughout) with no bridges, it is possible to sail in the initial country reaches, motoring in the later industrial built-up reaches where the wind is more erratic. Traffic is lighter than on the Nieuwe

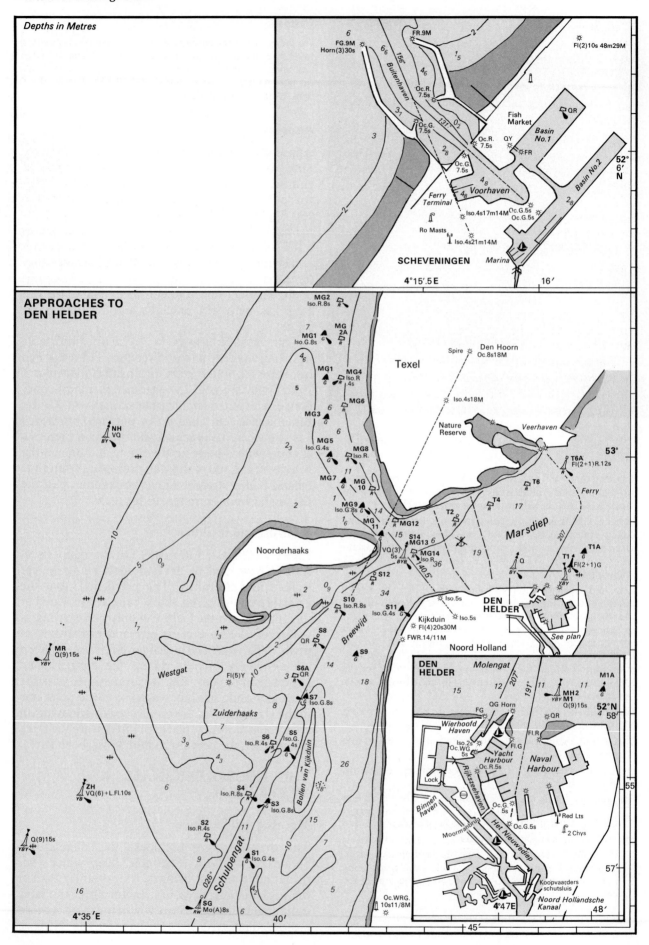

Depths in Metres

FR.9M

FG.9M
Horn(3)30s

Fl(2)10s 48m29M

156

Buitenhaven

Oc.R.
7.5s

137°

Oc.G.
7.5s

Fish
Market

QR

Basin
No.1

**52°
6′
N**

Oc.R.
7.5s

QY

FR

Oc.G
7.5s

Voorhaven

Basin No.2

*Ferry
Terminal*

Oc.G.5s
Iso.4s17m14M   Oc.G.5s

Ro Masts
Iso.4s21m14M

SCHEVENINGEN   *Marina*

4°15′.5E   16′

**APPROACHES TO
DEN HELDER**

MG2
Iso.R.8s

MG
2A

MG1
Iso.G.8s

Den Hoorn
Spire   Oc.8s18M

Texel

MG1

MG4
Iso.R.
.4s

MG6

MG3

Iso.4s18M

Veerhaven

NH
VQ
BY

MG5
Iso.G.4s

MG8
Iso.R.

Nature
Reserve

T6A
Fl(2+1)R.12s

53°

MG7

MG
10

Marsdiep

T6

Ferry

MG9
Iso.G.8s

MG
11

MG12

T4

T2

17

207

T1A

S14
MG13

T1
Fl(2+1)G

Noorderhaaks

VQ(3)
5s
BYB

MG14
Iso.R.

Q
BY

S12

**DEN
HELDER**

MR
Q(9)15s
YBY

S10
Iso.R.8s

S11
Iso.G.4s

Iso.5s

Kijkduin
Fl(4)20s30M
FWR.14/11M

Iso.5s

See plan

Noord Holland

S8
QR

S9

Westgat

Fl(5)Y

S6A
QR

**DEN
HELDER**

Molengat

M1A

Zuiderhaaks

S7
Iso.G.8s

12   207   191°

MH2
YBY M1
Q(9)15s

**52°N
4 58′**

FG   QG Horn

ZH
VQ(6)+L.Fl.10s
YB

S5
Iso.G.
4s

Bollen van Kijkduin

S6
Iso.R.4s

Wierhoofd
Haven

QR

Fl.R

Iso.2s
Oc.WG
5s

Yacht
Harbour

Fl.G

Naval
Harbour

55

S4
Iso.R.8s

Oc.R.5s

Lock

S3
Iso.G.8s

*Binnen
haven*

Oc.G.

S2
Iso.R.4s

Red Lts

2 Chys

57′

Q(9)15s
YBY

Schulpengat

Moormanbrug

Oc.G.5s

S1
Iso.G.4s

Koopvaarders
schutsluis

16

Oc.WRG.
10s11/8M

*Noord Hollandsche
Kanaal*

4°7E   48′

SG
Mo(A)8s
RW

4°35′E   40′   45′

Waterweg or Westerschelde but still requires great caution, particularly nearer to Amsterdam with its many ferries.

5M along to starboard Zijkanaal C leads via three opening bridges to Spaarndam and the Mooie Nel lake where there are several clubs and marinas, but outside the scope of this guide.

1M further on to port is Zijkanaal D with Jachthaven Nauerna, ☎ (02987) 17 22, less than ½M along to port again; a pontoon marina with an interesting club in a large barge, showers, toilets and lifting/repair facilities. This is quite a convenient resting-up base before making the long crossing to England.

Like the Nieuwe Waterweg the Noordzeekanaal to Amsterdam (*het* IJ, pronounced 'ai' as in train, is its name near the city) should be navigated with extreme caution, as there is considerable barge, deep-sea and ferry (there are many ferry crossings) traffic, in both directions and coming out of the many side docks and canals.

The main port offices are: sea waterways, Gem. Havendienst voor de zeehavens, Havengebouw, Amsterdam, ☎ (020) 22 15 15; inland waterways, Havengelddienst, James Wattstraat 84, Amsterdam, ☎ (020) 94 45 44.

There are a large number of mooring places of which I will list only a selection of the more easily accessible near the centre and the Oranjesluizen (the locks out into the Buiten IJ and the IJsselmeer). Moving along the canal from W to E:

De Ruyterkade to the W of the central station and in front of the harbour building. Pontoons on the river front but uncomfortable from the wash of passing ships.

WVDS Sixhaven, 1021 HG, Amsterdam, ☎ (020) 37 08 92 and 32 94 29. On the N side of the river opposite the central station, to which there is a frequent nearby ferry service. A small R buoy lies off its S corner and the entrance is a tree-overhung gap through its E wall. It is a pontoon marina (stern to posts) with showers, toilets, bar and lifting/repair facilities. It is a peaceful spot ideally situated for the ferry, then a short walk through the central station straight into the centre of the city.

ZV Aeolus, a small marina in the Johan van Hasseltkanaal-Oost, 1M E of the central station on the N bank with toilet and washing facilities.

Jachthaven Twellegea on the N bank near the Oranjesluizen, and ¾M up the zigzag Zijkanaal K just to the E of Nieuwendam.

For those yachts intending to lock into the IJsselmeer, the Oranjesluizen (3 locks) is approached along a buoyed channel at the E end of the canal, so make sure to keep to starboard. There are pilings to moor to, there is a megaphone to instruct you as well as traffic lights, and you can

Amsterdam – the harbour   *Photo* Amsterdam Tourist Office

make VHF contact. Beyond the locks is the Schellingwouderbrug, ☎(02904) 2 22, a fixed bridge (NAP +8.7m) but with one lifting span which is constantly available, and then the open IJsselmeer is ahead.

There are a number of other canal routes coming into the Noordzeekanaal on both N and S banks, within and outside Amsterdam, but beyond the scope of this book.

Amsterdam itself is very well provided indeed with entertainment, restaurants, museums (the Rijksmuseum and Van Gogh museum are most well known), and you may even take, dare I say, a busman's holiday tour of the concentric canals on one of the waterbuses.

DEN HELDER

Port radio (VHF)

Den Helder	VHF Ch	Hrs listening
☎ (02230) 1 12 34	14, 16	24
Moormanbrug	18	
Koopvaardersschutsluis	22	
Den Oever lock	20	
Kornwerderzand locks	18	

Entry signals

Contact should be made first on VHF.

From the signal station on the root of the west pier (Harssens) the following lights are shown.

Incoming traffic

R
W   no entry, stay beyond 200m of entrance

R
W   no entry, no traffic allowed within the harbour.
R

Outgoing traffic

R
  W   no traffic allowed in Marinehaven

R
W   no traffic allowed in Nieuwe Diep N of Moor-
      manbrug

R
W W no traffic allowed in Marinehaven and Nieuwe
      Diep N of Moormanbrug

R
W
R   no entry, no traffic allowed in Marinehaven and
      Nieuwe Diep N of Moormanbrug

Customs

Clearance can be obtained in the yacht haven. Het Nieuwe
Diep 23, ☎ (02230) 1 51 81 or 3 49 56.

Approach and entrance

Approach to Den Helder is very easy both in daylight and at night. Petten nuclear power station, a
square building with two 45m chimneys, is 6M
SSE of the Schulpengat offing buoy, SG (RW,
Mo(A)8s). The 55m pencil-like Kijkduin Lt tower
in the narrows opposite Noorderhaaks island is
also visible well out to sea. The Schulpengat channel is particularly well lit with R and G isophase Lt
buoys and 9M-range daylight (18M at night) Ldg
Lts on Texel island, the rear being a church spire.
The Molengat approach from the N between
Texel and Noorderhaaks is similarly well provided; 5M-range daylight Ldg Lts (8M at night)
on the mainland, isophase channel Lt buoys, and
an E cardinal Lt buoy off the end of Noorderhaaks
at the junction of the two channels with the
Marsdiep. The Marsdiep leading to Den Helder is
steep-sided and has a number of Lt buoys near the
harbour entrance which has 2 further sets of Ldg
Lts.

Tidal streams run fast across the entrance of
Marinehaven Willemsoord (4.7 to 9m least
depth), so once again it pays to use *Stroomatlas j* on
approach.

Nearing the entrance keep a sharp lookout for
ferries coming out of the Veerhaven and heading
for 't Horntje on Texel. The main harbour entrance has two successive sets of port and starboard
entrance lights. The westernmost of the outer set
of entrance lights is on the wall projecting out
from Harssens island (an island no longer!), and
the E one is a beacon MH4 on the N corner of a

dangerous drying patch extending N from the E
wall. Having passed between both sets of lights,
the yacht harbour is hard to starboard into the first
dock behind the island and signal station.

Mooring and facilities

Den Helder is the Netherlands' major naval port,
used by merchant vessels only for shelter, repairs
and provisions. Yachts and fishing vessels, however, have proved far more difficult customers to
keep out, particularly as both pastimes have infil-
trated the Navy itself.

Koninklijke Marine Jachtclub is part of the
Dutch Navy, and usually a few berths can be
found at the pontoons. There is a small clubhouse
with showers and toilets.

If you intend to take yet another inland route to
Amsterdam and the Noordzeekanaal via the
Noordhollands Kanaal then you can continue on
down the Nieuwe Diep along the W side of the
harbour, have the Vice-Admiral Moormanbrug
opened (2.7m height unopened) and lock through
the Koopvaardersschutsluis and its opening
bridge (both are in service throughout the day).
Hard to starboard after the lock and ½M N in the
Binnenhaven are two more marina/clubs: WSOV
Breewijd, another naval club, and WV Helder-
Willemsoord-Nieuwe Diep, both with showers
and toilets. There is a laundrette nearby (Binnenhaven 12). At the S, Westoever end of Industriehaven, is yet another marina with showers and
toilets, Jachthaven Den Helder, and this haven is
entered through yet another opening bridge over
an entrance through the W side of the Binnenhaven, S of the above two clubs.

The town of Den Helder is a 15 to 30-minute
walk W of the yacht harbour, and has a good mod-
ern shopping precinct, but little of historic
interest is left, even though the place has been
associated with the Dutch Navy since at least the
16th century. The main attraction is that it is a
good stepping stone to the Waddenzee and Frisian
islands, and also the IJsselmeer via two deep-
water channels to the two locks in its northern
dam.

APPENDIX

I. RYA NATIONAL CRUISING SCHEME

The RYA's training booklets *G15/83 National Cruising Scheme (Sail) – Syllabus and Logbook* and for the motor yachtsman, *G18/83 National Cruising Scheme (Motor) – Syllabus and Logbook* are obtainable from the Royal Yachting Association, Victoria Way, Woking, Surrey, GU21 1EQ, ☎ (04862) 5022. The booklets give a complete explanation of this scheme. Below is a brief, selective summary of the aims and requirements for the sail certificates. The *Power Driven Craft* endorsed certificates are similar to those for sail without the requirement for practical tests of ability under sail.

1. *RYA Competent Crew Certificate*
   Seatime required: 5 days, 100M, 4 night hours.
   Practical course: 5 days mainly seamanship and sailing.
   Shorebased course: Basic seamanship techniques and elements of theory of navigation and meteorology.
   Examination: None. Certificate awarded on satisfactory completion of course.

2. *RYA Day Skipper/Watch Leader Certificate*
   Seatime required: 10 days, 200M, 8 night hours.
   Practical course: 5 days seamanship, sailing and navigation.
   Shorebased course: Same as in entry 1 above.
   Examination: None. Certificate awarded on satisfactory completion of course.

3. *RYA/DTp Coastal Skipper Certificate*
   Seatime required: 20 days, 400M, 12 night hours.
   Practical course: 5 days practical skippering.
   Shorebased course: Navigation, meteorology and signals to the level required for the navigation of a yacht on coastal and offshore passages not requiring astronavigation.
   Examination: Oral for shorebased and practical course completion certificate holders. Practical for others.
   Aim to ensure a certificate holder has the knowledge necessary to skipper a cruising yacht on coastal cruises but not necessarily the experience needed to carry out longer passages.

4. *RYA/DTp Yachtmaster Offshore Certificate*
   Seatime required: 50 days, 2500M, 5 passages over 60M, 2 as skipper and 2 overnight passages.
   Practical course: None.
   Shorebased course: Same as in entry 3 above.
   Examination: Practical examination. Aim to ensure a certificate holder is an experienced yachtsman, competent to skipper a cruising yacht on any passage which can be completed without the use of astronavigation.

5. *RYA/DTp Yachtmaster Ocean Certificate*
   Open only to Yachtmasters Offshore.
   Seatime required: Ocean Passage.
   Practical course: None.
   Shorebased course: Astronavigation and world meteorology.
   Examination: Assessment of actual sights taken at sea plus written exam if no shorebased course proficiency certificate is held. Aim is to ensure that certificate holder is an experienced yachtsman, competent to skipper a yacht on passages of any length, in all parts of the world.

II. DEPARTMENT OF TRANSPORT REQUIREMENTS AND RECOMMENDATIONS FOR THE EQUIPMENT OF YACHTS

Pleasure yachts of 13.7m in length (45ft) or over are statutorily required to carry certain equipment as detailed in two government statutory instruments, which are published and obtainable from H.M.S.O.

The Merchant Shipping (Life Saving Appliances) Rules 1965
The Merchant Shipping (Fire Appliances) Rules 1965

The DTp makes 'recommendations' to owners of pleasure craft of less than 13.7m and divides these into two categories above and below 5.5m (18ft). Most vessels using this pilot book are likely to fall into the category from 5.5 to 13.7m and a brief summary of these recommendations is as follows (*Reed's Nautical Almanac* contains additional information).

General

- Adequate navigation lights to conform to the International Collision Regulations.
- Means of giving sound signals to conform to the Internatonal Collision Regulations.

Personal safety equipment

- 1 lifejacket for each person aboard; worn whenever danger exists of falling into the water, and kept in an immediately accessible place.
- 1 safety harness for each person aboard; worn in bad weather or at night when danger exists of falling overboard. Inadvisable to wear at speeds of 8 knots or more.

Rescue equipment

- 2 lifebuoys, minimum; if proceeding by night 1 of these should have a self-igniting light, 30m of buoyant line in reserve and be positioned in reach of helmsman.

Other flotation equipment, if going beyond 3M from land

- Either 1 inflatable liferaft of DTp accepted type, capable of carrying all members of the crew. Carried on deck or in locker opening directly to deck. Must be serviced annually.
- Or 1 solid dinghy carried on deck. May be collapsible, but should have permanent, not inflatable, buoyancy and oars and rowlocks secured to it. May be towed if tow secure.
- Or 1 inflatable dinghy with 2 compartments, one of which is fully inflated. Carried on deck, with oars and rowlocks secured. May be towed if tow secure.
- Or 1 inflatable dinghy with 2 compartments, which need not be carried on deck if the yacht has enough permanent buoyancy to float when swamped with 115 kilos (250lbs) added weight. May be towed if tow secure.

Fire fighting appliances

- 2 buckets with lanyards.
- Craft up to 9m (30ft), 1 fire extinguisher of at least 1.4 kilos (3lbs) capacity (dry powder) for cooker fires, plus 1 similar extinguisher for engine fires. Carbon dioxide or foam extinguishers of equivalent extinguishing capability acceptable. BCF and BTM extinguishers are also acceptable, providing owners realise fumes may be dangerous in a confined space.
- Craft over 9m and those with powerful engines carrying large amounts of fuel, additional extinguishers of at least 2.3 kilos (5lbs) dry powder or equivalent, and in some cases a fixed fire extinguishing installation will be necessary.

Other equipment

- 2 anchors, each with warp or chain of appropriate size and length. If warp, at least 5.5m (3 fathoms) of chain should be used between anchor and warp.
- 1 bilge pump.
- 1 efficient compass + 1 spare.
- Charts covering intended area of operations.
- 6 distress flares, of which 2 are rocket parachute type.
- Daylight distress (smoke) signals.
- Appropriate length of buoyant tow rope.
- 1 first-aid box with anti-seasickness tablets.
- 1 radio receiver for weather forecasts.
- 1 water-resistant torch.
- 1 radar reflector of adequate performance, as large as can conveniently be carried; mounted, if possible, at height of at least 3m (10ft) above sea level.
- Line suitable for use as inboard lifelines in bad weather.
- 1 suitable engine tool kit.
- Name, number or generally recognised sail number painted in prominent position on vessel or dodger in letters or figures at least 22cm (9in) high.

III. EXTRACT FROM *INTERNATIONAL REGULATIONS FOR PREVENTING COLLISIONS AT SEA, 1972*

Rule 10 'Traffic Separation Schemes'

a. This Rule applies to traffic separation schemes adopted by the Organisation.

b. A vessel using a traffic separation scheme shall:
i. proceed in the appropriate traffic lane in the general direction of traffic flow for that lane;
ii. so far as practicable keep clear of a traffic separation line or separation zone;
iii. normally join or leave a traffic lane at the termination of the lane, but when joining or leaving from either side shall do so at as small an angle to the general direction of traffic flow as practicable.

c. A vessel shall so far as practicable avoid crossing traffic lanes, but if obliged to do so shall cross as nearly as practicable at right angles to the general direction of traffic flow.

d. Inshore traffic zones shall not normally be used by through traffic which can safely use the appropriate traffic lane within the adjacent traffic separation scheme. However, vessels of less than 20m in length and sailing vessels may under all circumstances use inshore traffic zones.

e. A vessel, other than a crossing vessel or a vessel joining or leaving a lane, shall not normally enter a separation zone or cross a separation line except:
i. in cases of emergency to avoid immediate danger;
ii. to engage in fishing within a separation zone.

f. A vessel navigating in areas near the terminations of traffic separation schemes shall do so with particular caution.

g. A vessel shall so far as practicable avoid anchoring in a traffic separation scheme or in areas near its terminations.

h. A vessel not using a traffic separation scheme shall avoid it by as wide a margin as is practicable.

i. A vessel engaged in fishing shall not impede the passage of any vessel following a traffic lane.

j. A vessel of less than 20m in length or a sailing vessel shall not impede the safe passage of a power-driven vessel following a traffic lane.

k. A vessel restricted in her ability to manoeuvre when engaged in an operation for the maintenance of safety of navigation in a traffic separation scheme is exempted from complying with this Rule to the extent necessary to carry out the operation.

l. A vessel restricted in her ability to manoeuvre when engaged in an operation for the laying, servicing or picking up of a submarine cable, within a traffic separation scheme, is exempted from complying with this Rule to the extent necessary to carry out the operation.

IV. COASTGUARD AND RESCUE FACILITIES IN THE SOUTHERN NORTH SEA

Access to all facilities is available on VHF Ch 16

British waters

Yarmouth, Thames (Walton-on-the-Naze), and Dover Coastguards keep watch on VHF Ch 16.

Offshore lifeboats	Inshore lifeboats*
Wells-next-the-Sea	Wells-next-the-Sea
Sheringham	Cromer
Cromer	Happisburgh
Gt Yarmouth	Gt Yarmouth
Lowestoft	Southwold
Aldeburgh	Aldeburgh
Harwich	Harwich
Walton-on-the-Naze	Clacton-on-Sea
Sheerness	West Mersea
Margate	Burnham-on-Crouch
Ramsgate	Southend-on-Sea
Walmer	Sheerness
Dover	Whitstable
Dungeness	Margate
	Ramsgate
	Walmer
	Littlestone-on-Sea

* Inshore lifeboats mostly operate in summer only.

French waters

CROSSMA, the French rescue organisation for the English Channel and southern North Sea, keeps watch on VHF Ch 16. Lifeboats and rescue tugs are stationed at Calais and Dunkerque.

Belgian waters

Offshore and inshore lifeboats are based at Nieuwpoort, Oostende and Zeebrugge.

Netherlands waters

Life-saving service based on two life-saving societies; the Royal South Holland and the Royal North and South Holland.

Offshore lifeboats

Breskens
Burghsluis
Stellendam
Hoek van Holland
Scheveningen
IJmuiden
Den Helder

Inshore lifeboats

Breskens
Burghsluis
Ouddorp
Stellendam
Ter Heijde
Katwijk aan Zee
Noordwijk aan Zee

Zandvoort
IJmuiden
Wijk aan Zee
Egmond aan Zee
Den Helder

V. WIND AND FOG OBSERVATIONS AT NORTH SEA PORTS

Note Wind directions are by quadrant e.g. NE means N–NE–E and % frequencies are of total winds for that month.

	Fog days	Gale days	NE[8]	SE[8]	SW[8]	NW[8]	Calm
January							
Gt Yarmouth[1]	2	1	17	30	61	40	3
Felixstowe[2]	1	1	17	26	59	38	3
Shoeburyness[1]	2	2	23	29	56	39	3
Dover[3]	3	2	23	27	59	40	2
Calais[4]	5	5[7]	24	28	60	26	6
Oostende[5]	6	2	24	30	45	35	9
Vlissingen[6]	4	2	26	37	56	28	3
Rotterdam[6]	4	1	31	41	54	25	1
Amsterdam[4]	5	5	28	47	52	25	1
Den Helder[4]	3	8[7]	25	32	55	26	3
April							
Gt Yarmouth[1]	1	rare	33	33	43	41	2
Felixstowe[2]	1	rare	33	32	42	38	4
Shoeburyness[1]	1	1	33	29	44	42	1
Dover[3]	1	1	37	27	47	37	2
Calais[4]	3	3[7]	34	18	47	34	4
Oostende[5]	2	1	32	28	43	36	6
Vlissingen[6]	3	rare	36	28	46	41	2
Rotterdam[6]	2	1	37	27	47	41	1
Amsterdam[4]	2	2	36	28	44	41	2
Den Helder[4]	1	5[7]	38	26	43	37	1
July							
Gt Yarmouth[1]	rare	rare	28	29	48	40	4
Felixstowe[2]	rare	rare	28	26	49	41	2
Shoeburyness[1]	rare	rare	25	28	51	40	2
Dover[3]	rare	rare	28	23	58	34	2
Calais[4]	2	1[7]	28	13	58	32	5
Oostende[5]	1	rare	26	23	56	40	4
Vlissingen[6]	1	rare	29	24	49	48	1
Rotterdam[6]	rare	rare	29	26	51	47	1
Amsterdam[4]	1	1	33	26	48	45	1
Den Helder[4]	1	2[7]	31	21	46	44	3
October							
Gt Yarmouth[1]	2	1	14	23	68	42	2
Felixstowe[2]	1	1	21	23	62	41	3
Shoeburyness[1]	2	1	23	25	53	43	2
Dover[3]	2	1	26	25	52	43	2
Calais[4]	1	5[7]	21	24	59	28	7
Oostende[5]	4	1	30	31	50	32	5
Vlissingen[6]	3	1	34	40	47	30	2
Rotterdam[6]	3	1	29	42	52	30	2
Amsterdam[4]	3	2	33	45	56	33	1
Den Helder[4]	3	6[7]	31	42	45	31	2

Wind observations

[1] Mean of 0900,1500
[2] Mean of 0700, 1300
[3] 0900
[4] Mean 0700,1300,1900
[5] Mean of 0600,1200,1800
[6] Mean of 0800,1400,2000
[7] Force 7 or over
[8] Quadrant, e.g. N to NE to E. Therefore winds from the four cardinal points are included in both relevant columns below.

VI. TIDAL STREAMS

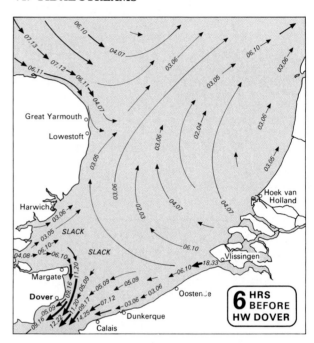

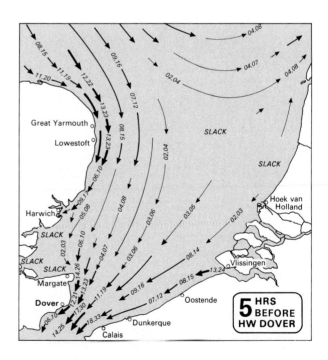

Figures shown against the arrows refer to the rates of stream.
E.g. *09,12*
0.9 knots at neaps
1.2 knots at springs

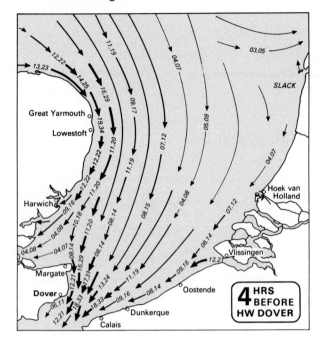

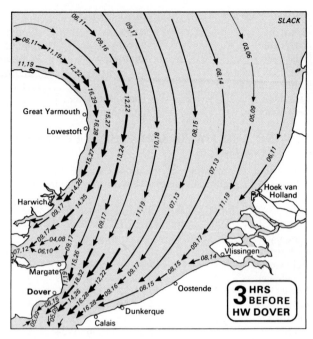

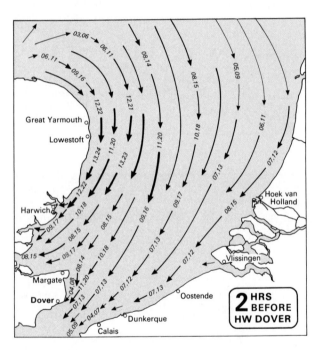

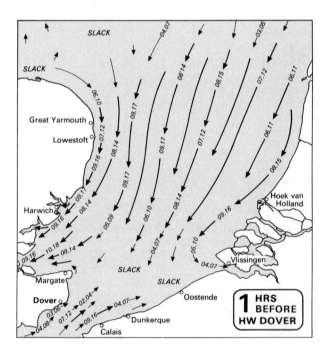

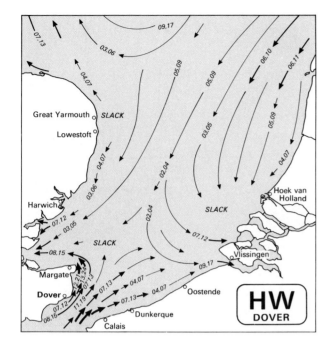

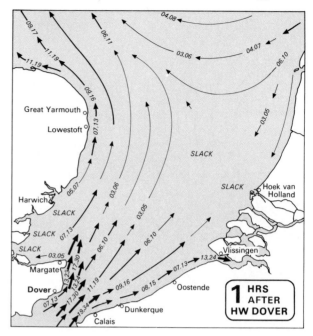

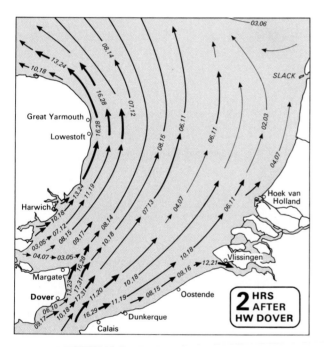

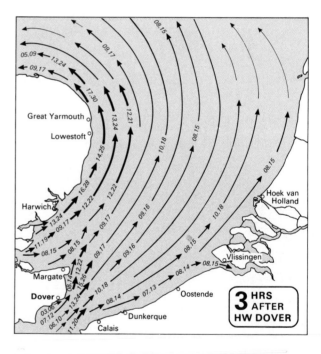

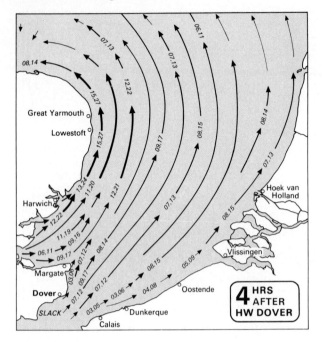

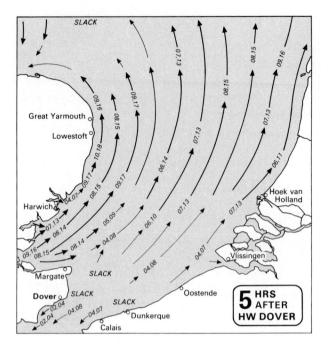

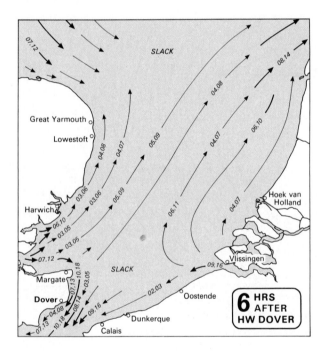

VII. SOUTHERN NORTH SEA CHARTS

BRITISH ADMIRALTY CHARTS
Passage charts

Chart	Title	Scale
1405	Texel to Helgoland	300,000
1406	Dover and Calais to Orfordness and Scheveningen	250,000
1408	Harwich to Terschelling and Cromer to Rotterdam	300,000
2182a	North Sea – southern sheet	750,000

Approach/harbour charts

Chart	Title	Scale
97	Zeebrugge	10,000
110	Oosterschelde – Westkapelle to Goeree	75,000
120	Westerschelde – Flushing to Zandvlietsluis	50,000
	Terneuzen	30,000
122	Approaches to Europoort	50,000
124	Noordzeekanaal	20,000
125	Approaches to Oostende	50,000
	Oostende	15,000
132	Nieuwe Waterweg and Europoort, Hoek van Holland to Vlaardingen	20,000
133	Nieuwe Maas and Oude Maas Vlaardingen to IJsselmonde and Dordrecht	20,000
139	Westerschelde – Valkenisse to Antwerp	25,000
191	Zeegat van Texel	50,000
	Den Helder Nieuwediep harbour	12,000
192	Oosterschelde	40,000
	Wemeldinge	10,000
	Tholen-Bergen op Zoom	25,000
323	Dover Strait – eastern part	75,000
325	Westerschelde – Oostende to Westkapelle	50,000
	Flushing	25,000
1183	Thames Estuary	100,000
1185	River Thames – Sea Reach	25,000
1186	River Thames – Canvey Island to Tilbury	12,500
1350	Dunkerque and approaches	25,000
1352	Ports on the north coast of France: Le Treport; Fecamp; Calais	15,000
1491	Harwich and Felixstowe	10,000
1504	Cromer to Orfordness	150,000
1536	Approaches to Great Yarmouth and Lowestoft	40,000
	Great Yarmouth haven: Lowestoft harbour	6,250
	Approaches to Lowestoft	20,000
1543	Winterton Ness to Orfordness	75,000
	Southwold harbour	10,000
1593	Harwich channel	10,000
1594	River Stour – Erwarton Ness to Manningtree	10,000
1605	Thames Estuary – Edinburgh channels	15,000
1607	Thames Estuary – southern part	50,000
1610	Approaches to the Thames Estuary	150,000
1698	Dover harbour	6,250
1827	Approaches to Ramsgate	12,500
	Ramsgate harbour	5,000
1828	Dover to North Foreland	37,500
1834	River Medway – Garrison Point to Folly Point	12,500
1835	River Medway – Folly Point to Maidstone	6,000
	Continuation to Maidstone	25,000

Chart	Title	Scale
1872	Dunkerque to Flushing	100,000
1892	Dover Strait – western part	75,000
1975	Thames Estuary – northern part	50,000
2052	Orfordness to the Naze	50,000
2151	River Thames – Tilbury to Margaret Ness	12,500
2322	Goeree to Texel	150,000
	Scheveningen	25,000
2449	Dover Strait to Westerschelde	150,000
2484	River Thames – Hole haven to London Bridge	25,000
2571	The Swale – Whitstable to Harty Ferry	12,500
	Whitstable harbour	5,000
2572	The Swale – Windmill Creek to Queenborough	12,500
2693	Approaches to Felixstowe, Harwich and Ipswich with the rivers Stour, Orwell and Deben	25,000
	Ipswich	10,000
2695	Plans on the east coast of England: Walton Backwaters	12,500
	Rivers Ore and Alde	25,000
	Southwold Harbour	7,500
3337	River Thames – Margaret Ness to Tower Bridge	12,500
	Thames Tidal Barrier	5,000
3371	Gabbard and Galloper banks to Europoort	150,000
3683	Sheerness and approaches	12,500
	Sheerness harbour	6,250
3741	Rivers Colne and Blackwater	25,000
	Brightlingsea harbour	12,500
3750	Rivers Crouch and Roach	25,000
	Burnham-on-Crouch	10,000

IMRAY CHARTS
Passage charts

Chart	Title	Scale
Y5	Yarmouth to Holland (IJmuiden)	343,000
C30	Thames to Holland and Belgium. Harwich and North Foreland to Hoek van Holland and Calais	182,000

Approach/harbour charts

Chart	Title	Scale
C1	Thames Estuary. Tilbury to North Foreland and Orfordness	122,000
C2	The River Thames. Teddington to Southend. Teddington to Vauxhall	17,000
	Vauxhall to Barking	14,000
	Barking to Southend	40,000
Y6	Thames Estuary – Northern Part	112,000
Y7	Thames Estuary – Southern Part	116,000
C8	North Foreland to Beachy Head and Boulogne	115,000
Y16	Walton Backwaters to Ipswich and Woodbridge	32,000
Y17	The Rivers Colne to Blackwater and Crouch	49,000
Y18	River Medway. Sheerness to Rochester with River Thames, Sea Reach	21,000
C28	The East Coast. Harwich to Wells	126,000
C29	East Coast of England. Orfordness to Blyth	261,000

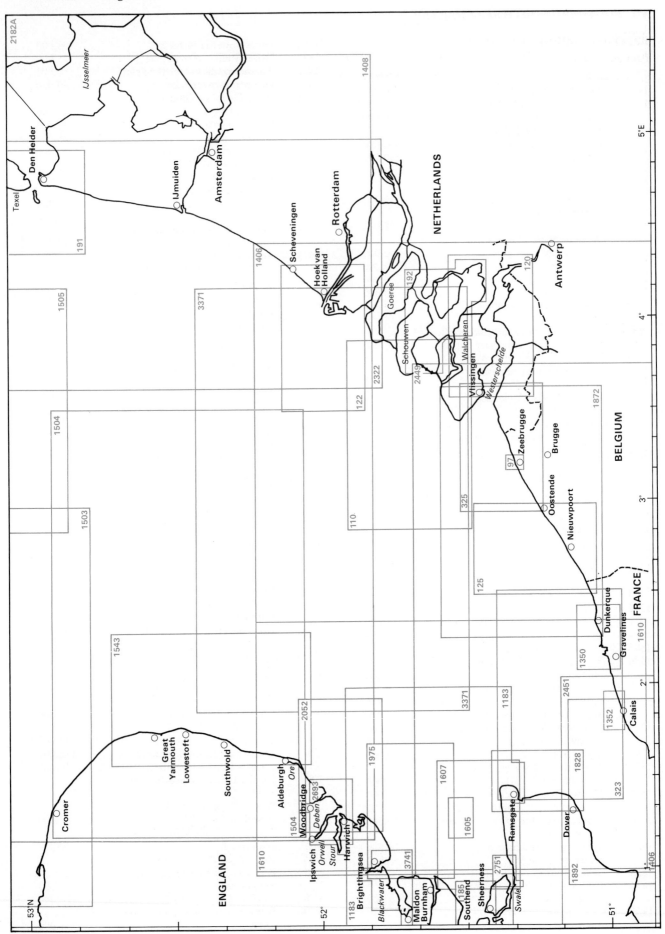

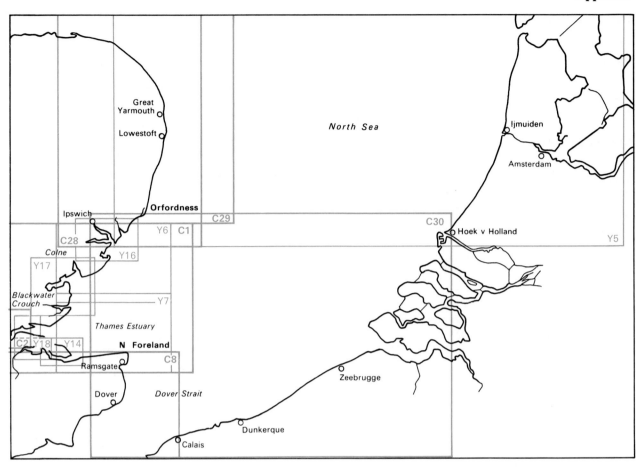

DUTCH HYDROGRAPHIC CHARTS

Chart	Title	Scale
1801	Noordzeekust, Oostende tot Den Helder	375,000
1803	Westerschelde (Vlissingen tot Antwerpen)	250,000
1805	Oosterschelde, en Veerse Meer	50,000
1807	Grevelingenmeer, Krammer, Volkerak en Haringvliet, Hollandsch Diep	50,000
1809	Nieuwe Waterweg, Nieuwe/Oude Maas, Spui en Noord, Dordtsche Kil, Brielse Meer	250,000
1810	IJsselmeer met Randmeren	210,000
1811	Waddenzee (Westblad) en Aangrenzende Noordzeekust	250,000

DUTCH STROOMATLASSEN

Note Figures in brackets show depth beneath surface for which average tidal effect has been measured.

a. Westerschelde (0–5m)★
c. Oosterschelde (0–5m)
e. Benedenrivieren (0–10m)†
f. Aanloop Hoek van Holland (0–5m)★
j. Zeegat van Texel, Aanlopen IJmuiden★ en Scheveningen (0–5m)
n. Noordzee, Zuidelijk deel‡

★ Available also in 0–10m and (in some cases) 0–15m for deeper draught vessels.
† Nieuwe and Oude Maas and Hollandsch Diep.
‡ Southern North Sea from Dover Strait to Denmark and Northumberland.

VIII. BIBLIOGRAPHY

Almanacs

The Macmillan & Silkcut Nautical Almanac. Macmillan Press Ltd

Practical Boat Owner Cruising Almanac. Practical Boat Owner

Reed's Nautical Almanac. Thomas Reed Publications Ltd

Hydrographer of the Navy

Admiralty List of Lights and Fog Signals

Vol A. British Isles and north coast of France. NP 74

Vol B. Southern and eastern sides of the North Sea. NP 75

Admiralty List of Radio Signals

Vol 1 Part 1. Coast Radio Stations in Europe, Africa and Asia. NP 281(1)

Vol 2. Radiobeacons, Radio Direction-Finding Stations, Radar Beacons. NP 282

Radiobeacon diagrams. NP 282(a)

Vol 3. Radio Weather Services. NP 283

Vol 5. Radio Time Signals, Radio Navigational Warnings and Position-Fixing Systems. NP 285

Vol 6 Part 1. Port Operations, Pilot Services and Traffic Management. NP 286(1)

Diagrams relating to Port Operations, Pilot Services and Traffic Management. NP 286(a)

Co-tidal and co-range chart – Dungeness to Hoek van Holland. Chart 5057

Co-tidal and co-range chart – Southern North Sea. Chart 5059

Dover Strait Pilot. NP 28

English Channel Passage Planning Guide. Chart 5500

The Mariner's Handbook. NP 100

North Sea (West) Pilot. NP 54

North Sea (East) Pilot. NP 55

Notices to Mariners – small craft edition (4 times per year). NP 246

Symbols and abbreviations used on Admiralty charts. Chart 5011

Koninklijke Nederlandse Toeristenbond ANWB

Almanak voor Watertoerisme, deel 1. Reglementen in vaartips (regulations and navigation tips)

Almanak voor Watertoerisme, deel 2. Vaargegevens (navigation data)

Ministerie van Verkeer en Waterstaat, Nederland

Getijtafels voor Nederlands (Netherlands tide tables).

Department of Transport

H.M. Coastguard. Coastguard, the co-ordinators, Safety at Sea, Small Craft Safety Checklist, Yacht and Boat Safety Scheme, Safety in the Dover Strait, etc.

H.M. Customs and Excise

Notice to Owners and Persons responsible for Pleasure Craft based in the United Kingdom. C1–15

Royal Yachting Association

Cruising Yacht Safety. C8/86

National Cruising Scheme (Motor) Syllabus and Logbook. G18/83

National Cruising Scheme (Sail) Syllabus and Logbook. G15/83

Planning for Going Foreign. Vol 1 (Belgium, N & W France, Holland, UK). C1/86

Pilots and Guides (see also under Hydrographer of the Navy)

Bowskill, Derek, *The East Coast. A Pilot-guide from Ramsgate to the Wash*. Imray, Laurie, Norie & Wilson Ltd.

Bristow, Philip, *Through the Dutch Canals*, Conway Maritime Press/Nautical

Coote, Jack, *East Coast Rivers. Southwold to the Swale*. Yachting Monthly

Coote, Jack & Delmar-Morgan, Edward, *North Sea Harbours and Pilotage. Calais to Den Helder*. Adlard Coles Ltd

Cove-Smith, C. *London's Waterway Guide*. Imray, Laurie, Norie & Wilson Ltd.

The Cruising Association, *Cruising Association Handbook*.

The Cruising Association, *Visiting Yachtsman's Guide to the London River*.

Tourism

Aquatic Sports in Belgium. Belgian National Tourist Office

Dover Yachtsman's Guide. Dover Harbour Board

Holland Watersports Paradise. Netherlands Board of Tourism

Ramsgate, Broadstairs and Margate, tidal and harbour information. Thanet District Council

Yachting Guide to Harwich Harbour and its Rivers. Harwich Harbour Board

IX. GLOSSARIES

Dutch Pronunciation

This a summary of the *major*, not all, differences in pronunciation between Dutch and English.

Most of the single consonants have similar pronunciations, some a little more abrupt than English. The ones which are significantly different together with some multiple consonants are:

ch similar to the 'ch' in the Scottish 'loch'

g as for 'ch' above

j similar to English 'y'

r rolled like the Scottish or guttural like the French

sch combined 's' and 'ch' (above) – the most difficult (e.g. Scheveningen)

sj similar to 'sh' in English

tj similar to 'ch' in cheese

v closer to English 'f' than to 'v'

w closer to English 'v' than to 'w'

Most of the single vowels also have similar pronunciations, again some a little more abrupt than English. The ones which are significantly different together with the multiple vowels/consonants are:

e, ee, i, ij often (not always) pronounced as the *a* in 'above' e.g. (*de* (the), *een* (a), *snedig*/witty *vriendelijk*/friendly)

aa as in (BBC) English 'bath'

au as 'ow' in 'how' (e.g. *blauw*/blue)

aai as for *aa* with 'ee' as in 'cheese' added (e.g. *saai*/slow)

ee as in 'train'

eeuw as for *ee* with 'oo' as in 'moo' at the end (e.g. *leeuw*/lion)

ei as in 'mail'

eu difficult. As for 'er' in 'her' but shorter e.g. *keus*/choice)

ie as in 'fleet'

ieuw combined *ie* as in 'fleet' (above) and *oe* (below) (e.g. *nieuw*/new)

oo as in 'boat' (the Dutch spelling is *boot*, pronounced as for the English)

oe 'oo' as in 'cool'

ooi combined *oo* and *ie* above (e.g. *mooi*/pretty)

oei combined *oe* and *ei* above (e.g. *moeilijk*/difficult)

ou (same as *au* above) as 'ow' in 'how' (*oud*/old)

uu/uw difficult. As for 'ee' in 'been' but with rounded lips, like 'u' in French 'lune' (*buur*/neighbour, *ruw*/rough)

ij not to be confused with the short *ij* at the end of some words (above). Same as *ei* above i.e. pronounced as in 'mail'. Sometimes spelt 'y', and always placed near 'y' in the alphabet. (e.g. IJmuiden, both letters are capitals at the beginning of a word)

ui close to *ou* above, as 'ow' in 'how' (e.g. IJmuiden)

ENGLISH TO DUTCH AND FRENCH

English	Dutch	French
aft	aaft, achter	arrière
anchor	anker	ancre
anchorage	ankerplaats	mouillage
anchorage prohibited	verboten ankerplaats	défense de mouiller
attendance	bediening	présence
bacon	spek	lard

English	Dutch	French
baker	bakker	boulanger
batten	zeillat	latte
beacon	baken	balise
beef	rundvlees	boeuf
bell	mistklok	cloche
binoculars	kijker	jumelles
black	zwart	noir
block	blok	poulie
blue	blauw	bleu
boat	boot	bateau
boom	giek	bome
bread	brood	pain
breadth (beam)	breedte	largeur
bridge	brug	pont
brown	bruin	brun
bulkhead	schot	cloison
bunk/berth	kooi	couchette
buoy	ton, boei	bouée
butcher	slager	boucher
butter	boter	beurre
cabin	kajuit	cabine
chain	ketting	chaîne
channel/fairway	geul	chenal
channel/waterway	vaarwater	chenal/canal
chart	zeekaart	carte maritime
cheese	kaas	fromage
chemist	apotheek	pharmacien
clew	schoothoorn	point d'écoute
closed	gesloten	fermé
cockpit	kuip	cockpit
current	stroom	courant
customs	douane	douane
cutter	kotter	cotre
dentist	tandarts	dentiste
depth	diepte	profondeur
diesel engine	dieselmotor	moteur diesel
dinghy	bijboot	prame
dinghy	jol	youyou
draught	diepgang	tirant d'eau
east	oost	est
ebb	eb	marée descendante
echo sounder	echolood	echosondeur
eggs	eieren	oeufs
ferry	pont	bac
ferry	veer	bac
fish	vis	poisson
fishing harbour	vissershaven	port de pêche
fishmonger	vishandel	marchand de poisson
fixed bridge	vaste brug	pont fixé
flood	vloed	marée montante
fog	mist	brouillard
foghorn	misthoorn	corne de brume
foot	onderlijk	bordure
fore	voor	avant
forecastle	vooronder	gaillard d'avant
foresail	voorzeil	voile de misaine
gale	storm	coup de vent
genoa	genua	gênois
grease	vet	graisse
green	groen	vert
greengrocer	groente handelaar	marchand de légumes
grocer	kruidenier	épicier
halyard	val	drisse
ham	ham	jambon
harbour	haven	bassin
harbour master	havenmeester	capitaine de port
hatch	luik	écoutille
head	top	point de drisse
height (air draught)	doorvaarthoogte	tirant d'air
high water	hoogwater	pleine mer
horn	nautofoon	nautophone
hospital	ziekenhuis	hôpital
immigration	immigratie	immigration
inner	binnen	intérieur
insurance	verzekering	assurance
jam	jam	confiture

English	Dutch	French
keel	kiel	quille
ketch	kits	ketch
leech	achterlijk	chute arrière
lifting bridge	hefbrug	pont basculant
light float	lightvlot	feu flottant
light vessel	lichtschip	bateau feu
lighthouse	lichttoren	feu
lights	lichten	feux
lock	sluis	écluse
locker	kastje	coffre
low water	laagwater	basse mer
luff	voorlijk	guidant
mainsail	grootzeil	grande voile
mast	mast	mat
mean	gemiddeld	moyen
meat	vlees	viande
mist	nevel	brume légère
mooring buoy	meerboei	coffre d'amarrage
mooring place	aanlegplaats	point d'accostage
mooring prohibited	verboden aan te leggen	accostage interdite
motor sailer	moterzeilyacht	bateau mixte
movable bridge	beweegbare brug	pont mobile
mutton	schapenvlees	mouton
neap tide	doodtij	morte-eau
no	ne	non
no, none	geen	pas de
north	noord	nord
office	kantoor	bureau
officer	beamte	agent
oil	olie	huile
open	open	ouvert
outer	buiten	extérieur
petrol engine	benzinemotor	moteur à essence
pork	varkensvlees	porc
port	bakboord	babord
post office	postkantoor	bureau de poste
prohibited	verboden	interdite
propeller	schroef	hélice
pulpit	preekstoel	balcon avant
pump	pomp	pompe
pushpit	hekstoel	balcon arrière
radio telephone	marifoon	telephone marin
radiobeacon	radiobaken	radiophare
range	verval	amplitude
red	rood	rouge
reed	mistfluit	trompette
rope	touw, koord	cordage
rudder	roer	gouvernail
sailmaker	zeilmaker	voilier
sausage	worstje	saucisse
schooner	schoener	goélette
sea level	waterstand	niveau
shackle	sluiting	manille
shop	winkel	magazin
shrouds	want	haubans
siren	mistsirene	sirène
sloop	sloep	sloop
south	zuid	sud
speed	snelheid	vitesse
spinnaker	spinnaker	spinnaker
spring tide	springtij	vive-eau
stand	stilwater	étale
starboard	stuurboord	tribord
stay	stag	étai
stem	voorsteven	étrave
stern	achtersteven	poupe
supermarket	supermarkt	supermarché
swing bridge	draaibrug	pont tournant
tack	hals	point d'amure
tiller	helmstock	barre
toilet	W.C.	toilette
tower	toren	tour
vegetables	groenten	légumes
water	water	eau
west	west	ouest
whistle	mistfluit	sifflet

English	Dutch	French
white	wit	blanc
withies (bound – leave to starboard)	steekbaken (gebonden)	osiers (reliés – laissez à tribord)
withies (loose – leave to port)	steekbaken (los)	osiers (non-reliés – laissez à babord)
yacht	jacht	yacht
yachtchandler	scheepsleverancier	fournisseur de marine
yawl	yawl	yawl
yellow	geel	jaune
yes	ja	oui

DUTCH TO ENGLISH AND FRENCH

Dutch	English	French
aaft, achter	aft	arrière
aanlegplaats	mooring place	point d'accostage
achterlijk	leech	chute arrière
achtersteven	stern	poupe
anker	anchor	ancre
ankerplaats	anchorage	mouillage
apotheek	chemist	pharmacien
bakboord	port	babord
baken	beacon	balise
bakker	baker	boulanger
beamte	officer	agent
bediening	attendance	présence
benzinemotor	petrol engine	moteur à essence
beweegbare brug	movable bridge	pont mobile
bijboot	dinghy	prame
binnen	inner	intérieur
blauw	blue	bleu
blok	block	poulie
boot	boat	bateau
boter	butter	beurre
breedte	breadth (beam)	largeur
brood	bread	pain
brug	bridge	pont
bruin	brown	brun
buiten	outer	extérieur
diepgang	draught	tirant d'eau
diepte	depth	profondeur
dieselmotor	diesel engine	moteur diesel
doodtij	neap tide	morte-eau
doorvaarthoogte	height (air draught)	tirant d'air
douane	customs	douane
draaibrug	swing bridge	pont tournant
eb	ebb	marée descendante
echolood	echo sounder	echosondeur
eieren	eggs	oeufs
geel	yellow	jaune
geen	no, none	pas de
gemiddeld	mean	moyen
genua	genoa	gênois
gesloten	closed	fermé
geul	channel/fairway	chenal
giek	boom	bome
groen	green	vert
groente handelaar	greengrocer	marchand de légumes
groenten	vegetables	légumes
grootzeil	mainsail	grande voile
hals	tack	point d'amure
ham	ham	jambon
haven	harbour	bassin
havenmeester	harbour master	capitaine de port
hefbrug	lifting bridge	pont basculant
hekstoel	pushpit	balcon arrière
helmstok	tiller	barre
hoogwater	high water	pleine mer
immigratie	immigration	immigration
ja	yes	oui
jacht	yacht	yacht
jam	jam	confiture
jol	dinghy	youyou
kaas	cheese	fromage
kajuit	cabin	cabine

Dutch	English	French
kantoor	office	bureau
kastje	locker	coffre
ketting	chain	chaîne
kiel	keel	quille
kijker	binoculars	jumelles
kits	ketch	ketch
kooi	bunk/berth	couchette
kotter	cutter	cotre
kruidenier	grocer	épicier
kuip	cockpit	cockpit
laagwater	low water	basse mer
lichten	lights	feux
lichtschip	light vessel	bateau feu
lichttoren	lighthouse	feu
lightvlot	light float	feu flottant
luik	hatch	écoutille
marifoon	radio telephone	téléphone marin
mast	mast	mat
meerboei	mooring buoy	coffre d'amarrage
melk	milk	lait
mist	fog	brouillard
mistfluit	reed	trompette
mistfluit	whistle	sifflet
misthoorn	foghorn	corne de brume
mistklok	bell	cloche
mistsirene	siren	sirène
moterzeilyacht	motor sailer	bateau mixte
nautofoon	horn	nautophone
ne	no	non
nevel	mist	broume legère
noord	north	nord
olie	oil	huile
onderlijk	foot	bordure
oost	east	est
open	open	ouvert
pomp	pump	pompe
pont	ferry	bac
postkantoor	post office	bureau de poste
preekstoel	pulpit	balcon avant
radiobaken	radiobeacon	radiophare
roer	rudder	gouvernail
rood	red	rouge
roundvlees	beef	boeuf
schapenvlees	mutton	mouton
scheepsleverancier	yachtchandler	fournisseur de marine
schoener	schooner	goélette
schoothoorn	clew	point d'écoute
schot	bulkhead	cloison
schroef	propeller	hélice
slager	butcher	boucher
sloep	sloop	sloop
sluis	lock	écluse
sluiting	shackle	manille
snelheid	speed	vitesse
spek	bacon	lard
spinnaker	spinnaker	spinnaker
springtij	spring tide	vive-eau
stag	stay	étai
steekbaken (gebonden)	withies (bound – leave to starboard)	osiers (reliés – laissez à tribord)
steekbaken (los)	withies (loose – leave to port)	osiers (non-reliés – laissez à babord)
stilwater	stand	étale
storm	gale	coup de vent
stroom	current	courant
stuurboord	starboard	tribord
supermarkt	supermarket	supermarché
tandarts	dentist	dentiste
ton, boei	buoy	bouée
top	head	point de drisse
toren	tower	tour
touw, koord	rope	cordage
vaarwater	channel/waterway	chenal/canal
val	halyard	drisse
varkensvlees	pork	porc
vaste brug	fixed bridge	pont fixé

Dutch	English	French
veer	ferry	bac
verboden	prohibited	interdite
verboden aan te leggen	mooring prohibited	accostage interdite
verboten ankerplaats	anchorage prohibited	défense de mouiller
verval	range	amplitude
verzekering	insurance	assurance
vet	grease	graisse
vis	fish	poisson
vishandel	fishmonger	marchand de poisson
vissershaven	fishing harbour	port de pêche
vlees	meat	viande
vloed	flood	marée montante
voor	fore	avant
voorlijk	luff	guidant
vooronder	forecastle	gaillard d'avant
voorsteven	stem	étrave
voorzeil	foresail	voile de misaine
W.C.	toilet	toilette
want	shrouds	haubans
water	water	eau
waterstand	sea level	niveau
west	west	ouest
winkel	shop	magazin
wit	white	blanc
worstje	sausage	saucisse
yawl	yawl	yawl
zeekaart	chart	carte maritime
zeillat	batten	latte
zeilmaker	sailmaker	voilier
ziekenhuis	hospital	hôpital
zuid	south	sud
zwart	black	noir

FRENCH TO ENGLISH AND DUTCH

French	English	Dutch
accostage interdite	mooring prohibited	verboden aan te leggen
agent	officer	beamte
amplitude	range	verval
ancre	anchor	anker
arrière	aft	aaft, achter
assurance	insurance	verzekering
avant	fore	voor
babord	port	bakboord
bac	ferry	pont
bac	ferry	veer
balcon arrière	pushpit	hekstoel
balcon avant	pulpit	preekstoel
balise	beacon	baken
barre	tiller	helmstok
basse mer	low water	laagwater
bassin	harbour	haven
bateau	boat	boot
bateau feu	light vessel	lichtschip
bateau mixte	motor sailer	motorzeilyacht
beurre	butter	boter
blanc	white	wit
bleu	blue	blauw
boeuf	beef	rundvlees
bome	boom	giek
bordure	foot	onderlijk
boucher	butcher	slager
bouée	buoy	ton, boei
boulanger	baker	bakker
brouillard	fog	mist
broume legere	mist	nevel
brun	brown	bruin
bureau	office	kantoor
bureau de poste	post office	postkantoor
cabine	cabin	kajuit
capitaine de port	harbour master	havenmeester
carte maritime	chart	zeekaart
chaîne	chain	ketting
chenal	channel/fairway	geul
chenal/canal	channel/waterway	vaarwater

French	English	Dutch
chute arrière	leech	achterlijk
cloche	bell	mistklok
cloison	bulkhead	schot
cockpit	cockpit	kuip
coffre	locker	kastje
coffre d'amarrage	mooring buoy	meerboei
confiture	jam	jam
cordage	rope	touw, koord
corne de brume	foghorn	misthoorn
cotre	cutter	kotter
couchette	bunk/berth	kooi
coup de vent	gale	storm
courant	current	stroom
défense de mouiller	anchorage prohibited	verboten ankerplaats
dentiste	dentist	tandarts
douane	customs	douane
drisse	halyard	val
eau	water	water
echosondeur	echo sounder	echolood
écluse	lock	sluis
écoutille	hatch	luik
épicier	grocer	kruidenier
est	east	oost
étai	stay	stag
étale	stand	stilwater
étrave	stem	voorsteven
extérieur	outer	buiten
fermé	closed	gesloten
feu	lighthouse	lichttoren
feu flottant	light float	lightvlot
feux	lights	lichten
fournisseur de marine	yachtchandler	scheepsleverancier
fromage	cheese	kaas
gaillard d'avant	forecastle	vooronder
genois	genoa	genua
goélette	schooner	schoener
gouvernail	rudder	roer
graisse	grease	vet
grande voile	mainsail	grootzeil
guidant	luff	voorlijk
haubans	shrouds	want
hélice	propeller	schroef
hôpital	hospital	ziekenhuis
huile	oil	olie
immigration	immigration	immigratie
interdite	prohibited	verboden
intérieur	inner	binnen
jambon	ham	ham
jaune	yellow	geel
jumelles	binoculars	kijker
ketch	ketch	kits
lait	milk	melk
lard	bacon	spek
largeur	breadth (beam)	breedte
latte	batten	zeillat
légumes	vegetables	groenten
magazin	shop	winkel
manille	shackle	sluiting
marchand de légumes	greengrocer	groente handelaar
marchand de poisson	fishmonger	vishandel
marée montante	flood	vloed
marée descendante	ebb	eb
mat	mast	mast
morte-eau	neap tide	doodtij
moteur à essence	petrol engine	benzinemotor
moteur diesel	diesel engine	dieselmotor
mouillage	anchorage	ankerplaats
mouton	mutton	schapenvlees
moyen	mean	gemiddeld
nautophone	horn	nautofoon
niveau	sea level	waterstand
noir	black	zwart
non	no	ne
nord	north	noord
oeufs	eggs	eieren
ouest	west	west
oui	yes	ja
ouvert	open	open
pain	bread	brood
pas de	no, none	geen
pharmacien	chemist	apotheek
pleine mer	high water	hoogwater
point d'accostage	mooring place	aanlegplaats
point d'amure	tack	hals
point de drisse	head	top
point d'écoute	clew	schoothoorn
poisson	fish	vis
pompe	pump	pomp
pont	bridge	brug
pont basculant	lifting bridge	hefbrug
pont fixé	fixed bridge	vaste brug
pont mobile	movable bridge	beweegbare brug
pont tournant	swing bridge	draaibrug
porc	pork	varkensvlees
port de pêche	fishing harbour	vissershaven
poulie	block	blok
poupe	stern	achtersteven
prame	dinghy	bijboot
présence	attendance	bediening
profondeur	depth	diepte
quille	keel	kiel
radiophare	radiobeacon	radiobaken
rouge	red	rood
saucisse	sausage	worstje
sirène	siren	mistsirene
sifflet	whistle	mistfluit
sloop	sloop	sloep
spinnaker	spinnaker	spinnaker
osiers reliés – laissez à tribord)	withies (bound – leave to starboard)	steekbaken (gebonden)
osiers non-reliés – laissez à babord)	withies (loose – leave to port)	steekbaken (los)
sud	south	zuid
supermarché	supermarket	supermarkt
téléphone marin	radio telephone	marifoon
tirant d'air	height (air draught)	doorvaarthoogte
tirant d'eau	draught	diepgang
toilette	toilet	W.C.
tour	tower	toren
tribord	starboard	stuurboord
trompette	reed	mistfluit
vert	green	groen
viande	meat	vlees
vitesse	speed	snelheid
vive-eau	spring tide	springtij
voile de misaine	foresail	voorzeil
voilier	sailmaker	zeilmaker
yacht	yacht	jacht
yawl	yawl	yawl
youyou	dinghy	jol

INDEX